UNDERSTANDING RADIOGRAPHY

UNDERSTANDING RADIOGRAPHY

Second Edition

By

STEPHEN S. HISS, R.T., B.S.

Radiology Manager
Lehigh Valley Hospital Center
Allentown, Pennsylvania

CHARLES C THOMAS · PUBLISHER
Springfield · Illinois · U.S.A.

Published and Distributed Throughout the World by
CHARLES C THOMAS • PUBLISHER
2600 South First Street
Springfield, Illinois 62717 U.S.A.

© *1978 and 1983, by* CHARLES C THOMAS • PUBLISHER

ISBN 0-398-04774-X

Library of Congress Catalog Card Number: 82-19231

First Edition, First Printing, 1978
First Edition, Second Printing, 1980
Second Edition, 1983

With THOMAS BOOKS careful attention is given to all details of manufacturing and design. It is the Publisher's desire to present books that are satisfactory as to their physical qualities and artistic possibilities and appropriate for their particular use. THOMAS BOOKS will be true to those laws of quality that assure a good name and good will.

Library of Congress Cataloging in Publication Data

Hiss, Stephen S.
    Understanding radiography.

    Bibliography: p.
    Includes index.
    1. Radiography, Medical—Image quality.   2. Radiography, Medical—Processing.
I. Title.   [DNLM: 1. Technology, Radiologic. WN 160 H673u]
RC78.H52   1983   616.07′572   82-19231
ISBN 0-398-04774-X

Printed in the United States of America
CB-1

*To my wife Patricia
and my daughters Kimberly and Laura*

FOREWORD

TWENTY years ago, many radiologic technology students graduated from their programs with only one or two textbooks and a well-worn, small pocket sized notebook in their possession. This notebook was full of their favorite techniques and hints on the trickier positions. When those x-ray technologists began work after graduation, automatic processing was still a big debate in the finance office, phototiming was still new, and stereoscopy was somewhat of a challenge on the manual locking tube tracks. If that new graduate looked around, he or she could probably find a kymograph that hadn't gathered too much dust, and one of the bigger technical challenges was remembering what to do if you came across one of the few high speed screen cassettes the department had. The only thing resembling a computer was the new "mag card" typewriter the radiologist's secretary had.

Now that young graduate of twenty years ago is a radiographer who wouldn't recognize the radiology department that housed the old school. About the only similarity between that department of twenty years ago and now is Room 3; it still "shoots light." Not only is that "new" department full of computers, so are the shelves of the old hi-fi and T.V. store down the street from our radiographer's home. Looking at the 16K model on the shelf of that store—it might make a nice high school graduation present for the radiographer's oldest child.

In less than one half of this radiographer's career, the world's storehouse of knowledge has quadrupled, giving rise to technological advances that were only a glimmer of an imagination just twenty years ago. What do the next twenty years have in store for this year's new graduates and for the twenty-year veterans? We know that the storehouses of knowledge and technology will quadruple again, and we know that the scientists and engineers who will spur and implement the changes are still in school.

A text, such as UNDERSTANDING RADIOGRAPHY, assumes a heavy responsibility in presenting the basic knowledge of radiography for both the new student and the experienced technologist. The basic knowledge of today's radiography cannot be limited to the techniques necessary to "take a good x-ray." Our veteran radiographers did that very well, and under circumstances that, though routine then, seem appallingly primitive today. Now those veterans, and new students alike, must move from one technological advance to another, often in the same working day. Anyone in radiography today works in a more cost- and quality-conscious health care world. The patients of today are more aware of value and quality than ever before, and the radiographers themselves are motivated to be more technologically mobile than was ever thought possible, to gain satisfaction from their careers. This text, in its Second Edition, adds even more to the radiographer's bank of fundamental knowledge, with chapters on the newest imaging techniques and modalities plus the basic soft computer controlled and enhanced imaging.

Twenty years ago, one of the most difficult theoretical challenges to an x-ray student was picking out a teacher's deliberate error in a schematic of a full wave rectification system. The word "rectification" isn't even included in the extensive glossary of this text. You will have to choose your own pet challenge from the more up-to-date matter given in this book. Proceed deliberately, but be excited. The steps on these pages may at first seem big, but master them. Twenty years from now, if you look back at this text, you may not recognize much at all, but you will understand why.

Robert L. Coyle, R.T., R., Ed.M.
Executive Director
Joint Review Committee on
    Education in Radiologic Technology

PREFACE TO SECOND EDITION

T**HE** First Edition of this work provided a firm base of information of radiographic imaging.

Four new chapters have been added which expand considerably the scope of this text.

It has been endeavored to provide, in a very practical format, a devotion to detail as this relates to the day-to-day clinical experience of the technologist. In this expanded edition, each of the four new chapters at the back of the book provides coverage of full and sufficient depth so that accurate insight may be obtained by the reader.

There has now been included a comprehensive chapter on radiation protection, covering complete and necessary details.

There is a complete chapter on radiographic tubes, x-ray production, and the nature and characteristics of x-radiation.

A chapter on the x-ray circuit utilizes a very clear and practical approach to this potentially confusing subject.

It has seemed important to include, in simple and concise terms, a chapter on T.V. cameras, image intensification, and digital fluoro subtraction.

S.S.H.

PREFACE TO FIRST EDITION

During the early planning stages of this text, a few important prerequisites were self-imposed in the firm belief that their absence would yield a publication so similar to those presently available that another text simply would not be justified. The information presented within the following pages is in some instances new ground for even the experienced technologist while, in other instances, old familiar concepts have been reassessed and aligned more closely with current data.

An important goal which had been set is that strict attention and ample time would be given to the many aspects of radiography which have, in the past, been treated perhaps too simplistically. Although complex physical formulae are not contained in this volume, an attempt has been made to not merely present these concepts of modern radiography for purposes of identification, but also to discuss and analyze each issue at hand from more than one perspective. Without this more rounded approach, much of the meaning is often lost, and as a result misconceptions and frustrations take the place of enlightenment.

It has been my intention from the outset that the information within these pages be presented in such a way that it can be readily understood, and that each concept discussed is covered thoroughly enough and with sufficient depth that an accurate insight can be gained to bridge the gap students often feel is present between classroom theory and its practical application.

In the end, it is often the concept of an idea that is most important to remember, because from it one can learn to answer many of his own questions.

The primary intention of this text is to provide those concepts and insights from which the technologist can grow into a competent professional.

S.S.H.

[xi]

ACKNOWLEDGMENTS

No undertaking of this type can be accomplished without a considerable amount of encouragement, conversation, advice, and reliable facts. Those individuals who have freely given their knowledge regarding radiography have my warmest and deepest appreciation: A great deal of thanks is heartily given to Mr. Thomas Callear, Mr. Edward Cook, Mr. Lee Erickson, and Mr. Robert Trinkle for their hours of time, valued information, and advice.

A number of commercial representatives also freely gave time from their busy schedules to help me obtain data I would not have otherwise been able to present in these pages. In appreciation for their individual efforts, I would like to thank Mr. Donald Becker, Mr. James Funk, Mr. Gary Goodridge, Mr. Harry Harter, Mr. William Orledge, Mr. Gene Oxley, Mr. Tony Passarelli, Mr. Lin Tiley, and Mr. James Wagner.

I would like to also express my thanks to Mr. B. Gilman Cutting for his help and suggestions.

Much appreciation is given to those who helped in proofreading the typed manuscript. Their opinions and comments weighed heavily in the final preparation of this text. For their effort, I would like to thank Ms. Rosann Allen, Ms. Cathy Blose, Ms. Veronica Brodovicz, Ms. Lynda Callaway, Mr. Ernest Griffith, Ms. Catherine Harbaugh, Ms. Mary Matas, Ms. Joann Newell, and Ms. Deborah Wolf.

I would also like to thank Mr. Michael Martinchick for his time in preparing the negatives used in the text, and for his skill, sincere encouragement, and concern for the success of this writing.

I wish to thank Dr. Albert Salzman, Dr. Sigmond Rutkowski, and the Associates in Radiology for making available important radiographs from the teaching library of the Division of Radiology at The Atlantic City Medical Center. I would also like to thank Mr. Charles Broomall, Administrator at A.C.M.C., for his help and consideration.

Sincere thanks must be given to Mr. Jonathan Law and Carmine Pierno for their interest and effort.

There are moments in one's career when a single choice must be made regarding the direction of one's career. Mr. Frank Horvath will never be forgotten for the opportunity he afforded by introducing me to radiography and for his unselfish guidance, good will, and trust.

Mr. Jack Cullinan's help and treasured advice have given me a wealth of knowledge and insight into radiography. His thoughtfulness, knowledge, and good will cannot be properly expressed in words, only felt within.

For the help and long hours spent printing the negatives to meet high standards, I very much want to thank Mr. Leroy Knupp, Ocean City Camera Shop, Ocean City, New Jersey.

The illustrations in this text were provided by Ms. Sue Criss. Her talent and cooperation in their preparation are indeed appreciated.

For the opportunity to work with a publisher who expresses sincere interest in the author's well-being without sacrificing quality, I wish to thank Mr. Payne Thomas. His attitude throughout this project was one of support, trust, and the provider of valued guidance. To Mr. Thomas and his staff at Charles C Thomas, Publisher, I wish to express my sincere appreciation for their help and effort.

Mrs. Patricia Donnon deserves a great deal of thanks for laboring through my first set of handwritten notes and putting them into the first typing. Untold hours must have been spent deciphering those scribblings.

In the long run, good will and trust often determine the success of an undertaking. For those gifts, unselfishly provided me throughout my early years, I wish to thank my parents, Nicholas and Sophia.

S.S.H.

CONTENTS

UNDERSTANDING RADIOGRAPHY

CHAPTER ONE

CHARACTERISTICS OF THE RADIOGRAPHIC IMAGE

T HE TERM RADIOGRAPH is most commonly used to identify a permanent image produced by x-rays; however, over the years, terms such as *roentgenogram* or *plate* have been used to identify the permanent image. *Roentgen*, of course, is taken from Wilhelm Roentgen's discovery of x rays, and *plates* was used because the first permanent images were on pieces of plate glass that had been coated with a silver bromide emulsion.

RADIOGRAPHIC BALANCE

The radiographic image must meet certain requirements to be of any medical value, and although the standards are considerably higher today than they were at some point earlier in time, the specific characteristics desired have not changed. Considering all the desirable properties an image should possess (see Fig. 1), technical balance is perhaps the most important. In a radiographic sense, balance is the relationship between contrast, density, and sharpness. It would be incorrect, however, to associate a specific contrast with a specific density, or sharpness. A balanced radiograph can have short or long scale contrast and can be light or dark. This is an important concept for the technologist to realize because if he can learn to identify a technically imbalanced image he will more easily know when to make technical adjustments or corrections. Figure 2 shows

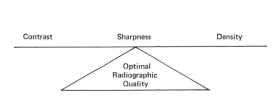

Figure 1. Contrast and density must compliment each other in order to produce optimal radiographic results. Too much or too little contrast for a given density or vice versa will destroy radiographic quality.

[3]

the diagnostic value of a well-balanced radiographic image as compared to one that is not. An imbalanced image may also be too flat or too light, and detail that one ordinarily expects to be present will be absent. It is

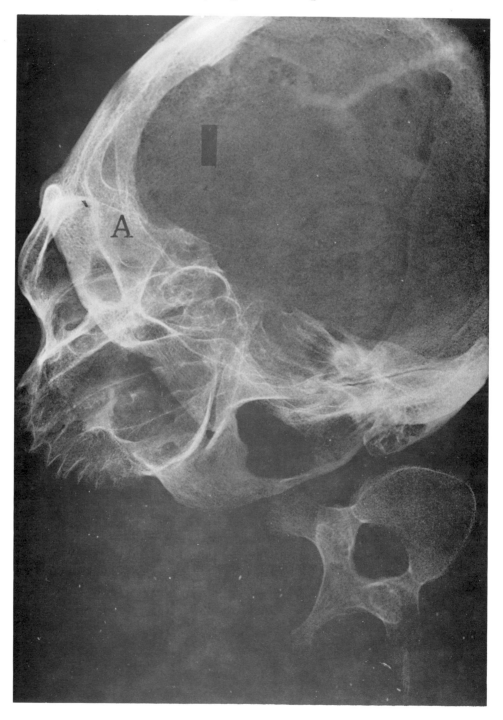

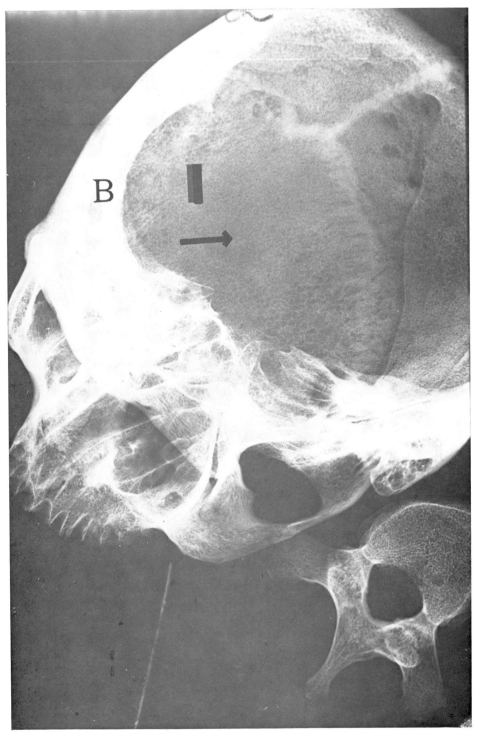

← Figure 2. Visibility of detail is much improved from *A* to *B* as a result of improved density.

important that such characteristics as these be identified as separate entities by the technologist so that he will have a basis from which corrections can be made. The author's feeling is that a technologist who cannot appreciate the quality or lack of it in a radiographic image will not be able to affect the appropriate adjustments necessary to correct the problem.

In summary, one can state that overall technical quality of a radiographic image is strongly dependent upon the compatibility that exists between contrast, density, and sharpness, and, if one is not dominant over the other, a certain technical balance has been successfully achieved. Later in the text, much discussion and evidence will be presented as to how such a balance can be obtained by using the various tools the technologist has at his disposal.

Radiographic Contrast

The amount of effort it takes to produce a radiographic image is of no importance if suitable radiographic contrast cannot be achieved; it is the

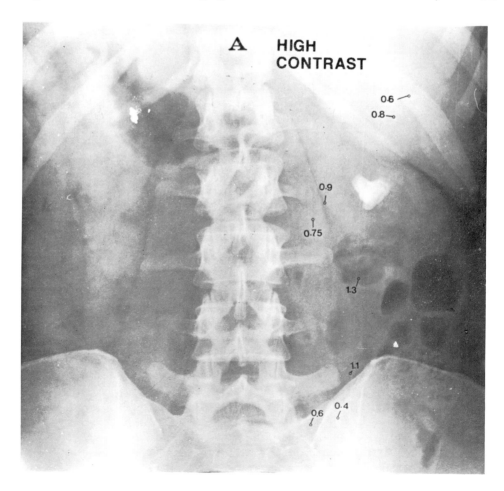

variations in density (contrast) that produce visibility of detail. Even in a nonradiographic sense, we cannot see any distinguishing parts if some type of contrast is missing. For example, upon looking at a solid green wall we will experience great difficulty in seeing any kind of detailed structures that are also green in color. However, if the various structures, cracks, curves, or chip marks are painted blue their presence is noted immediately. Similarly, if a radiograph exhibits only one tone, it is impossible to distinguish any body structures in the image. Fortunately, the radiographic image is composed of millions of tiny black silver crystals that, when viewed with the naked eye, form a pattern of different densities. Figure 3 shows a radiograph of an abdomen in which we see many dif-

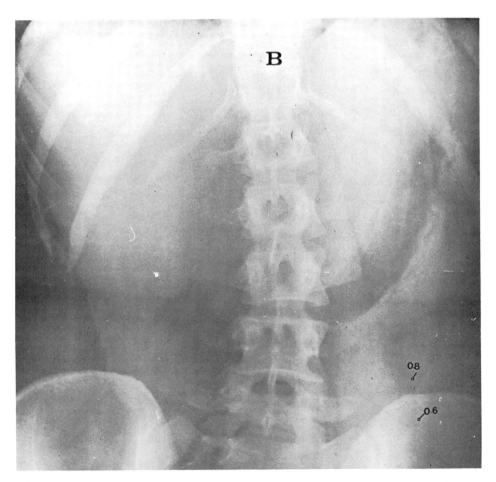

← Figure 3. In *A* the radiographic contrast is sufficient to easily demonstrate the various body parts. The overall density allows for optimal visualization as well. If the densities in *A* and *B* are compared in the pelvic region, the value of radiographic contrast can be understood. There is good sharpness as the borders of the various structures are well defined. *A* is indeed a well-balanced radiograph.

ferent shades of gray, each representing a body structure. Some of the
shades (or tones) have been given actual numerical values. The kidneys,
for example, can usually be seen, and it should be pointed out that the
only reason we can distinguish their presence in the radiograph is that
they display a slightly different density than do the structures in their
immediate surroundings. As we scan the abdomen and look at the inferior
areas of the liver where the gallbladder is located, no such distinction
can be made because the gallbladder has the same density as the tissue
around it. The pelvic bone is easily distinguished in the radiograph because the space it occupies in the film is a different density than its surroundings. In general, one can say that it is the difference in radiographic
tones (densities) that produces contrast, and the ease with which a body
part can be seen in the radiographic image is dependent on the *degree* of
contrast present between its own tone or density and that of the body part
or area around it. The various densities (see Fig. 3) in film B are closer
in relationship than those in film A. The simple formula noted below is
used to illustrate actual contrast values.

$$2^D - 1^D = \text{Contrast}$$

In Figure 4, the contrast values for A and B in the areas measured are
noted; compare with C and D. The reason why the various densities can
be produced in the image is that as x-ray photons pass through the body
they are randomly absorbed by the various body parts. This differential
absorption within the body causes the *exiting* photons to carry that absorption pattern on to the film as demonstrated in Figure 5. The x-ray
film in this regard acts something like a mirror: it responds to the various
intensities of the remnant photons emerging from the patient as a result
of random tissue absorption. This aspect of radiography is known as subject contrast, which is extremely important and will be given proper
attention later in the text. So far we have said in essence that radiographic
contrast is the difference in density between two or more tones and that
such contrast can be measured by subtracting density A from density B.
Also, it was pointed out that unless sufficient contrast exists in an image,
no diagnostic information of value can be seen. We will now go on to
cover some of the important factors that contribute to overall radiographic
contrast.

Elements of Radiographic Contrast

There are actually two major elements that make up radiographic
contrast: subject contrast and film contrast. Subject contrast is the amount
of differential absorption that has taken place among the various body
structures lying in the path of the x-ray beam. The more pronounced these

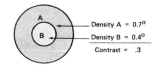

(As With The X-Ray Image These Densities Were Produced Only
With Changing The Number Of Black Specks Per M.M. Area)

Figure 4.

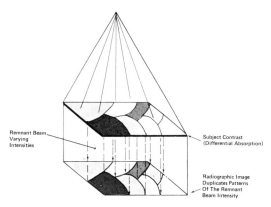

Figure 5. There is a direct relationship between subject contrast, film contrast, and radiographic contrast.

absorption differences are, the greater the subject contrast will be (see Fig. 6, which shows that the radiographic contrast is greatest between A and B). Other things remaining constant, the degree of differential absorption (subject contrast) will determine radiographic contrast. As subject contrast is increased, radiographic contrast will increase; as subject contrast decreases, radiographic contrast will decrease (assuming all the factors are held constant).

A certain harmony or compatibility must exist between subject contrast and film contrast. Film contrast denotes the inherent sensitivity a given emulsion has to variations in the intensity of remnant photons striking its surface. This sensitivity is manufactured into the film and can be increased and decreased by the manufacturer according to the requirements of the radiologist and technologist. Figure 7 shows two radio-

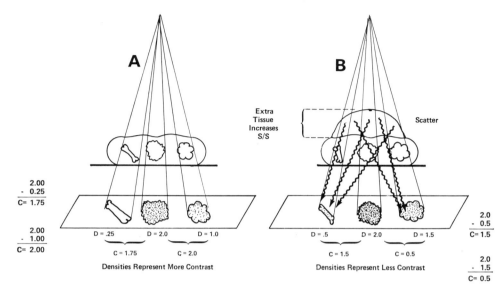

Figure 6. As x-rays are absorbed to different degrees, a corresponding pattern is produced by the film.

graphs of the same patient taken with the same exposure but using different film; by reviewing the radiographs carefully, you will note that film (A) has produced a greater degree of contrast than (B). The determining factor in this case is the film's own ability to emphasize the varying intensities of remnant photons striking the film. Thus, it can be correctly stated that if subject contrast (S.C.) and scatter (S/S) are held constant, film contrast controls radiographic contrast. This can be summarized as follows:

$\uparrow$ S.C. + (fixed film contrast) = $\uparrow$ radiographic contrast
$\downarrow$ S.C. + (fixed film contrast) = $\downarrow$ radiographic contrast
(Fixed S.C.) + $\uparrow$ film contrast = $\uparrow$ radiographic contrast
(Fixed S.C.) + $\downarrow$ film contrast = $\downarrow$ radiographic contrast

Why Different Film Contrasts are Used

Students may ask why film is manufactured with different sensitivities to the remnant beam. The answer is that a harmony must exist between film and subject contrast before a good radiographic result can be gained. There is indeed an optimum radiographic contrast for different body parts which will best demonstrate *that* particular area, and it is the responsibility of film contrast to help produce this optimum condition for the viewer.

For example, the chest region has a great degree of subject contrast; if a high contrast film were used to record this area of the body, the re-

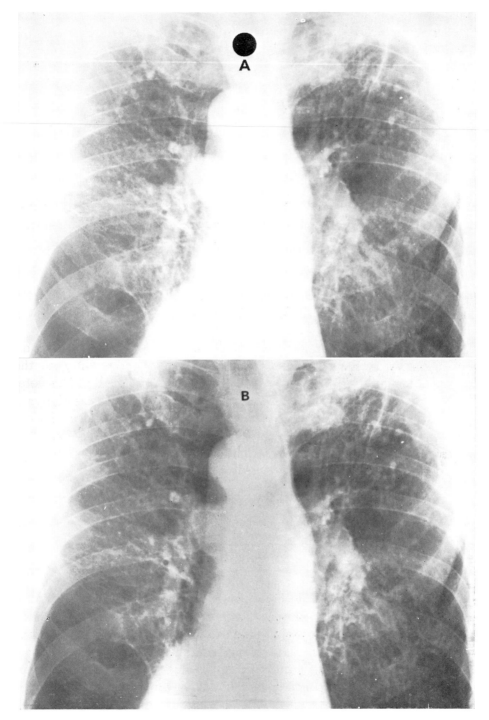

Figure 7. Given the same conditions, the film's inherent ability to emphasize intensities of the remnant beam will determine radiographic contrast. The film used for *B* was a low contrast film.

sulting radiographic contrast might be outside this optimum range (too much radiographic contrast). Likewise breast tissue has an extremely low subject contrast because the differential absorption of the structures is very low; if a low contrast film were used to record that particular body part, an extremely low radiographic contrast would result and would be below the optimal range for diagnosis. With this in mind, one can more readily understand the purpose of manufacturing various film contrasts: Various body regions cause different absorption abilities, and x-ray film is a sufficient variable to complement subject contrast or modify its effect to produce the desired radiographic contrast for good visualization of detail.

There are many factors that in turn contribute to film contrast and subject contrast. For example, it was stated that film contrast is the inherent ability of a film to record varying intensities of the x-ray photons exiting the patient, but it is generally held that processing is an important factor in film contrast. It will be discussed in Chapters Two and Three how different chemical activities and developing solution temperatures affect film contrast. When film contrast is mentioned, one should mentally include processing as well.

Subject contrast in itself is complex and involves such factors as body thickness, density, atomic number of the structure under examination, and kilovoltage.

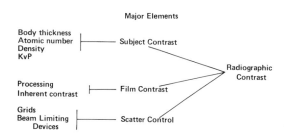

Radiographic Density

There must be an overall density present in the radiographic image to produce optimal detail visualization. Density is obtained when a sufficient *accumulation* of black metallic silver crystals is present in the film; this accumulation of silver is directly related to the number of remnant photons that struck the film during an exposure. If a small number of x-ray photons strike the film, the accumulation of black silver crystals will be sparse and when the film is placed on the x-ray illuminator, the eye will have great difficulty seeing the intended detail. However, if a remnant beam is sufficiently intense, the overall accumulation of silver would be evident along with associated detail. In Figure 8 two circles have been drawn with different image densities. Circle A produces an image of low density as compared to B. The only reason these circles appear different is that there are more dots per square inch in B (the dots themselves are the same size); they represent deposits of black silver.

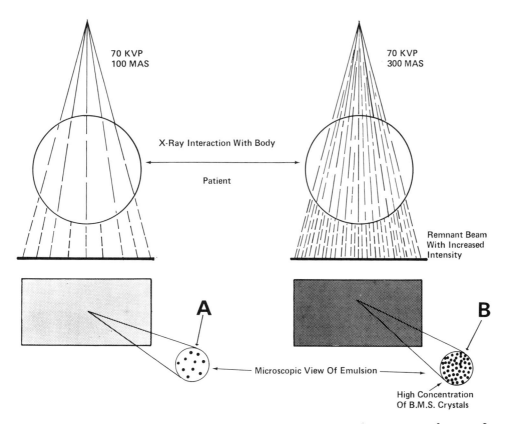

Figure 8. The degree of blackness (density) in a radiographic image is the specific result of how concentrated the black metallic crystals in the film's emulsion are after processing.

	A	B
Exposure =	2,000 photons	4,000 photons
Absorption =	1,000	2,000
Remnant photons =	1,000	2,000
Radiographic Density =	may be 1.0	may be 2.0

Elements of Radiographic Density

Radiographic density like radiographic contrast is the result of a combination of factors. The two major factors are film speed (which includes processing) and sufficient quantities of remnant photons. X-ray film man-

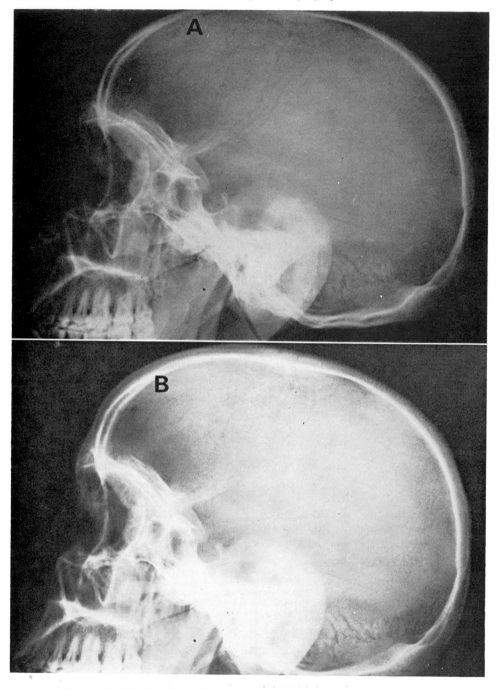

Figure 9. The density in *A* is approximately twice the density in *B*.

ufacturers have developed an ability to "give" x-ray film not only the sensitivity to record *varying intensities* (film contrast) but also an overall sensitivity to x-rays. The sensitivity a film has to x-ray photons or light from intensifying screens is known as film speed, and it has a strong influence on overall radiographic density. In the above example, the radiographic density in A was given as 1.0. However, if another type of x-ray film were used with the same exposure, a density perhaps of 2.0 may result because the inherent sensitivity of the second film to a given intensity of x-ray photons is greater and will respond more dramatically to the exposure. Stated more concisely, for a given intensity of remnant photons, the second film will produce more black silver because of its increased sensitivity. Figure 9 shows two different speed films taken under equal exposure and processing conditions. Film speed is primarily affected by the manufacturers ability to control emulsion thickness and crystal size.

In summary, radiographic density is the degree of blackness over the area being viewed and it is determined by the film's speed and the quantity of *remnant* radiation striking the film.

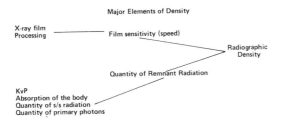

Major Elements of Density

X-ray film
Processing ———— Film sensitivity (speed)

Radiographic
Density

Quantity of Remnant Radiation

KvP
Absorption of the body
Quantity of s/s radiation
Quantity of primary photons

VISIBILITY VERSUS SHARPNESS OF DETAIL

In addition to radiographic contrast and density, the desired result is to see the various body structures with maximum sharpness. Sharpness or definition is the third major image characteristic. The radiographic image actually has two types of detail: (1) visibility of detail and (2) sharpness of detail. If the technologist can develop a clear distinction between these forms of detail, he will have a good start in reaching his goal of understanding radiography. Figure 10 is a schematic drawing of an x-ray tube emitting a primary beam. As the primary beam passes through the object under examination and proceeds to the film as remnant radiation, it produces a geometric pattern much the way a shadow of a tree is produced by the sun's rays. You will note that x-ray film records a "geometric impression" into its emulsion as shown, and also a pattern of contrast and densities. Thus, the radiographic image is actually a composite of two distinctly different entities, one being responsible for producing sharpness

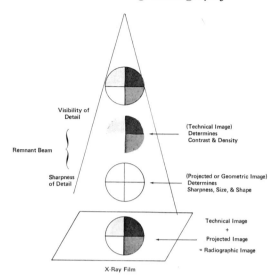

Figure 10. A remnant beam is actually composed of two distinct entities, the technical image (contrast and density) and the geometric image (sharpness). These components can be altered by the technologist independently of each other.

and the other being responsible for providing sufficient density and contrast. Those items that have an effect on sharpness or definition are said to control definition of detail and those items that effect radiographic contrast and density have control over visibility of detail.

Sharpness of Detail

The projected x-ray beam (or projected image) is primarily responsible for producing sharpness of detail, and anything that alters the projected beam will alter sharpness of the image. There are many things that influence definition of detail, and they will be covered in Chapter Six. In short, they are focal spot size, focal film distance, object film distance, screens, and motion.

Visibility of Detail

Visibility of detail is something quite different. As sharpness of detail is controlled by the geometry of the projected beam, visibility of detail is how well we can see the structures that have been transferred to the film's emulsion by the quantity and variation of remnant photons.

In Figure 11, the overall radiographic detail is poor because the image is too dark and because contrast is poor, so visibility of detail is poor. An analogy can be used to help explain this concept: If one looks at a tall building at a distance, it can be seen rather easily; the sharp borders of the building are also distinguished. However, if a fog bank would move into the area, the geometric dimensions of the building would not be

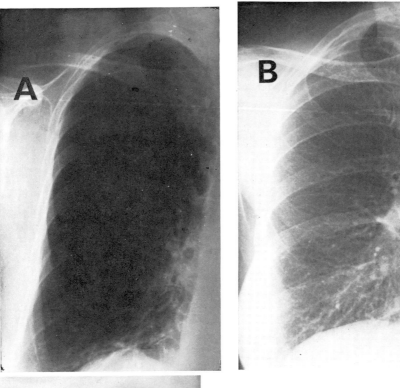

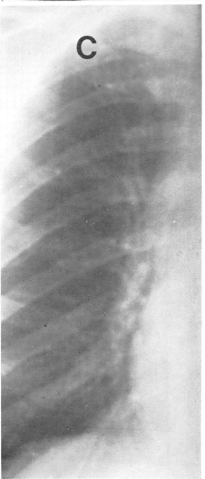

Figure 11. Visibility of detail (contrast and density) must be adequate to *see* various body structures. *A* has optimal sharpness which cannot be fully appreciated because of excessive density. *B* shows correct density. *C* shows poor sharpness but good density and contrast.

visible. This certainly is not because the structure is no longer present, but because something occurred that obstructed its *visibility.* If the same building is viewed on a clear night, the building once again is not visible, although the geometric structure of the building is present. In this instance, the building is not visible because of blackness of the night. It is important that the technologist can distinguish these two forms of detail. If one recognizes a radiograph that exhibits poor visibility of detail he need only work to correct those factors that can affect visibility of detail.

If radiographic detail is poor because of a geometric problem, only the factors that control the projection are to be corrected and not those related to visibility of detail.

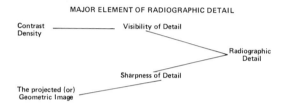

MAJOR ELEMENT OF RADIOGRAPHIC DETAIL

Contrast
Density ——————————— Visibility of Detail

Radiographic Detail

Sharpness of Detail

The projected (or) Geometric Image

SENSITOMETRY

Sensitometry is the process of giving quantitative values to the image characteristics of contrast, latitude, and density and also the arrangement of their values into a logical form or chart that can be easily analyzed. Figure 12 shows two gray scales with actual density values indicated. The numbers were obtained from readings made by an instrument known as a densitometer. The device is very sensitive to various levels of light and will give numerical readings to each density tone it is presented with. From the general appearance of the two scales, it is quite obvious they are different, yet unless we use the densitometer, it is difficult to put these scales into workable terms for comparison purposes.

A graph could be used to display these values, which would furnish much more information. In Figure 13, the vertical scale measures the degree of density a particular film was able to produce and the horizontal scale gives the amount of radiation that was required to produce that density. Figure 14 shows how the two gray scales would appear if translated into a chart known as an H & D curve. The dots indicating density levels are then connected to form a curve that when interpreted shows some important characteristics of the film being tested. Further, it should be noted that H & D curves are broken into three major sections with dividing points shown below.

$$\text{Toe} = 0^D \text{ to } 0.25^D$$
$$\text{Body} = 0.25^D \text{ to } 2.0^D$$
$$\text{Shoulder} = 2.1^D \text{ to maximum reading}$$

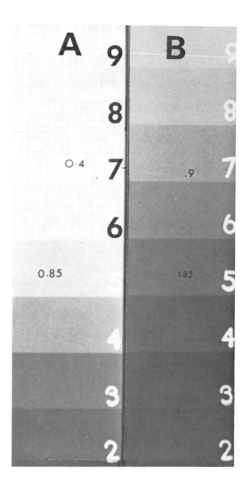

Figure 12. A gray scale of densities is depicted: *A* is a high or short scale, *B* is a low or long scale.

As you view the curves in Figure 14 it is evident that they have various shapes and also occupy different areas of the chart. These two factors in essence provide a technologist with all necessary information regarding the film's contrast, speed, and latitude characteristics.

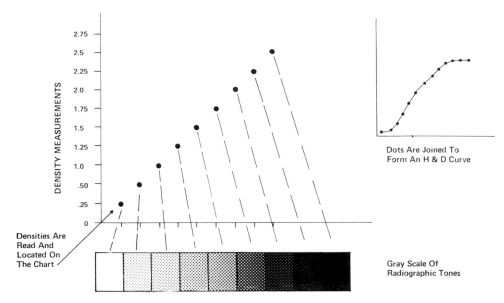

Figure 13.

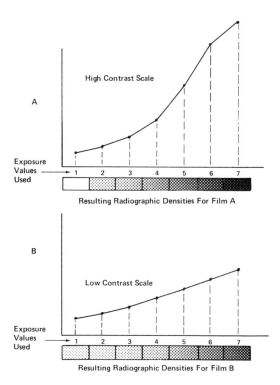

Figure 14.

The Location of the H & D Curve

Location refers to how far toward the left or right the body of the curve appears on the chart. In Figure 15, we see two curves at various locations. Curves are tangent lines which can present a problem for interpretation because density readings vary along the curve from one point to another. With this in mind, 1^D was agreed upon to be used as a fixed reference point. For example, in the body in curve A, 1^D is located far to

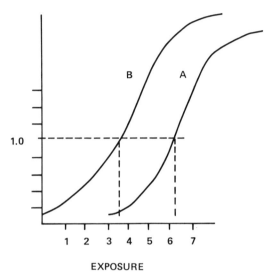

EXPOSURE

Figure 15. The left to right location or position of the curve, at D^1, is used to determine speed.

the right side of the chart. Keeping in mind that the numbers along the horizontal scale indicate quantity of exposure, we can see that much more exposure was needed before film (A) produced enough black metallic silver to reach a density of 1^D. On the other hand, in film (B), less exposure was needed to make a density of 1^D. Thus, by using the location or lateral position of a curve, the technologist knows that (A) is much less sensitive to x-rays and therefore is termed a slow film, certainly slower than film (B). If you examined the relative exposure values that were used to produce a density of 1, you would find that one-half the exposure was required to produce the same amount of film density in film (B), showing that (B) is twice as fast as film (A).

The Shape of the H & D Curve

Within reason, the position or location of the curve can fall anywhere on the chart, but it should be pointed out that location has no relevance to its shape (shape means the overall appearance or form of the curve).

Figure 16 shows three curves at approximately the same position on the chart; from a theoretical point of view, all three films have the same speed because 1ᴰ was obtained with the same exposure values, yet the curves are different in other ways. Film (1) has a very vertical body with a sudden fattening at the shoulder and toe. Film (2) produced a curve that has a more moderate angle and with a more subtle change at the shoulder and the toe. Film (3) has a flat curve but the shoulder and toe areas show a more obvious transition than in Film 2.

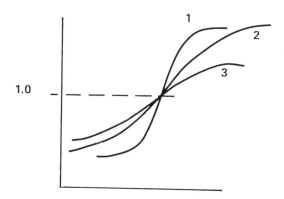

Figure 16. All three curves represent the same speed but different contrasts.

It can be stated that the shape of the curve tells the technologist two important film characteristics: contrast and latitude. Film contrast has already been defined as the ability of a film to record varying intensities of remnant photons. Film latitude is the same, but when the term latitude is used it specifically refers to how well the film can record varying photon intensities at the transition point of the curve where the shoulder and the toe meet the body.

Film Latitude

Film latitude is an important factor because valuable diagnostic information is often lost in these transition points. For example, when a radiograph is made of the chest, much of the lung field is in the upper

---

Figure 17. Film *A* represents a wide latitude (low contrast) film while film *B* represents a narrow latitude (high contrast) film. Actual H & D curves were made with these films; note that curve *A* takes a more gradual turn at the shoulder compared to curve *B*. By reviewing these radiographs it can be seen that film contrast plays an important role in the ultimate diagnostic image, since subject contrast can be altered by film contrast. The H & D curves were drawn to represent the contrast of the films; all other factors were held constant.

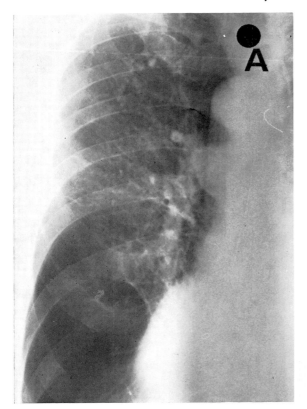

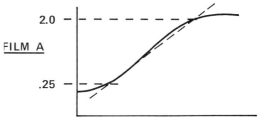

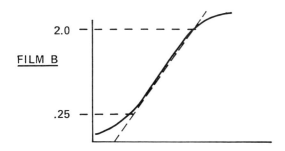

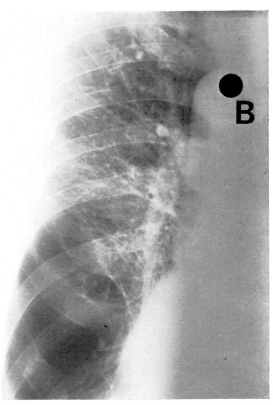

density region of the curve; if the film being used produces a very abrupt transition from body to shoulder, the film is known as a narrow latitude film. Because of the film's ability to make a quick transition at these points, some detail could be lost. However, a film with wide latitude characteristics would have little or no difficulty recording detail in the transition areas because of the gradual and smooth transition the film makes from the body to the shoulder or toe areas. Figure 17 (A and B) shows a radiograph of the same chest made with a narrow latitude film and a wide latitude film. If one is looking for detail at the densities at the lower end of the curve, a wide latitude film would again be more helpful because of its gradual transition characteristics from the body of the curve to the toe. The probability of seeing diagnostic information in this case is greater if a wide latitude film is used.

In short, it can be correctly stated that a wide latitude film is able to expand the useful range of diagnostic densities beyond what could be seen with a narrow latitude film, thus providing the viewer with more *patient information* (detail). It is generally true that wide latitude is desirable except possibly when examinations using contrast materials are performed or when subject contrast is especially low. It should be pointed out that film latitude and film contrast are dependent on one another to the extent that a high contrast film will almost always have a narrow latitude and a low contrast film will almost always have wide latitudes.

Film Contrast

Film contrast is the sensitivity of the film to various intensities of the remnant photons. Figure 5 shows a remnant beam emitted from an object has abrupt variations of intensities. In general, x-ray film can be manufactured to record these variations in three ways: to produce accumulations of silver quantities commensurate to the beam's intensities; to produce accumulations of silver that will in effect exaggerate these intensities; and to produce accumulation of silver that will result in an image which would deemphasize the beam's intensities.

Review the two gray scales in Figure 14. In doing so, we would see that the density differences in film (A) are more abrupt and obvious. As the densities this film produced are plotted on a chart, we note the body of the curve becomes very vertical, which is evidence that film is very sensitive to variations of intensity. Film (B), which was made using the same factors the gray scale produced, is more *moderate in gradiations* of densities from black to white; this is evidence that this film is not as sensitive to the intensities of the remnant beam. Thus, film (A) is known as a high contrast film and film (B) is known as a low contrast film. As will be pointed out in Chapter Two, all types of film have their place in radiography and one should be aware of the importance when making a choice. (Film contrast should compliment subject contrast.)

Determining Film Contrast from the H & D Curve

The term average gradient is used most commonly to determine radiographic film contrast. The average gradient for a film is shown in Figure 18. The average gradient is used because the true gradient or slope of all H & D curves change depending on which part of the curve you are viewing as it makes its tortuous route from the toe to the shoulder. Two horizontal lines are drawn covering a suitable diagnostic density range within 0.25^D and 2.0^D: These lines show the *average* gradient of diagnostic densities along the body of the curve. From these and with the formula shown below an actual film contrast value can be obtained.

$$\frac{D^1 - D^2}{Exp^1 - Exp^2} = \text{Average Gradient}$$

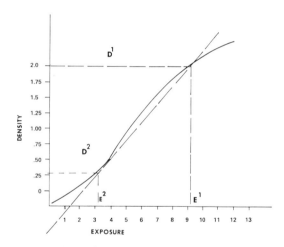

Figure 18. The average gradient for this hypothetical curve is 3. Actual film gradient values range from approximately 1.5 to 3.5. A film that has an average gradient over 1.0 will tend to exaggerate the differences between the various intensities of the remnant beam. Because of the low inherent subject contrast, film usually emphasizes rather than deemphasizes subject contrast.

From radiographs and curves shown in Figure 17, it is evident that one can choose from a wide range of film types. There are several terms used to describe the characteristic curves, but H & D and sensitometric curves are the most common. The body of the curve is referred to by other terms than those used in this text (two of these are straight line and gamma), however, these terms refer to actual angle or slope and not the average.

Base Plus Fog

If a film is processed without any type of exposure, it is not perfectly clear. This faint density, termed base plus fog or simply base fog, is caused by two major factors. First, the polyester material used in making the film's base is given a pale blue tint during manufacturing, to improve film contrast characteristics. Second, even though unexposed, a chemical reaction between the unexposed silver bromide crystals and the developing solution causes a low level of density. Base fog is always present in diagnostic film, but, under normal circumstances, it should never exceed 0.2^D. The presence of excessive base fog (over 0.2^D) causes a moderate to dramatic decrease in radiographic contrast, as well as an increase in radiographic density (see Fig. 19).

Figure 19

THREE MAJOR CAUSES OF INCREASED BASE FOG

1. IMPROPER FILM STORAGE
   - Storage Area Too Warm
   - Held Beyond Expiration Date
   - Unprotected from Fumes and
     Scatter Radiation
   - Excessive Background Radiation

2. IMPROPER SAFE LIGHT CONDITIONS
   - Too Many Lights
   - Too Large a Bulb Size
   - Scratched Filter
   - Safe Light Too Close to
     Work Area
   - Improperly Sealed Windows
     and Doors

3. IMPROPER DEVELOPMENT
   - Solution Temperature Too High
   - Solution Too Concentrated
   - Contaminated Developer Solution
   - Too Much Replenishment

CHAPTER TWO

RADIOGRAPHIC FILM

THE NEED for an x-ray film was not to produce a visible image, because crude fluorescent crystals were already in use by 1895. However, the fluorescent image was very poor in detail and certainly not permanent (an image that could be easily stored for future reference was needed). The earliest types of x-ray film were actually glass plates which were coated with silver bromide crystals. These crystals changed chemically after being exposed and developed to yield a visible permanent image. One will occasionally hear the term "flat plate" being used, and this term recalls those days when the radiographic image was made of plate glass. One can easily imagine how difficult it must have been to handle these highly breakable plates.

A need for something more manageable was clear, and about 1924 the first film using a flexible base similar to what we know today was manufactured. The base was made up of a cellulose acetate material, and its usage was a truly revolutionary advance in medical radiography.

In this chapter, the basic composition of film will be discussed, and an effort will be made to relate how the components contribute to the desired result of a high quality radiographic image. The various physical and chemical processes that take place during the film's exposure to x-ray and processing sequences will also be introduced in this chapter.

FILM MANUFACTURING, AN ART

To classify the manufacturing of x-ray film as an art may at first seem presumptuous. There is no doubt that a high degree of technical information and scientific research weighs heavily in designing and manufacturing film. Yet, this raw scientific information has to be arranged and applied in such a way that the product will be of highest quality, much the way the composer uses the science of notes and harmonic chords to produce a symphony. The film manufacturers must work with the physical laws that govern the transfer of energy from the remnant x-ray beam to a latent image and eventually to a visible image. The laws of physics governing the production of a good radiographic image are common to

[27]

all manufacturers. However, the ability of the manufacturer to work with these laws and their skill in doing so is truly an art.

DIAGNOSTIC X-RAY FILM COMPOSITION

Emulsion

Subcoating

Base

Subcoating

Protective Layer

Emulsion

Emulsion

Silver Bromide Crystals

Gelatin

Figure 20. Basic components of a typical double emulsion x-ray film.

Various Types of Film Composition

As complex as film composition can be, Figure 20 illustrates its basic parts. All x-ray film is similar in its basic design and composition. How-

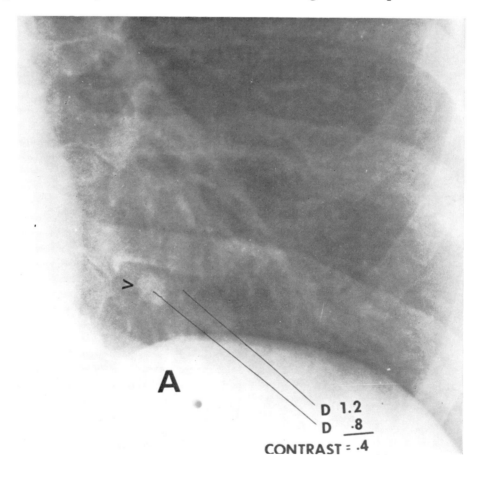

A

D 1.2
D .8
CONTRAST = .4

ever, there are some films that contain only a single emulsion and are used much less frequently in medical radiography as compared to double emulsion film. Today's needs for film are certainly more expansive than those of 1924 when almost any quality of permanent image was warmly greeted. It should be pointed out that x-ray film is also manufactured for industrial purposes. Our primary interest in this chapter will, of course, be aimed at the medical type. The most widely used type of diagnostic film is screen film. This film is made especially to be used with intensifying screens. Nonscreen film, also known as direct exposure film, is used much less often in recent years and is, of course, designed to be used with direct exposure only.

Radiographic film must produce an extremely high quality image, that is to say, it must produce an image that makes visible as much of the body part under examination as possible. As more and more body structures can be made visible radiographically, the higher the image quality

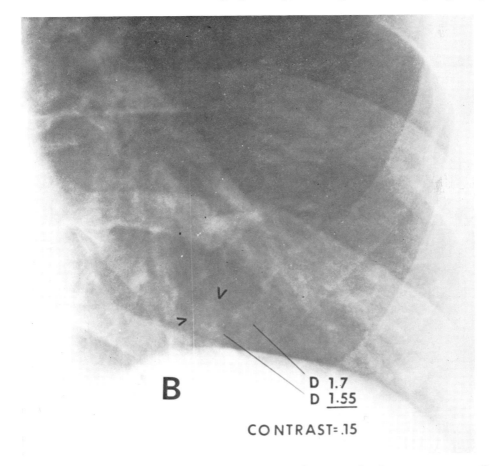

← Figure 21. *A* shows increase in visibility of detail as a result of proper contrast; *B* shows poor contrast and visibility of detail.

and the more valuable that particular film becomes. The types of diagnosis today's radiologists are trained to make are often so finite that only film which yields the best image possible should be used. Figure 21 shows some of these subtle but very important radiographic findings. One can easily appreciate how nothing but the best image should be used for diagnostic purposes. In a true sense, x-ray film equals the maximum that will ever be seen by the radiologist. During the author's research with major film manufacturers, it has become more evident that each component part of the film produced has been researched to the limits of reason and the substances used in manufacturing are of the highest quality available. The technologist should make every effort to position the patient properly and use the best combination of exposure factors possible since a high quality film cannot compensate for the selection of poor exposure factors.

The Film Base

The base is the support material of the x-ray film. Its primary purpose is to support the emulsion with sufficient firmness so that it can be handled easily. The emulsion goes through some important changes while the film is being transported through the various processing solutions, and here is where the film's base often meets its toughest test of quality. As mentioned earlier, the base was originally made of glass. Later, a plastic material (cellulose acetate) was used until the early 1960s when it was found that cellulose acetate caused some problems with film transport in automatic processors.

A polyester material is now used in its place. Modern x-ray film is sometimes referred to as safety film because of its resistance to fire. X-ray film can burn, however, but it requires much more heat and will burn much more slowly than the type that was used earlier.

Film Base Characteristics

The base must be fairly hard yet flexible; it must also be clear. (This does not mean totally transparent, but rather clear of any foreign particles that might detract from information held by the emulsion.) Light areas of the film appear as such because the emulsion has been dissolved away and light from the x-ray illuminator is transmitted easily to the observer. If the base was poorly made it might not transmit light properly or evenly and the light areas of the image would suffer. This could be especially important when viewing small structures of low film density such as lung markings in a radiograph of the chest. If a film is processed without being exposed to light or x-rays, all of the emulsion will be dissolved away by the fixer solution and you will see a bluish tint in the base. The purpose of this tint is to reduce the harsh light emitted by the illuminator. In short, a glare would be evident if the base was perfectly transparent and this would be very objectionable to the radiologist. By reducing the glare, the radiologist is better able to interpret the subtle density tones in the radiographic image. This might be similar to what happens as we walk down a street during a bright sunny day and try to see something located toward the sun. The glare and harsh light of the sun would make it quite difficult to see the object well. However, if you used a pair of sunglasses, the blue tint of the sunglasses would help reduce the harsh light and would increase the visibility of the object. Thus, the radiographic value of having a blue tint in the base is that it allows one to see the various tones of the radiograph more easily.

Base Thickness and Parallex Distortion

Base thickness has an effect on the sharpness of the radiographic image (see Fig. 22). You can see that each side of the film has its own separate image. The viewer, of course, looks at both sides simultaneously as the light from the view box transmits through the film. One can easily see that if the two images are not perfectly superimposed over each other a slight degree of blurring can result. Under normal conditions, this blurring is not excessive and is not a noticeable problem; however, one can see that the thickness of the base can have an effect on this type of image sharpness. As the base is made thicker, sharpness decreases. This type of blurring is commonly referred to as parallex distortion. Single-coated emulsion film does not possess this problem.

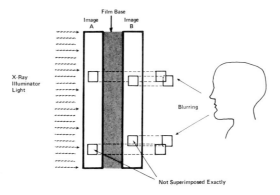

Figure 22. When *two different layers* of emulsion are used, the thickness of the base will determine the degree of parallex distortion.

A more practical problem that relates to base thickness is that regarding transport through an automatic processor. It can be seen that the tolerance in spacing between the transport rollers of the automatic processor is very important. Once set, the film must be able to pass between them with just enough pressure to assure transport yet not so tightly that excessive pressure would cause artifacts or a delay in transport (jamming). In general, the thickness of modern film bases is approximately .008 inches thick.

Base Plus Fog

It was earlier mentioned that the base had a slight blue tint, and in fact, this accounts for a low degree of film density when viewed. However, there is an additional density in the base that is caused by other factors besides the blue coloring. The base material is somewhat affected by the processing chemicals. During development, the base absorbs some developer solution causing a low level of chemical reaction which, in turn, causes a slight fogging effect. The third contributing factor to base plus fog (base fog) is low level radiation exposure to the emulsion of the film. The film is actually exposed to background radiation while in storage. Although this at first may seem to be inconsequential, considerable testing and troubleshooting throughout the country has shown this to be a more significant problem than originally thought. Many reports of so-called "bad film" have revealed the actual cause to be background radiation.

Base fog, then, is simply a low level density that can be detected on a film if it is processed without any exposure whatever. Quantitative standards have been established for acceptable base fog limits which should not exceed 0.2^D. If a reading is obtained greater than 0.2, radiographic quality could be in jeopardy. It should be remembered that very often important body structures to be seen are those of very discreet density variations: If they cannot be reproduced because of base fog, diagnostic value of the radiographic image is unnecessarily limited.

Silver Bromide Crystals

The silver bromide crystal is the heart of the radiographic image. It is what actually causes the density on the x-ray film. Manufacturing these crystals is very complex, and there is no purpose in describing it here. However, some characteristics of silver bromide crystals are important and these will be described here.

Crystal Size, Quantity, and Shape

One of the most striking things that come to mind is the extremely small size of the crystals used in producing x-ray film. For general purpose, screen type film, the silver bromide crystals are approximately 0.1 to 0.5 microns, and it is important to know that crystal size affects the quality of the image as shown in Figure 23. Although the size of these

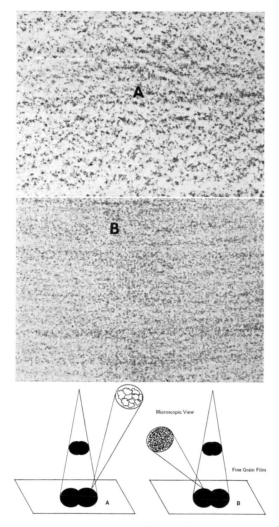

Figure 23. *A* represents a coarse grain film and *B* represents a fine grain film.

crystals has been enormously enlarged, the radiographic effect is representative of what actually occurs in a real silver bromide crystal. It can be seen quite easily that as larger crystals are used they individually cover a larger area. With such an increased area they are individually less able to duplicate body detail. Figure 23B illustrates an ideal situation as extremely small crystals are used. However, in Figure 23A one can see that these larger crystals influence greater areas of the image and decrease their ability to represent individual finite body structures. Crystal size has been increased to a point where one crystal represents too many body structures and cannot produce maximum detail. Since it is impossible for the same crystals to produce different densities, the film's ability to reproduce detail would decrease with larger crystals. With this in mind, one can begin to appreciate that an ideal situation is where each crystal influences only its miniscule area of the body part and can thus better represent the structure in detail. In other words, if crystals are produced beyond a certain size they will record neighboring structures as well as their own. When this occurs, radiographic detail dramatically decreases and produces an image that is commonly referred to as grainy.

Crystal Size and Radiographic Contrast

Crystal size can be varied by the manufacturer with a process known as digestion. It is very important that manufacturers control the size of the crystal because it affects the film's ability to produce contrast and density. As the size of the crystals vary with each other throughout the same emulsion, the film will tend to deemphasize intensities of the remnant radiation which, in turn, will produce a lower contrast image. It is the ability of the manufacturer to mix the various sized crystals throughout the film that produces various film contrast characteristics. Crystal size also affects the overall speed (sensitivity). As will be explained later in more detail, once a crystal has been exposed by the light of the intensifying screen or by direct x-ray exposure, it will change in composition and become black metallic silver via the developing process. It is important to understand that the more black metallic silver grains there are on the film after processing, the darker the overall image will be. If two crystals, one small and one large, are exposed equally, the larger crystal will indeed block more light emitted by the x-ray illuminator, giving that area of the film a dark appearance. A smaller crystal, of course, would allow more illuminator light to pass through the viewer and that area of the film would appear less dense. A film made with larger crystals will produce more overall blackness (density) than a film made with small crystals if both were exposed equally. In fact, changing crystal size is a common method used by film makers to produce film with various speed characteristics. The film with a fast speed will produce more density per exposure because its larger silver bromide crystals will intercept more x-ray photons, thus causing a larger black area in the film's emulsion.

The thickness with which silver bromide crystals are coated also strongly influences film speed for reasons that are outlined in Figure 24. It can be seen here that with equal exposure values, the film with more crystals (thicker emulsion) will absorb more photons, thus producing more black metallic silver during processing.

There is no typical shape of silver bromide crystals. A micrograph of a typical crystal used in manufacturing x-ray film would show that these crystals are irregularly shaped with many variations. The shape of the crystal has very little apparent effect on radiographic quality.

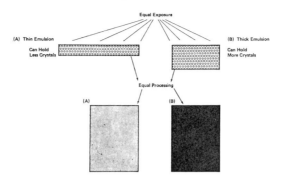

Figure 24.

Film Contrast and Latitude

Film contrast is influenced by using different crystal sizes in the film's emulsion. As the crystals become more varied in size, contrast will decrease; as the size of the individual crystals become more consistent throughout the emulsion, the film will have high contrast characteristics. Film latitude, as was already described, is the film's ability to record *diagnostic* densities toward the extreme light and dark density regions. It is generally held that a low contrast film will have a wide latitude characteristic and vice versa.

The Gelatin

The gelatin contained in the x-ray film's emulsion plays a very important part in the film's overall quality. Gelatin and silver bromide crystals are mixed to produce what is known as the emulsion. The mixing process is quite a delicate operation because any variation in consistency or in how evenly the emulsion is spread over the film's base will cause very obvious radiographic changes. If this *coating* process is not properly controlled it will affect radiographic contrast as well as density. The practical problem that this would cause to the technologist is that the same type of film would produce different densities and contrast values from one month to the next. With this kind of inconsistency, it would be very difficult to produce predictable radiographic results.

As refined and controlled a substance as gelatin is, it can simply be described as a clear gel which is refined very meticulously from calf's skin. Its properties, however, are much more important: (1) Gelatin must be perfectly transparent; (2) It must be able to hold the silver bromide crystals firmly in place, especially during processing; (3) Gelatin must be able to swell and contract during the processing cycles. One can imagine a radiographic image whose crystals have "slipped" out of position. Of course, this is not likely to happen, but it *does show any degree* of crystal shifting would be disastrous to the quality of the radiographic image. The task of holding the silver bromide crystals firmly in place is made considerably more difficult once the exposed film begins its route through an automatic processor. In addition, the gelatin must be able to resist the various chemicals with which it comes in contact during processing.

There is a moderate degree of friction against the film surface caused by the processor's rollers and, as noted above, it is during processing that the gelatin becomes most important in the radiographic image. As soon as the film is immersed in the developer solution, the gelatin begins to swell beyond its original thickness. During this expansion, the silver bromide crystals must be held in exactly their original position. The importance of the swelling lies in the fact that a situation must be created whereby the developer solution can easily seep into the depths of the emulsion to begin the developing action on those crystals located along the film's base. If the gelatin could not swell sufficiently during developing, underdevelopment would occur and the radiographic density and contrast would be decreased.

After this very important emulsion swelling period, the film is transported to the fixer and then into the wash where clean fresh water is used to "bathe" the gelatin of the residue chemicals from the developing and fixing cycles. The gelatin must undergo its last important change, shrinking and hardening, as the film is transported through the drier section of the processor. The film can then be stored for future reference.

The Effect of Double Film Coating

X-ray film can be manufactured with single or double coating; double coated film is the most common. The main reason is that intensifying screens are used in pairs as illustrated in Figure 25. It can be easily seen that a film with emulsion on both sides of its base would take advantage

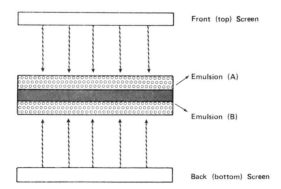

Figure 25. Double coating is essential to general screen radiography to make effective use of screen light emission from both screens.

of the light emitted by both screens. Two thinly coated emulsions are more practical than one thick single coating, because the light from the intensifying screens would not be able to penetrate the thick emulsion and expose the crystals lying against the film base. Also the developing solution would certainly have more difficulty in reaching the exposed crystals lying close to the base.

Film Storage

Film storage is a simple problem if one keeps in mind that the emulsion in x-ray film is sensitive to circumstances other than exposure by light or by the x-ray beam. The basic rule is to store film in a cool, dry area (50 to 65% humidity at a temperature from 40° to 70°). As the atmospheric conditions reach the upper limits of these regions, the shelf life* of x-ray film is reduced dramatically. Another factor in film storage besides the atmospheric conditions is that x-ray film is sensitive to various types of fumes and gases. Certain strong gases will produce a low level of ionization which will cause fogging. If film is stored beyond its expiration date, some very definite problems could result. The most likely would be decreased radiographic contrast and increased base fog.† The extent of these is directly related to how badly the film storage requirements were violated. It is not a good economic practice, however, to throw out film immediately upon the arrival of its expiration date. Upon the realization that a supply of film is beyond its expiration date, from a practical point of view a sample film can be pulled and tested for contrast and speed. This can be done by exposing a current and old film equally with a sensitometer and reading the base fog and a density at approximately 0.5^D. If both films produce equal densities the old film can probably be used.

LATENT IMAGE FORMATION

Although proper film composition is crucial to a good radiographic image, a process known as latent image formation is the mechanism by which the film's component parts work together with the exposure to produce the eventual visible image. The strange thing is that latent image formation process is only a theory, that is to say, it has not been totally proven. The reason for this lies in the infinitesimal size of the structures

---

* A period of time when something can be stored without danger of deterioration.

† Because there is already some density on the film due to base fog, the film will produce, depending on the degree of fogging, slightly more density than would otherwise be expected for that exposure value.

involved. In fact, the latent image itself cannot be detected by any means so far known.

In reality, there are many so-called "images." There is the radiographic image, the visible image, and the projected image, and all refer to slightly different aspects of radiography. In short, the latent image is the ionization of the exposed silver bromide crystals in the emulsion as *compared* with those crystals that have not been exposed and therefore are *not ionized.*

Before going into a detailed discussion of this important topic, it might be well to reflect for a moment on a more general level. Figure 26 shows that a structure can be made visible by making a number of black dots on a sheet of paper. Upon inspection of a routinely exposed and processed x-ray film, we should imagine that the various black and grey tones present are the result of accumulated or concentrated tiny spots of black silver. For example, if we closely inspect the composition of a newspaper picture we will begin to see that the image being viewed is made up of an accumulation of very tiny dots. You will also note that where the dots have become concentrated, the picture looks darker and the lighter areas are caused by a more sparse distribution of these dots. Thus, the radiographic image is nothing more than a series of random concentrations of tiny black metallic specks. As was noted earlier, the latent image is caused by the ionization of silver bromide crystals as a result of an exposure by x-rays or intensifying screen light. The greater the exposure, the more silver and bromide ions will be liberated within the crystal and will result in a greater latent image formation. In short, the latent image is the mixture or pattern of ionized and nonionized silver bromide grains as a result of the exposure.

ANATOMY OF A RADIOGRAPHIC IMAGE

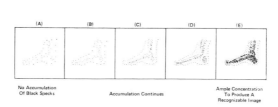

No Accumulation
Of Black Specks   Accumulation Continues   Ample Concentration
                                                                 To Produce A
                                                                 Recognizable Image

Figure 26. As the number of black metallic particles present in the exposed and processed film's emulsion increases, more body parts will become visible. However, *too many* black silver particles will decrease visibility due to excessive radiographic density and reduced contrast.

The Crystal Lattice

In order to more clearly understand the latent image, we will now introduce a structure commonly referred to as the crystal lattice. The lattice is a combination of atoms made of silver (+) ions and bromine (−) ions arranged in such a way that, when struck by an x-ray photon, the positive silver ions and negative bromine ions begin to move within the crystal in rather predictable patterns. The crystal lattice may be thought of as a single molecule of silver bromide.

In order to describe the process of latent image formation, we must gain some additional insight into the crystal of the film. The overall chemical composition of the film's crystal was described above and is illustrated in Figure 27. The lattice indicates that each correctly formed crystal has a specific and crucial arrangement of silver and bromine ions in addition to several small specks of sulfur compounds. The sulfur compounds are located in the outer edges of the lattice arrangement and are commonly known as sensitivity specks. Their purpose is to trap the free-floating, negative bromine ions that will be released when the crystal is ionized by absorption of light or x-ray photons. In a very general way, the lattice cube may be thought of as a closed environmental area in which these ions may roam after the exposure. When the exposure puts these negative ions into motion, the sensitivity specks trap them.

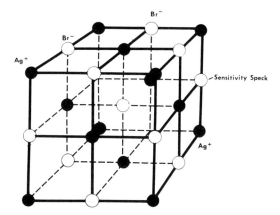

Figure 27. A schematic drawing of a silver bromide crystal is shown here before exposure.

The sensitivity speck is often referred to as a defect in the silver bromide crystal. The reason for this is that the crystal lattice provides a route by which the ions can travel after being liberated by the exposure. However, the sensitivity speck is an area of the crystal lattice that prevents easy movement of the negative bromine ions and in that sense are defects in the crystal. The defect has been described by Christensen, Curry III, and Nunnally. They compare the crystal lattice to a brick wall within which one row of improperly sized bricks were laid. The result of this incompatibility of structure is an increase of pressure on that defective row of bricks. Similarly, the crystal lattice has built into it an incompatibility in structure (the sensitivity speck) which will not allow the roaming negative bromine ions to pass by without entrapping them. With the above description of a typical silver bromide crystal, we may proceed with what actually causes the latent image.

We can begin the description of latent image formation by describing one random photon as it is emitted by the patient. Of course, many of these photons are not absorbed by the silver bromide crystal and no ionization or latent image formation results. However, those photons that do expose the crystal will cause ionization. Thus, the more photons of intensifying screen light or of x-rays there are to strike the crystal, the more photons absorption (by the crystal) takes place, and more latent images will result. When a photon is absorbed by the crystal, the negative bromine ions immediately begin to move about the lattice structure (see Fig. 28). The ions will continue to travel for some distance until they approach an area of increased chemical pressure caused by the sensitivity speck. The negative ion (see Fig. 29) causes the once neutral sensitivity speck to take on a negative charge. Simultaneously, the newly liberated negative ion may collide with other ions and release others from their

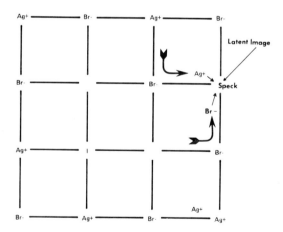

Figure 28. A schematic drawing of the same crystal after exposure. As a result of crystal ionization, silver and bromine ions are formed and begin to move about in the crystal lattice toward the sensitivity speck.

position in the crystal lattice as well. These additional bromine ions eventually accumulate at the sensitivity specks located throughout the silver bromide crystal. The positive ions are also released in the ionization process. Once the sensitivity speck traps a negative bromine ion, it acquires a negative charge and the more slowly moving positive silver ions are attracted to it. When the positive silver ions begin to accumulate at the sensitivity speck, a chemical change takes place resulting in the production of a piece of black metallic silver. The enlarged sensitivity speck once again has a neutral charge but soon begins to impede the flow of other negative bromine charges. It soon takes on the negative charge and begins to attract the positive silver ions, producing a larger piece of black metallic silver. This cycle continues until all of the negative ions in the crystal have been attracted to the sensitivity speck or until the film is placed in the fixer.

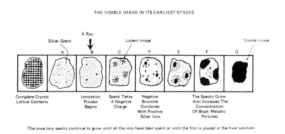

Figure 29.

The latent image is, in summary, formed by a process of ionization between silver bromide crystals of the emulsion. The once stable crystal lattice picks up energy from the incident photons as described and produces black metallic silver at the site of the sensitivity speck. A few last points should be made regarding the sensitivity speck before closing this discussion. (1) Without this chemical defect (sensitivity speck), the process of latent image formation could not occur. (2) There is usually more than one sensitivity speck in each crystal. (3) The number of photons striking the silver bromide crystal determines how many of the ions will be released and thus controls the extent to which the speck can grow. (4) The greater the accumulation of black metallic silver in the film's emulsion, the greater will be the film's overall radiographic density (see Fig. 30).

We must now direct our attention to transforming the latent image into a visible image, and this is accomplished by the reducing agents of the developer solution. Reducing agents, by definition, are substances that are able to give up electrons to another substance very easily. Actually, the overall effect of the developing solution is to simply act as a catalyst to the activities described above. It should be pointed out that the basic differences between latent image and the visible image is merely the size to which the sensitivity speck has grown as the result of the ionization process.

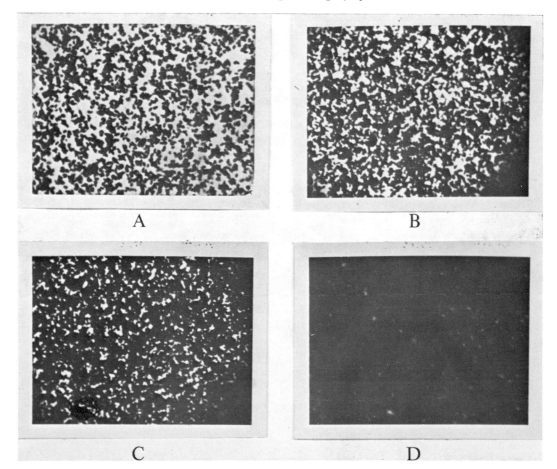

Figure 30. Four radiographs were exposed to various densities (A at 0.42, B at 0.65, C at 0.98, and D at 1.56), and tiny sections were magnified × 800 and photographed, in order to demonstrate that radiographic density is caused by the accumulation of black metallic silver grains in the emulsion. *Courtesy of* E. I. DuPont.

During development, the sensitivity speck becomes much larger and does so very quickly. For example, in a ninety-second processor, all of the silver and bromine ions are attracted to the sensitivity speck in twenty seconds. It is interesting to point out that a film could produce a visible image without the use of developer reducing agents—however, this would take an extremely long period of time, perhaps thousands of years. Thus, the developer only acts as a catalyst to the *natural activity* of the silver bromide crystal after the exposure.

Basically, what occurs is that the developer solution has an abundance of bromine ions to repeatedly give the sensitivity speck a negative charge, thereby attracting more quickly all of the free silver ions that are available. The sensitivity speck also acts as a doorway through which the reducing agents may enter with their abundance of negative bromine ions, as noted in Figure 31.

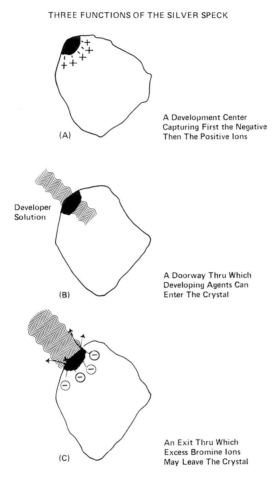

THREE FUNCTIONS OF THE SILVER SPECK

(A) A Development Center Capturing First the Negative Then The Positive Ions

Developer Solution

(B) A Doorway Thru Which Developing Agents Can Enter The Crystal

(C) An Exit Thru Which Excess Bromine Ions May Leave The Crystal

Figure 31. Without the presence of at least one sensitivity speck in a crystal, no latent image would be possible. No negative bromine ions would be trapped as a result of an exposure.

DEVELOPER AND FIXER SOLUTIONS

A textbook can be written on just the material involving the latent image, film manufacturing, and processing; the level of chemistry one needs to understand such a text is beyond the scope of our discussion. There are some general points that should be made to give a little insight into the composition of the x-ray processing solutions. Each manufacturer of developing solution has a slight variation in the chemical composition. However, the basic generic agents of developing solutions and fixing solutions are similar (see Fig. 32). The major difference in the composition of x-ray processing solutions is determined by whether or not they will be used manually or in automatic processors. Hand processing solutions are slightly different than those of automatic processing. The chemicals used in automatic processing have additional additives which prevent chemical fog, which is very likely to result from high processing temperatures. Additives that will aide in film transport are also included in the developing solution designed for automatic processing. This will be covered in more detail later.

Figure 32

GENERAL FUNCTION	CHEMICAL	SPECIAL FUNCTION	
	developer		
Reducing Agents	Phenidone	Quickly builds gray tones	Converts exposed grains to black metallic silver.
	Hydroquinone	Slowly builds black tones for contrast	
Activator	Sodium carbonate	Swells and softens the emulsion so that the reducing agents may work more effectively.	
Hardener	Glutaraldehyde	Controls emulsion swelling to allow better transport through automatic processor.	
Restrainer	Potassium bromide	Helps prevent reducing agents from causing chemical fog.	
Preservative	Sodium sulfite	Prevents rapid oxidation of the developing agents.	
Solvent	Water	Liquid for dissolving chemicals.	
	fixer		
Fixing	Ammonium thiosulfate	Clears away unexposed silver bromide crystals.	
Acidifier	Acetic or sulfuric acid	Stops and neutralizes developer activity.	
Hardener	Aluminum chloride or sulfide	Shrinks and hardens the emulsion.	
Preservative	Sodium sulfite	Maintains chemical balance of fixer.	
Solvent	Water	Liquid for dissolving chemicals.	

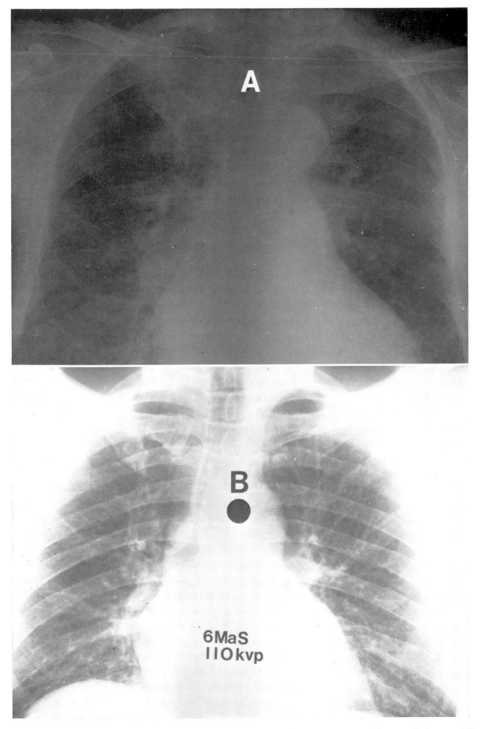

Figure 33. *A* shows decreased visibility of detail due to extreme chemical fog, while *B* shows a properly processed film.

Chemical Fog

Figure 33 shows the effect of severe chemical fog radiographically. Chemical fog, as you can see, increases density and reduces contrast substantially. Chemical fog results when reducing agents *invade* the *unexposed* silver bromide crystals. These crystals would ordinarily not be reduced and would be washed away from the surface of the film. The radiographic image, of course, depends solely on differences between one tone and another, which means that the white or lighter areas of the image are as important to maintain as are the black. If reducing agents invade the unexposed crystal (which should appear white on the film) the white areas would begin to diminish and darken slightly as a certain amount of black metallic silver is produced. Thus a buildup of density in these *inappropriate areas* gives rise to an overall grey flat look. Chemical fog is caused by reduction (see Fig. 34) of unexposed silver bromide crystals causing a decrease in radiographic contrast and an increase in radiographic density.

There are only a few things that can commonly cause reducing agents to attack and reduce unexposed crystals; they are (1) too high a developing solution temperature; (2) using a highly concentrated solution; (3) too long a developing time, and (4) contamination of solutions.

Figure 34.

COMMON CAUSES OF CHEMICAL FOG

1. ↟ Developing Temperature
2. ↟ Developing Time
3.   Chemical Contamination
4. ↟ Replenishment
5.   Chemical Imbalance;
   —Weak Restrainer

Note: Chemical fog occurs when *unexposed* silver bromide grains are converted to black metallic silver.

The fixing solution has, by comparison, a rather sedentary role in film processing, and its chemical balance in general is not as delicate as that of the developer. However, a certain degree of control must be maintained for good processing. The basic purposes of the fixer are that it dissolve the unexposed crystals from the film's emulsion and harden the emulsion so that the film can be put into storage. In summary, it is the fixer that clears the film so the light areas of the film can be produced. It should be pointed out, further, that the fixer will dissolve the black metallic silver from the film, causing a bleaching effect *if* fixing time is extended for several hours. Thus, the fixer solution can reverse the effect of the reducing agent. Under normal conditions, once the film is placed in the fixer all developing activity ceases.

CHAPTER THREE

AUTOMATIC PROCESSING

INTRODUCTION

ONE OF THE most important pieces of equipment in a radiology department is a reliable processor. Over the past ten years automatic processing has come a long way in design and dependability. Proof of this is in the total dependency all departments have placed over recent years in automatic processing units. In this chapter, the fundamental design and its functioning systems will be discussed with some emphasis on troubleshooting, malfunctions, and breakdowns. It has been the author's opinion that overall radiographic quality is not possible until there is at least a general working knowledge regarding a processor's component parts.

Figure 35 is a photograph of a conventional darkroom that was used up until the late 1950s. With some imagination one can begin to see some of the more obvious problems common to the manual method of x-ray

Figure 35. *Courtesy of* Eastman Kodak Company.

film processing. Each film had to be individually racked on metal frames and placed in various solution tanks. This became a problem when large quantities of film had to be handled at the same time. The "racked" films were placed in the developing tank for three to five minutes, then into the rinse and fixer tanks. In the meantime, the darkroom attendant's hands were wet and, before another batch of film could be put into the developing tank, had to be dried. Manual timers had to be set for each developing, fixing, and washing phase, and sometimes two different batches of films were placed in the same tank at different times. The darkroom attendant was often literally in solutions up to his elbows. In addition to all this, it was impossible for anyone to view the film for quality until at least eight to ten minutes had elapsed after the film was sent to the darkroom for processing. Needless to say, replenishing the solutions was a messy affair; in addition to all the other duties, the darkroom attendant had to maintain a fairly accurate log relative to the number of films processed so that he would know when the solutions needed to be replenished with fresh solution.

As a result of all this film handling and general activity, a lack of control occurred; coupled with long periods of time wasted waiting for film to progress through the various solution tanks, there was much interest in replacing this method with something more manageable and quicker. Thus the strong interest in automatic processing was to (1) increase department efficiency by making the finished radiographs available more quickly (hopefully to the extent where it would be practical for a technologist to keep a patient on the table until the film was viewed); (2) reduce the mess of wet darkroom floors and walls, a result of open developer fixer tanks evaporating into the room air; (3) reduce artifacts by limiting the number of times the film's delicate surface had to be handled; and (4) help stabilize radiograph quality.

Centralized Versus Dispersal Processing

Indeed the use of automatic processors has proliferated over the past fifteen years to the point where they are now taken for granted. Depending on the configuration of the department, automatic processors can be used in one of two basic ways: centralized and dispersal. Centralized processing is when one or two processors are centrally located in the department. The result of this is that very often technologists have to transport cassettes long distances to the darkroom. Also, a large backlog of cassettes to be processed is very common, resulting in long waiting times. The dispersal processing system is utilized when several processors are strategically located throughout the various sections of the department to serve just a few rooms. The result of this is reduced transport of the exposed cassettes through the halls, and because the processor is responsible for fewer radiograph rooms there is less waiting. This arrangement is generally considered to be more efficient.

Early Automatic Processors

The first automatic processor was installed about 1959. It measured approximately ten feet long, five feet high, and thirty inches wide. Figure 36 shows the relative size of the first Kodak M processor compared with the new Kodak M6A-N®. The M completely processed a film in seven minutes (dry to dry). This was less time than could be done in just the developing and fixing alone using the manual method.

With automatic processing the darkroom has taken on a new look and function. It has become much smaller and more pleasant to work in.

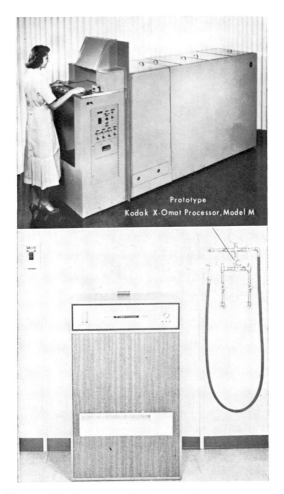

Prototype
Kodak X-Omat Processor, Model M

Figure 36. *Courtesy of* Eastman Kodak Company.

Figure 37. *Courtesy of* Eastman Kodak Company.

The prerequisites for a modern darkroom are much simpler by comparison. A floor drain is needed. A simple hot and cold water supply is connected to the now familiar mixing valve, and an appropriate arrangement of pass boxes (depending on whether the dispersal or centralized system is used) are installed. Sufficient counter space, 220 volts of electrical power, a film bin, and a good ventilation system are needed as well. There must be, of course, an ample supply of safe light fixtures. The types of safe light most commonly used are shown in Figure 37. It has a reddish lens and is known as 6-B wratten filter. It projects a reddish light to which x-ray film is not sensitive, thus causing no exposure on the final image if used properly. Other types of filters may be necessary depending on the film used. A wratten 2 filter is used for film sensitive to ultraviolet light. The term *safe light* instills a sense of false security, because if improperly used some light rays do escape to expose the film. Such an exposure to safe light is known as postexposure. Postexposure produces an unwanted increase in density and decreased radiographic contrast. A 10-watt bulb should be used and the fixture should be no closer to the film than two feet. Under these conditions, the film can be handled safely using the safe light for a period of no more than thirty seconds (or at three feet for sixty seconds). An indirect type of safe light fixture can also be mounted in the ceiling. Although more expensive, it often produces a much brighter light throughout the entire darkroom. Safe lighting should not be used at all with ceni film as well as other films with a *photographic* type of emulsion.

SOME IMPORTANT CONSIDERATIONS
FOR AUTOMATIC PROCESSING

As was pointed out earlier, manual processing was a terribly inefficient operation relative to space consumption, mess, and wasted time. Typically, a film had to be placed in the developer for an average of five minutes, rinsed for thirty seconds, fixed for ten more minutes, washed for at least thirty minutes, and dried by a variety of methods that could take approximately thirty minutes additional time. Thus the total time taken for this operation was approximately seventy-five minutes, or almost two hours. The chemicals and the film used were, of course, especially balanced and designed for such a slow system, which featured slow activity of the developing solution and relatively low solution temperatures that ranged between 68 and 75° Farenheit. The advent of automatic processing (dry to dry) systems, however, placed new demands on the developing and fixing solutions because of the new requirement for extremely short developing times and the use of developing temperatures of approximately 90 to 95° Farenheit.

Also with automatic processing, new demands had to be made of the film itself. Manufacturers had to produce a film that (1) could build a visible image very quickly; (2) would be able to be transported through a roller system without damage; and (3) had an emulsion able to resist relatively high developing temperatures without melting or accumulating chemical fog.

One of the important changes that was made in the film to be used with automatic processing was a reduction in the thickness of its emulsion. Figure 38 illustrates why this was done. Since the gelatin is not visible in the finished radiographic image, and because it inhibits the developer solution's progress to the exposed silver crystals, a reduction of its bulk made the silver bromide crystals more vulnerable to the processing solutions, and thus development could occur more quickly. Also, fewer silver bromide crystals were used in the emulsion. This might be some cause for concern since it is the silver that eventually produces the image we see, but careful calculations had been made regarding the overall quantity of silver that must be initially present in the emulsion to produce optimum results, and considerable care was taken by the manufacturers to stay well within the limits. The first important consideration regarding rapid processing is that a compatible film (thin emulsion) is used so that the developer solution can more readily penetrate through its bulk and

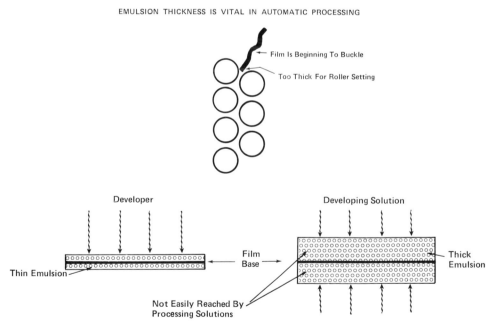

EMULSION THICKNESS IS VITAL IN AUTOMATIC PROCESSING

Film Is Beginning To Buckle

Too Thick For Roller Setting

Developer

Developing Solution

Thin Emulsion

Film Base

Thick Emulsion

Not Easily Reached By Processing Solutions

Figure 38. Emulsion thickness has not been reduced to the extent of detracting measurably from radiographic quality.

attack the exposed silver crystal. Also, of course, much care was taken in the design of the emulsion, to allow it to withstand the high developing temperatures of automatic processing so the gelatin would not liquify.

In summary, the primary changes in film composition were (1) to reduce the thickness of the emulsion (less gelatin primarily) and (2) to make the film's emulsion more resistant to higher developing temperatures. With regard to the processing solutions themselves, their composition had to be changed for several reasons: (1) a properly exposed radiographic film would otherwise be overdeveloped by the increased temperature and hyperactivity of the processing solution; (2) a more effective preservative had to be added to the solution to retard oxidation caused by high solution temperatures; and (3) hardening additives had to be added to the developer to prevent the film's emulsion from softening too much and gumming the processor's transport rollers.

MAJOR SYSTEMS IN AUTOMATIC PROCESSORS

The Transport System

In manual processing, the darkroom attendant moved film from one tank to another; with automatic processing an interesting series of rollers and gears do the work. In Figure 39, we can see a unit of rollers known as a rack. In the early M model they were so large a ceiling mounted crane was used to remove them from the processor for repair and cleaning. In Figure 40, we see the arrangement of racks and crossovers used in more modern automatic processors. The accompanying schematic illus-

Figure 39. *Courtesy of* Eastman Kodak Company.

tration shows how the units are able to transport a film from the entrance rollers to the bin of the processor.

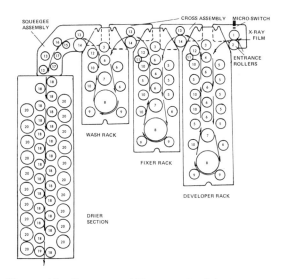

Figure 40. *Courtesy of* Eastman Kodak Company.

Thus the transport system's function is to move the film through the various processing solutions and drying tubes without damage to the film's surface. Another important characteristic of a processor's transport system is that it must move the film at a very consistent and controlled rate of speed. All other things being equal, the speed of the film moving through a processor determines the developing, fixing, washing, and drying time.

The Margin of Error

With manual processing, developing time was so long that a thirty-second error would not have had an especially damaging effect on the film's image. By comparison, with automatic processing a five-second error in developing time would cause a notable change. The automatic processor is much like a high powered racing car: It will go faster but will tolerate very little errors in fine tuning. With automatic processing, the total developing time, for example, is approximately twenty seconds. This is certainly fast and a five second error (plus or minus) can cause a 20 percent variation in density.

In Figure 41, we can get a better idea of how all the individual rollers function. A small motor turns a shaft which has "worm gears" set along its length. In turn, each rack has a complementary set of rollers that fit firmly into the worm gears of the main drive shaft. The rack also has a main drive and its gear is made to fit a chain which turns all of the smaller rollers of the rack. This general design is common to most automatic processors.

Figure 41. A—transport drive motor; B—chain drive to transport drive shaft; C—main drive shaft; D—worm gear; E—belt drive to dryer transport rollers. *Courtesy of Eastman Kodak Company.*

The Crossover Assembly

As the film begins its upward travel through the rack, there must be some way to eventually guide the film over into the next solution tank. Figure 40 shows how the transport system accomplishes this. The crossover, as you have probably already imagined, is simply a bridge that allows the film to cross from one tank into the other. You will also note a long metal plate which is located at all critical turning points. These devices, known as guide shoes, assure the film will maintain proper alignment between the next set of rollers.

As with most chemical reactions, some type of by-product is produced. In the case of processing x-ray film, *reaction particles* form on the film's surface as it is immersed in the developing solution; these are basically bromide and gelatin deposits. These reaction particles form tiny clumps on the surface of the film and, when allowed to accumulate in large numbers, act as a barrier to the developing solution (see Fig. 42). If a film had a coating of reaction particles on its surface, the developing solution would experience great difficulty in reaching the exposed silver bromide crystals in the film's emulsion and underdevelopment would occur. However, slight agitation of the film is sufficient to shake these reaction particles loose from the film. This agitation is accomplished by arranging the transport rollers in such a way as to cause a moderate bending of the film, as shown in Figure 44B.

In short, the transport system is an arrangement of racks placed in

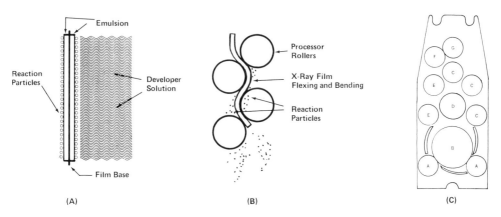

Figure 42. *A* shows an enlargement of film layers; *B* shows how the film is flexed back and forth to loosen the reaction particles; *C* shows the profile of an entire rack with the actual placement of various rollers.

each solution tank, crossovers, an assortment of gears, and an electric motor. It has two functions: (1) to move the film through each solution tank without causing artifacts and (2) to prevent reaction particles from accumulating on the film.

Artifacts Commonly Linked to Transport Problems

The most common artifact linked to transport problems is scratches. The direction in which the scratches run is important because it can help pinpoint the problem. Most, if not all, scratches caused by transport problems run the direction the film traveled through the processor. A dark scratch is common when the film is abrased after the exposure but before developing (see Fig. 43A). A light scratch usually occurs before exposure or the fixing or wash cycle (see Fig. 43B). If scratches (see Fig. 44) are noted approximately one and one-half inches apart running in the direction of film travel, they are usually caused by guide shoes that have come out of adjustment. Pi lines are white lines that run *across* the direction of film travel; they are frequently formed by new *processors* but this problem ceases shortly after installation.

Another type of artifact that causes plus density markings is known as pressure marks. It is important to know that x-ray film is very sensitive to pressure, especially after exposure. Any type of extraneous pressure against the emulsion, epecially during the processing, will cause the silver bromide crystals to overreact to the reducing agents. The result of this is an increase in density at the point of the pressure. Figure 45 illustrates this problem. With this in mind, one can imagine what might happen if the surface of the rollers is not smooth. Small but disturbing pressure (density) marks can be seen on the finished radiograph; they appear as faint, dark blotches. Keeping the rollers clean by scrubbing the crossovers daily and the deep racks weekly will help to reduce the possibility of pressure marks.

The Replenishment System

If a large number of paper sheets were continually dipped into a tank of colored water and the paper was so treated that it would have a neutralizing effect on the strength of the solution, the solution would eventually lose its color and potency. A similar situation occurs in the developing and fixing solution in automatic processing. As you know, the developer solution has a strong alkaline content, and the fixer has a strong acid content. As a steady stream of film is transported through these solutions in the automatic processor, the film's emulsion absorbs a certain amount of this liquid. Along with this simple absorption, an additional chemical reaction takes place in the film's emulsion, and soon the developing and fixing solutions tend to break down in strength. If this situation were allowed to continue, the resulting radiographic image would exhibit a marked decrease in contrast and density: The solution would simply become "exhausted." Figure 46 shows a radiograph process under this exhausted solution condition.

Built into all automatic processors is a replenishment system; its sole function is to maintain a predetermined level of chemical strength (activity) in the developer and fixer (Fig. 47 schematically shows its more important component parts). When a film is pushed between the two entrance rollers, they separate slightly to set a pair of delicate microswitches; these switches close an electrical circuit which energizes the replenishment motor which drives two pumps (one for the developer and one for the fixer solutions). As illustrated, each pump is connected to its perspective solution tank by plastic tubing. The fresh replenishment solution flows through the tubing past a flow gauge of the processor and into the appropriate tank. As you would expect, the flow of solution continues only while the entrance rollers engage the microswitches. With this system operating properly, the processor can then automatically compensate for the lost solution spent in developing each film. This, of course, stabilizes the chemical activity and strength of the developer and fixer solutions and prevents the possibility that they would become weakened and eventually exhausted. It is possible for a processor to pump too little or too much solution into the developer and fixing tanks, however.

Overreplenishment

Overreplenishment occurs when too much developer or fixing solution is pumped into the processing tank, thereby increasing the strength and activity of the solutions in excess of the prescribed limits. Typically, what

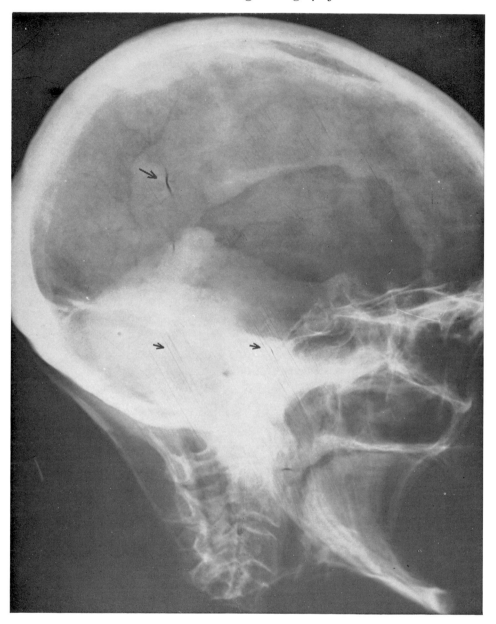

happens in this situation is that the developer solution becomes so active it indiscriminately begins to develop the unexposed silver bromide crystals in the film's emulsion. Thus, the unexposed silver bromide crystals are in a sense "forced fed" with reducing agents and begin to take on some density. Figure 48 illustrates how this would effect the image: the radiograph was properly exposed but processed in a developer solution that had been ovrreplenished. There is a sharp increase in density and a notable decrease in contrast. The film is too dense because the film was developed

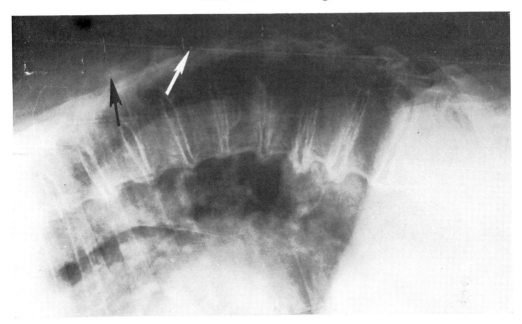

← Figure 43. In *A*, along with the black scratches, note the black "crescent" mark caused by the film being bent over a sharp edge such as a fingernail. *B* shows scratch lines that appear white, depending on when the scratch occurred.

Figure 44. As this unexposed film demonstrates, scratches may be caused by poorly adjusted guide shoes.

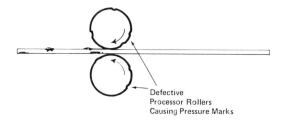

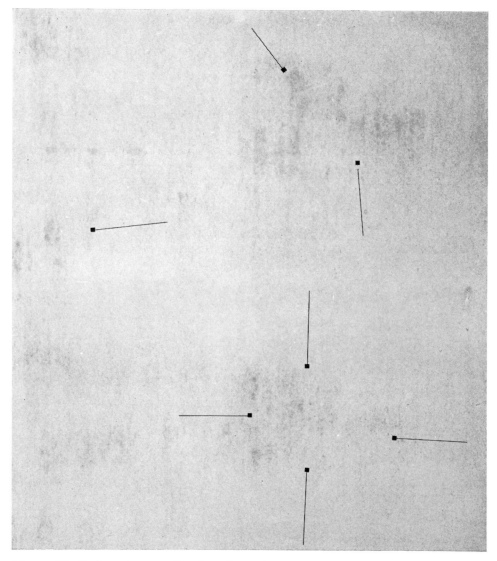

Figure 45. It is quite possible that these densities would appear more subtle than those shown here. The extent to which this occurs is dependent upon the degree of pressure by the rollers and in some cases film type. Certain film types are more sensitive to pressure.

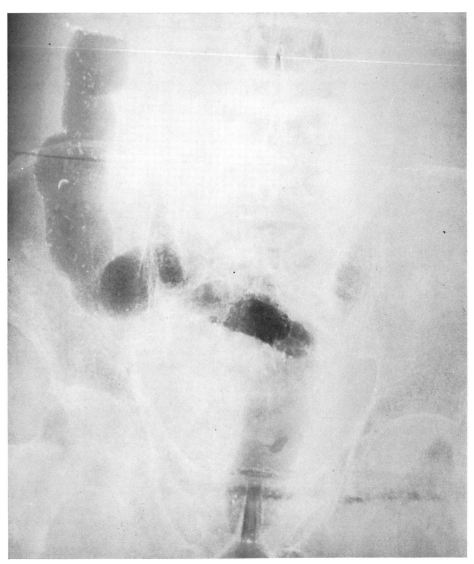

Figure 46. As a result of decreased density and contrast, visibility is poor.

beyond the correct level, and the contrast is reduced because the unexposed crystals (which should represent the white areas) have taken on
some density. Therefore the overall relationship between the black and
white areas of the film is diminished. The most common causes for overreplenishment are (1) the flow rate is too high and too much solution
from the replenishment tank; (2) the microswitches are set too closely
and spontaneously engage the motor of the replenishing pump.

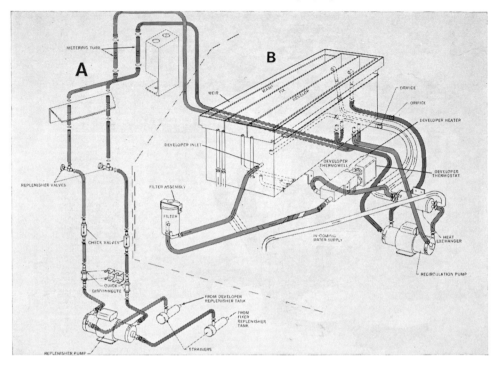

Figure 47. A composite of the recirculating system and the replenishment system. *Courtesy of* Eastman Kodak.

Underreplenishment

The result of an inactive or exhausted developer solution is decreased density and reduced contrast. Also, because the hardener additives in the developing solution are weak, the film's emulsion becomes soft and sticky, making transport difficult. With underreplenishment, not enough fresh solution is drawn into the processing tanks. Figure 46 shows a properly exposed radiograph processed in an exhausted developer solution. The strength of the developer chemicals are below the prescribed level to the extent that not enough silver bromide crystals have been reduced to black metallic silver particles, resulting in a marked decrease in density and contrast.

There are several causes for underreplenishment. One is that the microswitches were "set" so loosely that they cannot close the circuit, even as the film passes through the entrance rollers. Secondly, the diaphragms on the pumps might be defective and unable to draw solution from the replenishment tanks. Third, the plastic tubing through which the solution flows to the processor from the large replenishment floor tanks could be pinched or blocked in some way. Fourth, an air pocket may have formed

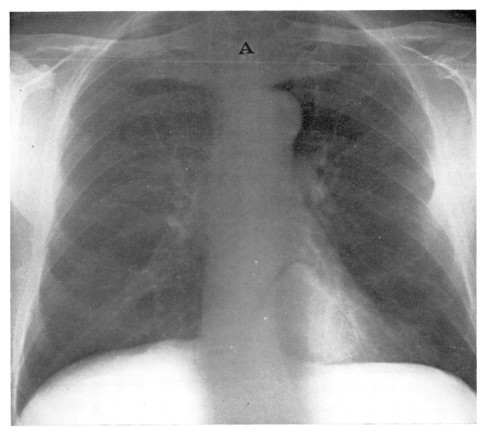

Figure 48. As a result of hyperactivity, developer reducing agents attack the unexposed silver bromide crystals and density begins to increase in what should be white or light areas of the film. Contrast decreases, producing an overall gray dense radiograph. A similar image can be obtained from excess scatter (S/S) reaching the film.

in the plastic tubing making it difficult for the pumps to draw the solution into the processing tank. Sometimes a pinhole, crack, or a poorly sealed connector in the tubing will allow such an air pocket to form.

The flow gauge can be used by the technologist to check what amount of solution is being pumped into the processor for each film put through the entrance rollers. Besides this, however, problems in transport or drying could be a clue to improper replenishment. First, it should be recalled that there is some hardener in the developer solution and certainly a great deal in the fixer solution. However, if replenishment is poor the hardener in the developer solution might be too weak and thus allow the emulsion to swell too much and become sticky and cause a jam. Also one might be suspicious of underreplenishment if an increase in abrasion marks are noted on the film.

There are many indicators to under– and overreplenishment. However, an alert technologist could prevent many of these problems by periodically looking at the processor's flow meters to see if the replenishment being pumped into the processor is the proper amount. Figure 49 shows the correct method for feeding film.

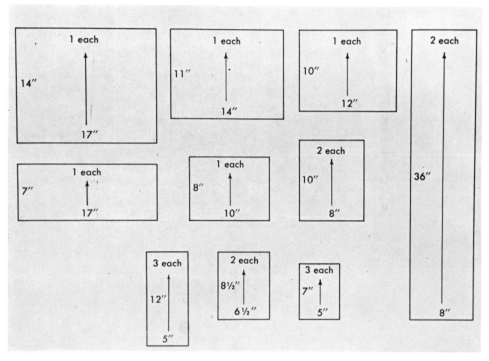

Figure 49. The correct method for feeding film into an automatic processor. If this is not done correctly, over development could result. *Courtesy of* Eastman Kodak Company.

Setting Replenishment Rates

The developer and fixer pumps can be adjusted to draw a specific amount of fluid into their respective processing tanks. Under normal working conditions, replenishment pumps work while the film is passing through the entrance rollers. If the flow valve is adjusted toward an open position, more solution would be carried into the processing tank and vice versa. Figure 50 shows such a flow gauge installed in a Kodak M6A-N Processor. The higher the metal balls rise with the solution during replenishment, the more solution is being pumped into the tanks. Figure 51 was supplied by the Eastman Kodak Company and outlines the correct flow rate for various conditions. You will note that the number of films one expects to process per day determines the prescribed flow rate.

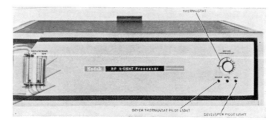

Figure 50. *Courtesy of* Eastman Kodak Company.

As the number of films processed per day increases, the replenishment rate decreases. Although this might seem contrary to logic, when solutions stand idle in a processor oxidation increases and breaks down and weakens the solutions. Thus with low volume processing, the tank must be slightly "overcharged" with fresh solution to compensate for the increased oxidation that takes place.

Figure 51. Correct replenishment rates. *Courtesy of* Eastman Kodak Company.

Average Number of Films Processed During 24 Hours of Processor Operation	Recommended Rates of Replenishment per 14-inch Length of Film Travel	
	Developer	Fixer
24	95 ml	135 ml
25–50	90 ml	135 ml
51–75	85 ml	135 ml
76–100	75 ml	135 ml
101–125	70 ml	120 ml
126–150	65 ml	120 ml
over 150	60 ml	100 ml

The Drying System

The drying system is designed to deliver hot air to the surface of the film after it has been adequately washed. It should be pointed out that the film is not considered totally processed until it has been dried. The drying system's major components consist of a heating element, a blower, a thermostat, and drying tubes. Figure 52 shows where these parts are located in the processor. The thermostat consistently monitors the air temperature being blown through the plastic tubing of the dryer section. If the air temperature goes below a prescribed level, the thermostat will close an electrical circuit, and the heater will automatically bring the air temperature to the required level.

Normal drying temperature is approximately 120 degrees Fahrenheit. Usually, there is little need to vary this range except possibly when a long run of roll type film is being processed when the temperature might be increased to 150 degrees. Drying temperature is certainly influenced by atmospheric humidity.

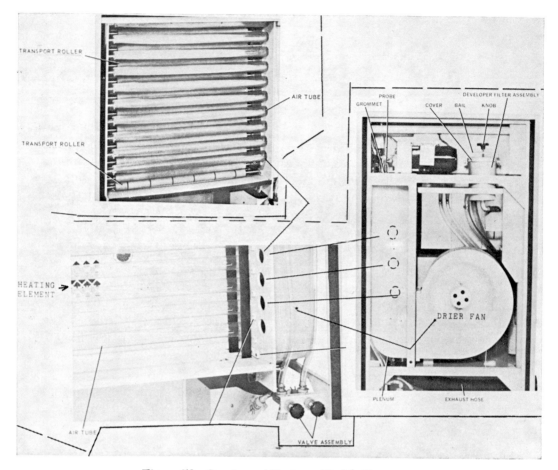

Figure 52. *Courtesy of* Eastman Kodak Company.

Drying Problems

Drying artifacts are not uncommon and are often indicators of problems that exist not in the drying system itself but rather with the strength of the developer and fixing solutions. If the replenishment system is under-replenishing the developer and fixer solutions, the film's emulsion might be too soft and unable to dry quickly enough in the dryer. Also, inadequate washing will often result in a greasy, shiny-looking film.

The Recirculating System

The recirculating system is possibly the most involved of the four major systems in the automatic processor. It has three basic functions:

(1) to filter out reaction particles and other chemical impurities in the developer solution and thereby help maintain its strength; (2) to agitate the developer solution so that chemical components will be equally dispersed throughout the processing tank; (3) to maintain and stabilize the temperature of the developing solution. The reader should recall the concept of time-temperature processing. It has also been pointed out that automatic processors do not allow a wide latitude for malfunctions before a deleterious effect on radiograph quality results and that maintaining temperature is a very important aspect of good quality processing. Figure 53 shows three radiographs that were equally exposed but processed at different developing temperatures. You will note the obvious density and contrast variation that ultimately result. For any given developer solution, as the temperature increases a condition known as chemical fog sets in. Chemical fog is the result of hyperactive solution or prolonged developing time. Chemical fog causes an increase in radiographic density and a notable decrease in radiographic contrast.

In Figure 47B, a schematic of the recirculating system is depicted. As the reaction particles fall from the film, a sludge accumulates in the developing tank. These reaction particles are eventually filtered out by a filter which is located in the recirculating system. The reader will also note the circulation pattern of both the fixer and developer solutions. The developer solution temperature is held constant by balancing it with the incoming cool water and the developer heater located in the heat exchanger. Optimal developing temperature is considered to be from 92 to 95° F. A developer thermostat (for the developer solution) is always monitoring the solution temperature as noted in the illustration. If the solution goes below a predetermined level, the thermostat will close an electrical circuit and the heater will begin to warm the solution to the desired level; once obtained, the thermostat will open the circuit and the heater will turn off until it is signaled by the thermostat once again that the solution temperature is too low. A small red light will brighten in the front of the processor panel when the developer heater is working. The developing solution passes through a device known as a heat exchanger, which is simply a heat conduction device. If, however, the developing temperature begins to rise too high, the heat exchanger could be used to lower the temperature to the desired level if the incoming water would be deliberately lowered by the technologist to perhaps 65° or 70°. As the cold water begins to pass through the heat exchanger, the metal casing cools down and the neighboring metal casing through which the developer solution flows would also begin to cool. In this manner the developing solution is refrigerated and could drop from 110° to 94° in approximately twenty minutes.

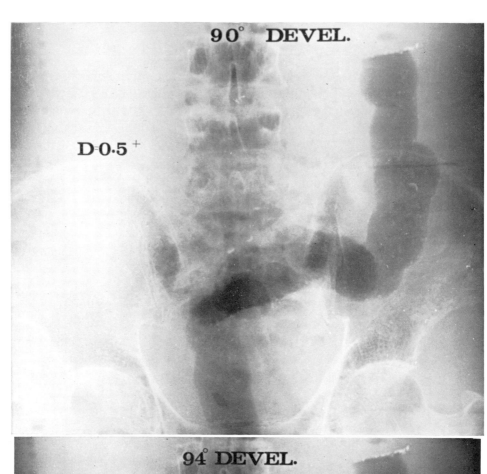

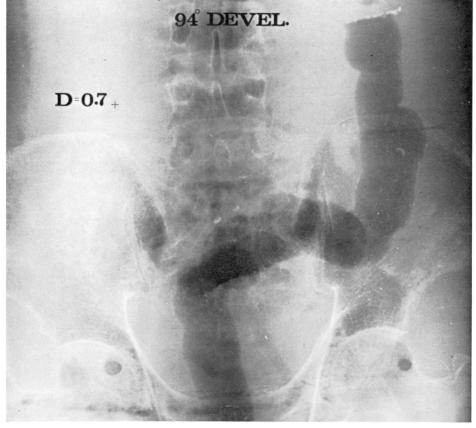

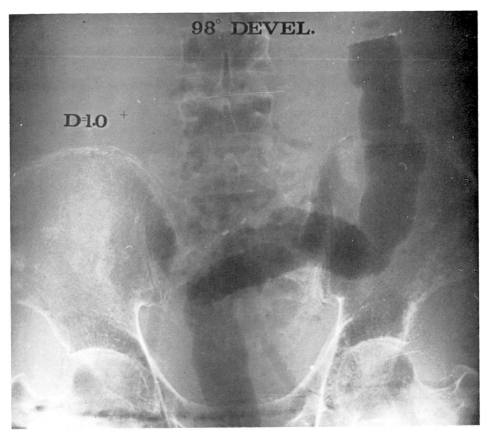

← Figure 53. Developer solution temperatures play an important role in obtaining maximum radiographic quality. As the temperature increases beyond its correct level, density increases and contrast decreases. Note the variations in film density ranging from 0.5^D to 1.0^D.

In short, the developer solution thermostat monitors the developer temperature and if necessary signals the heater to warm up the solutions. The cool incoming water is used to reduce the developer temperature if necessary and a small filter traps impurities so they will not accumulate in the developing solution and thus break down the strength (activity) of the solution. In general it is good practice to keep the incoming water temperature slightly lower than the desired developer temperature level. For example, the water temperature might be held at 86 to 88° and the developer thermostat set at 92°.

TROUBLESHOOTING PROCESSOR PROBLEMS

A great deal of effort has been expended in seminars dedicated to the sole purpose of uncovering some of the typical problems that can arise with automatic processing. The intent of this section is simply to point out some day-to-day problems and offer some suggestions that might help in finding the cause. Many have already been discussed. In the author's own experience, the following lists the most common problems, in order of frequency, and their corrections.

Trouble	Probable Cause	Correction
Dark films	Incoming water too hot	Adjust mixing valve
	Developing thermostat out of adjustment	Adjust downward
	Overreplenishing developer	Adjust downward, check microswitches
	Contamination (might have an ammonia smell)	Drain contaminated tank, clean thoroughly, and use fresh chemicals
	Light leak in darkroom	Repair
Light films, poor contrast	Poor replenishment	Adjust microswitches
		Repair leaking pump, broken pump or motor
		Check for air in lines
		Check for pinched lines
	Developer temperature	Adjust incoming water temperature
		Adjust thermostat
		Adjust defective thermostat
Films have a brownish look	Inadequate developer, replenisher	Adjust replenishment
Films have a milky appearance	Inadequate fixer replenishment	Adjust fixer flow rate
Films have a greasy appearance, often accompanied by a high contrast	Inadequate water supply	Turn on water or increase flow rate
	Wash tank might be empty	Fill tank
Jamming	Transport rollers not turning properly	Check gears and drive mechanism
	Poor replenishing rates	Adjust system
	Guide shoes out of adjustment	Adjust guide shoe
	Feed tray not properly aligned	Align
Scratches	Guide shoes not adjusted	Adjust shoes
	Raised or roughened areas of a transport roller	Replace roller
	Nicks in feed tray	Replace
Black, flaky marks	Accumulation of algae	Clean tank and roller
		Replace filters in developer and water
Increased fog	Old film	Eliminate
	Contamination of developer	Replace solutions
	Increased developer temperature	Adjust
	Overreplenishment	Adjust downward
	Fogged film	Replace

Explanation

Dark Films

Dark films are usually the result of overactive developer. Developer chemical activity raises with temperature. As explained earlier, developer solution temperature is somewhat dependent on the incoming water temperature via the heat exchanger. Lowering the water temperature may be the only necessary adjustment.

Occasionally, the thermostat itself will malfunction and will unnecessarily engage the developer heater. Adjusting the thermostat down will often correct the problem.

Overreplenishing could easily occur if the aforementioned microswitches go out of adjustment and engage the replenishing pumps too often. The pumps will continue to pump new solution into the tanks and eventually the reducing agents will become too strong.

Contamination of the developer tank with fixer can occur with only 3 to 4 ounces of fixer. Under these conditions, the tank and rack must be rinsed thoroughly, the old solution drained, and new solution be put in its place. When the developer solution becomes contaminated with fixer, a strong ammonia odor will result.

Light Films, Poor Contrast

Light films are usually caused by a developer solution with low chemical activity. This may be caused by simple oxidation or by developing too many films compared to the volume of fresh replenisher being pumped into the processing tank. Poor replenishment can be caused by a number of situations: (1) Microswitches not engaging the replenishment pump when film passes through the entrance rollers. A simple adjustment can correct this. (2) Faulty pump or motor. Here one might note leaking of solution from the diaphragms of the pump indicating the solution is escaping through the pump before it can reach the tanks. (3) Air in the replenishment lines. Air can be sucked into the lines through a tiny pin hole or a crack in the tubing or in a connector. Air in the lines will prevent the pump from making an adequate vacuum and little, if any, solution will move through the lines into the tank. One can usually see the air pocket in the tubing and with some imagination and manipulation of the lines it can be removed without dismantling the system in any way. The pump must be on during the attempt, however. (4) Pinched replenishment lines will prevent the solution from moving when the pump is pumping. (This is a very common situation.) To prevent this from happening, a two-inch plastic pipe may be purchased and used as a conduit through which all the replenishment lines are placed.

It is, of course, entirely possible that the developer temperature is too low. It was discussed earlier what happens to the reducing agents when the developer temperature is low. Low developer temperature could be the result of a defective developer heater. The thermostat could also be defective or simply set too low. In the latter case, an easy adjustment is all that is necessary. It is also a possibility that the incoming cold water pressure is too high and has overridden the hot, causing the developer temperature to drop via the heat exchanger.

Films Have a Brownish Look

Oxidation of the developer is usually responsible when a low density, low contrast, brownish image results on the film. Oxidation is when air mixes with solutions and eventually breaks down their potency.

To correct this problem, the developer tank should be drained, cleaned, and refilled with freshly mixed solutions. In addition, the replenishment system should be checked and adjusted so the problem will not recur. Often the problem involves a low volume processor and the most likely correction would be a very high flow rate setting (see Figure 51).

Films Have a Milky Appearance

It should be kept in mind that although the same electric motor is used to drive the replenishment pumps, there are two distinct replenishment lines and it is very possible the developer could be working well and the fixer poorly. Basically the same regimen described in the section on light films should be followed, placing attention on the fixing side.

Films Have a Greasy Appearance

The processed film must be washed thoroughly of all the developer and fixer solutions. The problem is usually very simply corrected by opening the water flow valve to attain a higher flow rate.

Jamming

If the reader can remember the drive mechanism, it will be recalled that each roller responsible for moving the film has a small set of gears attached. The gears of each roller mesh snuggly under normal conditions. With this arrangement, all the rollers turn at the same rate; however, if one of the nylon gears is stripped, the roller pauses. If the film reaches that roller while it is slipping, it could cause enough of a delay in transport to cause a jam.

On the other hand, the film's emulsion might become too sticky, which would cause the film to adhere to the roller. This condition would cause a temporary delay in transport sufficient to cause a jam as the following film catches up with the first. Occasionally, one can find a film completely wrapped around one of these rollers. Such a condition could be corrected by examining the replenishment system carefully and making the necessary adjustment. You will recall with poor replenishment the hardener in the developer and fixer solutions would not be very effective and the film would become too soft and sticky.

Finally the guide shoes must be set so that the film will pass between the rollers at crucial points along its route. This is a difficult problem to correct because it is important to find the one guide shoe that is out of adjustment.

Scratches

Although the film's emulsion is firmed by the hardener, it is still sensitive to deliberate abrasions. The guide shoes, if out of adjustment, will make white scratch marks one and one-half inches apart in the direction of film travel. A simple adjustment will usually correct this problem. Often, the shoes are set to turn the film too sharply causing unnecessary friction on the film surface resulting in scratches.

Raised or roughened areas of the transport rollers cause unnecessary friction on the film's surface as it passes between the rollers. This situation causes indiscriminate areas of abrasions. Some rollers may be sanded smooth, but soon after immersion in the solution, the roller will probably become defective in the same manner; replacement is the best initial recourse. Pi lines are common in new processors and disappear after a short break-in period.

Black, Flaky Marks

City water can be extremely dirty to the point where algae will grow and actually accumulate in the wash and developer tanks and on the racks. This condition will advance to the point where black flakes of algae will float in the water tank and eventually become "pressed" onto the surface of the film by the rollers. Nothing can be done except drain the tank, wash out the racks with hot water, and thoroughly clean the wash tank. In addition to this, a new filter should be put on the incoming water lines before the mixing valve. In areas with dirty water, filters should be replaced weekly. To help obviate this problem, it is sometimes recommended that the water tanks be allowed to drain over the evening hours. Algae filters and algae sticks should also be connected to the incoming water lines.

Increased Fog

The causes listed occur with equal frequency. Old film can certainly be eliminated by a conscientious method of inventory checks. Contamination rarely happens spontaneously. It often follows a jam when film, rack, and crossovers are pulled too carelessly, causing fixer to drip into the developer tank or vice versa. Keep in mind that because the processing tanks are small, only slight amounts of contaminants are needed to cause this problem. Increased developer activity will cause a situation where unexposed silver bromide crystals are "force fed" with reducing agents. One can easily imagine that the unexposed crystals are as crucial in producing light areas of the film (which in turn make up radiographic contrast) as the exposed crystals are important in blackness in the image. Obviously if the white areas of the film become gray through improper developing, optimal differences between the white and dark tones would be diminished, and reduced detail would result.

In summary, processing problems are often caused by poor quality checks and maintenance programs. In general, automatic processors are made for long service. Basically, student technologists are not expected to troubleshoot an automatic processor, yet a working knowledge of the major systems is important. There are, of course, many more problems that could arise, but those presented above are the more common and can often be corrected by the supervising technologist or a trained darkroom attendant.

ROUTINE MAINTENANCE

Although the processor can take care of itself in many ways, there are some points that should be followed. First, the deep racks should be cleaned with an abrasive material like Scotch Brite® weekly. The crossovers should be similarly cleaned daily to prevent reaction particles from excessive accumulation. The developer filter should be replaced monthly or every 5,000 films, and, if the water supply is dirty, the hot and cold water filters should be replaced weekly as well. The replenishment should be checked throughout the day. Some processors have a very convenient method of checking rates as was noted earlier regarding the Kodak M6A-N Processor. It was mentioned that the water tank should be drained each night to prevent the accumulation of algae. Most hospitals have a solution company that performs a monthly maintenance check. With this inspection, rollers, gears, guide shoes, lines, pumps, belts, etc. are checked for wear. Any part showing more than a moderate degree of wear should be replaced.

Sensitimetric Strips

A method of checking consistency of developer activity should be devised, since this directly relates to the appearance of the radiographic image. The author uses a method that involves the use of a sensitometer made by E. I. DuPont which sensitizes (exposes) an x-ray film to varying intensities of light so that a gray scale will result after processing. After the film is exposed with the sensitometer, it is developed and the developing temperature is recorded on the film. A densitometer is then used to read the densities on the gray scale at two predetermined points. The densitometer reads the tones of the densities and they are recorded.

The base fog is also read and under normal conditions the reading should not go beyond a density of 0.2. A base fog reading above this accepted level indicates a potential problem. Usually, however, it is a result of the developer temperature (which might be caused by an elevated incoming water temperature) or simply too much fresh replenishment solution being pumped into the processor's developing tank. It should be pointed out that an elevated base fog reading will certainly indicate the presence of chemical fog and possible malfunction.

CHAPTER FOUR

INTENSIFYING SCREENS

ONE CAN EASILY lose sight of the importance intensifying screens have on radiograph quality. In early radiography, screens, as we know them today, were simply not available. As a result of this, and because the type of film that had been used was much less sensitive to x-rays, exposure times were unbelievably long. An abdomen exposure would require several minutes and the exposure time for a hand or a wrist examination was perhaps thirty to sixty seconds. As demands for higher quality images were made during the early 1900s, methods of making faster exposure systems (film-screen combinations) were being tested. Also, the early pioneers of x-ray technology had to cope with x-ray generators that were not able to produce very penetrating rays or adequately high milliamperage values. It was clearly understood, however, that even with faster and more powerful x-ray equipment, other avenues had to be explored that would provide a faster exposure system. These early pioneers soon realized that Roentgen's discovery of x-rays was made with help of luminescent phosphors. Phosphors are substances that can give off light when they absorb certain electromagnetic wavelengths such as visible light, ultraviolet light, and x-rays. The color of light given off varies with the particular phosphor being excited. But generally the colors emitted by the phosphors used in radiography are green and blue and ultraviolet. Further experiments with these phosphors showed that perhaps luminescence (the process of phosphors giving off light) could be put to good use especially because the phosphor materials are more sensitive to the x-ray photons than to the film itself. This is because certain phosphors can absorb (capture) more radiation from a given exposure than x-ray film, and so phosphors will yield a greater response to an equal exposure. Figure 54 illustrates this point. With 500 x-ray photons, approximately 150 latent images will be produced compared to only five latent images when directly exposed. Thus, the phosphors will produce an effect that is thirty times greater for conventional fast screens (sixty for rare earth screens).

The color of light these phosphors produce for radiography is usually

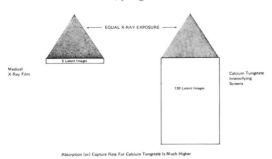

Figure 54. Exposure time, patient motion, and patient dose are decreased because of calcium tungstate's ability to absorb x-radiation as compared to the x-ray film. The actual ratio of "latent images" formed between a screen and nonscreen exposure is dependent on the speed of the screen being tested.

blue. Screen type x-ray film is, therefore, manufactured to be blue-sensitive, which is to say that a blue-sensitive film will respond better to blue light than it will to any other color. If the type of screen used gives off ultraviolet or another type of emission, one must be careful to use a compatible film, one that will respond well to that particular frequency. Thus the color of screen light used must be matched to the sensitivity of the particular type of film being used.

Another fact that inspired the use of screens is the deleterious effects of radiation to human tissue. Thus screens had another advantage, that of reduced exposure to patients. With the factor of reducing exposure time, and the added advantage of decreasing patient exposure, research on intensifying screens continued in earnest. We will see later to what point these advantages extend when modern intensifying screens are used. In summary, one might say the purpose of using intensifying screens is to reduce patient exposure and patient motion through more efficient absorption of x-rays. Screen radiography is possible because manufacturers can produce x-ray film that is more sensitive to the screen's emitted light than to the remnant x-ray photons. There is one important disadvantage with using screens, however, and although it will be discussed in some detail later, it must be noted at this point that screens greatly decrease radiographic sharpness.

GENERAL CONSTRUCTION

The gross anatomy of an intensifying screen is quite easy to comprehend. Figure 55 illustrates a cross-sectional view of a typical intensifying screen.

The Backing

The purpose of the backing is to act as a rigid support for the phosphor materials spread upon it. Originally, the backing was made of a card-

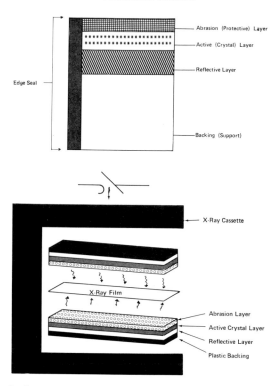

Figure 55. Although there is some variation between manufacturers, *A* is representative of most screens. *B* depicts a pair of screens positioned in a cassette.

board of high quality, but today a plastic material has become popular among some manufacturers and technologists. It is very important that screens and film surfaces make absolute contact or image sharpness will be substantially reduced. In order to assure good contact, the screens must be perfectly smooth and flat. With this point in mind, cardboard could be a disadvantage because as atmospheric humidity increases and decreases, a cardboard backing could expand and then contract. These changes would reduce film screen contact and decreased image sharpness can result. In contrast to this, plastic backing is not affected by moisture in the air and thus more consistent contact is likely to result.

The Reflective Layer

The reflective layer is the next component in the construction of a typical intensifying screen. As shown in Figure 56, its function is to reflect or bounce back light that has been produced by the phosphors. This has a direct effect on how many light rays will strike the film and be absorbed into the film's emulsion. Obviously the more light rays that are absorbed by the film, the more black silver crystals will be present in the visible image which would produce darker radiographic images.

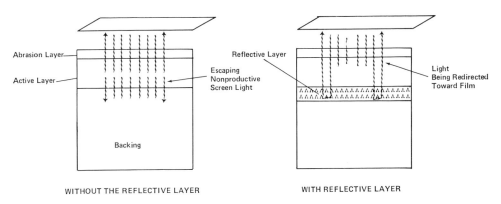

Figure 56. The reflective layer is one of the primary factors controlling overall screen speed—the efficiency with which it can bounce escaping screen light back to the film.

You have already come to realize that it is the light from the intensifying screen that exposes the film and not the x-ray photons themselves. In fact, depending on the screens, only 5 percent or less of the density seen is caused by x-rays directly exposing the film. Figure 57 illustrates the concept that light is emitted by the photons in all directions similar to a lightbulb; if the stray light rays (green or blue) can be captured and redirected back, the film will experience a greater exposure. This is, however, accompanied by decreased sharpness. One can correctly say that, as

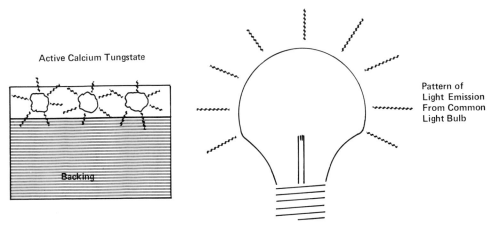

Figure 57. The root cause of screen unsharpness is the pattern of emission of the crystal. If the light from each crystal would travel in vertical direct pattern toward the film, screen unsharpness would be greatly improved.

the reflective layer becomes more efficient, image sharpness decreases but the exposure system increases in speed.

The Active Layer

The active layer is the heart of the intensifying screen. It has two components, the phosphor itself and the suspension material. The phosphor most commonly used is calcium tungstate. In the past, a phosphor known as barium lead sulfate was more efficient at absorbing x-rays and thus emitted more light. More recently, calcium tungstate phosphors were further developed and have improved to where their absorbing abilities are greater than barium lead sulfate. The active layer is approximately 4 to 6 $\times$ 10^{-3} of an inch thick, depending on the speed of the screen.

The suspension material that surrounds each calcium tungstate crystal and holds it in place is made of a plastic substance, thus its primary function is to suspend each crystal in place. Later we will see how these elements of the active layer can influence the quality of the radiographic image.

The Edge Seal

By reviewing the cross section noted above, it might be possible to imagine the radiographic effect if these components would begin to separate. The result in decreased sharpness would surely be great. This would most likely occur if moisture in the air would settle between the components and cause buckling and separation. To prevent this from happening, a sealing agent is used to cover the edges of a screen sandwich. Often, technologists will use too much liquid when cleaning screens—the edge seal will prevent excessive moisture from seeping between the screen's various layers unless a small crack is present in the edge seal (penetrating moisture could inevitably damage the entire intensifying screen).

THE PROCESS OF LIGHT EMISSION

Light emission is a phenomenon in which x-rays are absorbed by the phosphor crystals and react by emitting light rays. In the simplest terms, it is a process of energy conversion. A high frequency energy beam is absorbed by a substance with a relatively high atomic number (calcium tungstate) and in turn a lower energy of visible light is produced. Within reasonable limits, there are two types of intensifying screen, light emission processes: phosphorescence and fluorescence. Each gives off a visible light but there are slight yet important differences between the two, and there is usually a color distinction in the emitted light. Fluorescence, the more common of the two, is the type of light emission that occurs in diagnostic cassettes, is usually a light blue, and is emitted only while the fluorescent crystal is being excited (irradiated) by the x-rays. With phosphorescence, the emission, usually yellowish, continues after the x-rays have stopped irradiating the phosphorescent crystal. Fluorescent crystals glow only while being irradiated, and phosphorescent crystals can continue to give off light after the exposure has terminated (also known as after glow or screen lag).

Fluorescence

Fluorescence is a process in which a phosphor becomes ionized by x-ray photons and gives off light. The atoms of these materials contain broadened outer orbits that can be thought of as bands. Although much wider, these bands are actually orbits in which electrons are moving about. Figure 58 shows how the outer bands contain three different zones, each having a slightly different energy. A detailed discussion of the physics involved is not necessary here. Suffice it to say that such bands do exist, and their presence is an important characteristic of the fluorescent material. It should be understood that the inner orbits are confined to specific energies. It is only the outer orbits of these materials that widen and contain bands.

The various zones of the outer orbit are known as the filled zone, the trapped zone, and the conduction band. We can now begin a brief ex-

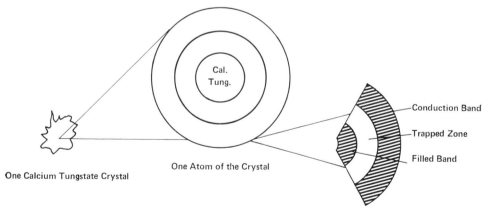

Figure 58. Without these lower energy valence bands, calcium tungstate crystals would have no practical value in diagnostic radiography.

planation into the mechanics of screen light emission. In Figure 59 we see an entering x-ray photon has evacuated an electron from an area of the filled band. The entire orbit becomes very unstable and attempts to improve the situation by drawing on an electron from the trapped zone. As the trapped electron moves to satisfy the vacancy, which is in a lower energy level of the orbit, the trapped electron gives up its excessive energy so it will be compatible with the energy of its new location. This "excess" energy is seen by us in the form of green or blue light and goes on to expose the film.

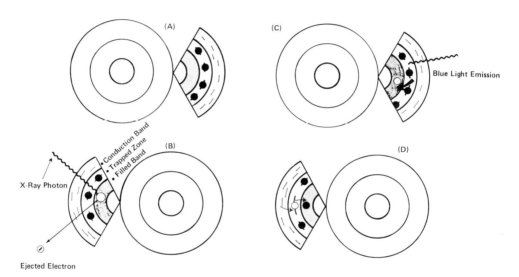

Figure 59. Screens produce light photons because of an ionization process. *A*, a stable atom of calcium tungstate. *B* shows an incident x-ray photon striking the atom's outer bands, causing a vacancy in the filled band, and resulting in the entire atom becoming unstable; in an attempt to become stable once again, one of the "trapped" electrons is pulled to fill the vacancy. In *C* the electron transfers to a lower energy band and must give up the excess energy to "fit" comfortably in the filled band; the excess energy given off is often the frequency of blue light and goes on to expose the film. In *D* the vacancy created in the trapped zone is filled by an electron from the conduction zone.

The amount of light emitted by a screen is primarily dependent upon (1) inherent ability of the crystals to absorb the x-ray photons, (2) the thickness of the active layer, and (3) the efficiency of the reflective layer. This process of light emission is caused by vacancies forming in the various bands of the outer orbit. As might be expected, there is a total reshuffling of the electrons located in this area as they begin to fill each other's vacancies. Also, with each reshuffling more light is emitted. This process goes on *only* as long as the exposure continues: Upon termination of the exposure, the ionization process and resulting light emission process cease immediately with fluorescent crystals.

Phosphorescence

Phosphorescent phosphors are very similar to fluorescent materials in general. However, the electrons are considerably slower in filling the vacancies created by the incident x-ray photons. In these crystals, the electrons in the trapped zone often do not have the total amount of energy needed to move directly into the vacancy in the filled band. During an x-ray exposure, the "trapped" electron travels to the conduction band where it gains additional energy to pass through the trapped zone and continue on to satisfy the vacancy in the filled band. It is possible, however, that an electron may not have sufficient energy to pass through the trapped zone even after it was somewhat revitalized by the conduction band. In this instance, the electron has to make several such trips to gain sufficient energy to pass through the bindings of the trapped zone. The procrastination or inability of the electron to immediately satisfy the vacancy in the filled band causes a continuation in the light emission process of the phosphorescent materials to the extent that they will continue to give off light after the exposure has terminated.

Phosphorescent materials are not often used in radiography today. They were, however, very popular up until the late 1950s and early 1960s when they were used by manufacturers to make fluoroscopic screens. Fluoroscopic screens were made with a phosphorescent material known as zinc cadmium sulphide. Since the brightness of the image produced was insufficient to satisfy current demands, an entirely new fluortechnology involving image intensifiers has evolved.

In summary, fluorescence and phosphorescence are light emission processes that are different in the color of light they emit, and whether or not they continue to produce light after the termination of the exposure. The light emission process is caused by phosphors absorbing x-ray photons and a resulting ionization process in which excess quantities of low energy emission (visible light rays) are emitted to expose the film.

Other Important Crystal Characteristics

It was thought by some that crystals themselves lost their light emission abilities with prolonged use. If the actual process of light emission described above is recalled, it can be seen quite easily that as long as there is an ample supply of electrons and vacancies, ionization and light emission will continue without change. Actually, the calcium tungstate crystals are quite stable (they can retain their electrons) and will produce equal light emission for hundreds of years of exposures.

Intensifying screen crystals have no particular shape and can vary considerably in both size and shape. With respect to size, however, most crystals used range from 4 to 8 microns, depending on screen speed. As crystal size increases, the amount of light it produces will increase when a similar exposure is used. Clearly, larger crystals will produce more light because they contain more atoms to become ionized and also will intercept more remnant x-ray photons than a smaller crystal would.

CLASSIFICATION OF SCREENS

The discussion up to this point has been concerned with a "typical" crystal and screen. Manufacturers can introduce certain chemical additives into the standard calcium tungstate crystal to slightly alter its emission characteristics. Other changes in the overall makeup of the screen can yield still further light emission differences. As a result, conventional intensifying screens can be categorized into three general groups based on their light producing efficiency. These are detailed (or slow), medium, and fast. Two of these types, exposed equally as shown in Figure 60, would yield strikingly different results. The screen that produced the darkest image would be known as a fast screen because it obviously produced much more light than the other two. By using a fast screen, the technologist can afford to reduce the exposure time by one-half and still maintain a comparable diagnostic image compared to medium speed.

The term *intensification factor* has important bearing because it tells the technologist to what extent a particular screen will intensify or amplify the x-ray exposure. Manufacturers can produce screens to amplify the beam by a certain factor. For example, if a medium speed screen was exposed using 100, Ma, ½ second, at 70 KvP, a certain density, such as 1^D, might be produced on the film. If a fast screen is chosen and the same exposure factors were used, the result in density would be doubled, equal to 2^D.

The significance of the intensification factor comes into play when new screens have been purchased for a department and new techniques have to be written or when the technologist knows he is about to use a faster or slower screen than was previously used. Generally it is unwise for a department to have more than one speed of screen available for routine diagnostic studies because of the changes in density that are bound to result in exposure errors. To find the extent to which a given screen can amplify the beam (the intensification factor), the exposure value needed to produce a given radiographic density with screens is

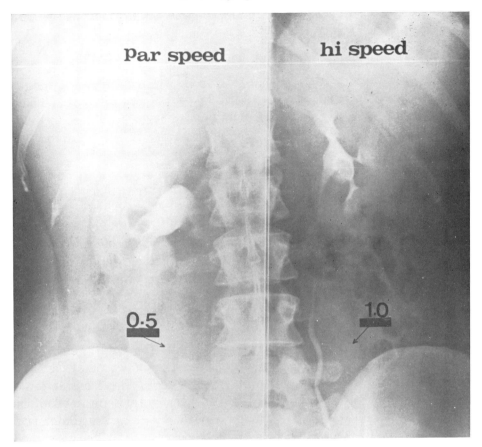

Figure 60. The variation in film density between fast and medium speed screens.

divided into the exposure value needed to produce the same density without screens.

$$\text{Exposure:} \quad \frac{(\text{MaS}) \text{ exposure without screen}}{(\text{MaS}) \text{ exposure with screen}} = \text{I.F. factor}$$

If a situation arises where 200 Ma, ½ second, at 70 KvP is used non-screen to produce a density of 1.0, screens at 200 Ma, ¼₀ second, at 70 KvP would yield a density of 1.0. With the formula above it can be found that the intensification factor for that particular screen is 20. The manner in which this information can be put to practical use will be thoroughly discussed in Chapter Eleven. The emphasis now will be placed on the concept that each category of screen has its own intensification factor and that I.F. tells the technologist the extent to which a particular screen will amplify the beam's effect on the film to produce the desired density. In screens with high I.F. factors, only about 5 to 6 percent of the remnant x-ray photons are put to use.

The Practicality of Having Different I.F.

In general, slow, medium, and fast screens account for a range of I.F. from 30 to 60 respectively. It has been pointed out that with conventional screens, detail decreases as faster screens are used. Slow screens are used for certain examinations simply because of the superior radiographic sharpness they provide. An example of this is the increased use of detail (slow) screens for extremity and, in some cases, mammographic work. Figure 63 compares the detail obtained when using fast and slow screens.

There are times when extreme detail can be realistically compromised somewhat in order to obtain a lower patient dose or to prevent patient motion. In such cases a faster screen is used because it can amplify the effect of the beam substantially over the slow screens. Examinations such as these might include pelivemetry and pediatric work. Examples for reducing patient motion include emergency work, portable work, and special procedures when very fast exposures must be made using automatic film changers. Also, many emergency patients are not able to hold still long enough to make a good exposure. Special procedures often require the run of rapid sequence films. Fast screens help considerably to make the exposure time as short as possible so more frames (exposures) can be taken while the contrast material is moving through the vascular system. Also there is a great advantage in special work in that fast screens reduce the heat load substantially on the x-ray tube. Between the two extremes of slow and fast screens, there is, of course, the medium speed screen which is in the view of many radiologists and technologists an adequate compromise between the advantages and disadvantages of slow and fast types. From this discussion it can be seen that there is a practical application for each group of screen speeds, and the proper selection is a matter of making an appropriate and realistic compromise between speed, patient dose, and sharp detail. Many more radiology departments are equipped with fast speed screens today than just a few years ago. The reason for this is that manufacturers have been able to produce improved detail characteristics of faster screens through constant research and product development.

HOW DO SCREENS AFFECT THE RADIOGRAPHIC IMAGE?

Contrast and Density

An image must possess only three things to be of high quality, but all three must be present: (1) sufficient contrast; (2) sufficient density;

and (3) sharply defined structures. The importance of screens in obtaining a high quality image cannot be overly emphasized, because they have a very pronounced effect on all three. The effect that screens have on radiographic contrast is important as well. Figure 61 shows two radiographs of the same body part. The exposure values were correctly adjusted so that density could be compared. Film B was exposed with a par speed screen and film A was exposed without any screens but with the same technique. Exposure for C was increased approximately forty-six times to produce a similar density to the screen film. Note the variation in contrast from B to C.

It should be clear by now that the light emitted by the screen is for the most part totally responsible for the diagnostic image. To illustrate this another way, two sheets of paper were fitted over one half of the screen so that the film would receive no light from the phosphors. The other half of the film was, of course, able to receive blue screen light. By viewing Figure 62 one can see the extent to which the unassisted x-ray exposure affects the film.

The reason why contrast is affected when using intensifying screens is not totally understood. It should be pointed out that radiographic contrast does not appear to change between the various speeds of screens but only in the initial nonscreen to screen conversion.

Screen Sharpness

With an understanding of the profound effect screens have on radiographic contrast and density, another interesting aspect of screen radiography may be discussed. It is vital that the structures radiographed be sharply delineated in the image. With all the advantages screens offer in radiography, they have an important disadvantage as they cause unsharpness. Whenever screens are used, radiographic sharpness decreases, and this trend worsens as faster screens are used (see Figure 63).

Modulation Transfer Function

In brief, modulation transfer function (M.T.F.) refers to the manner in which the phosphors transmit their light to the film. The film, of course, is responsible for receiving the pattern of emitted light intensities and makes a permanent image of them (see Figure 64). Here we see how screen phosphors react to varying intensities of the remnant beam and produce proportionate or representative amounts of light, which in turn will produce proportionate or representative densities on the film. The varying amounts of x-rays striking the phosphors are caused by variations in tissue absorption. M.T.F. is the process of transferring the information of the remnant beam to the emulsion of the x-ray film.

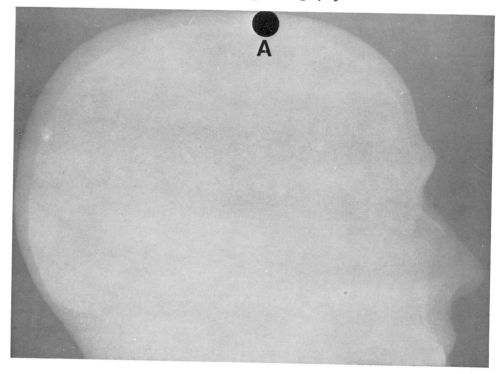

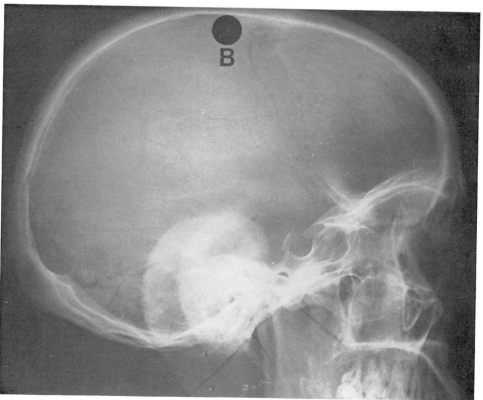

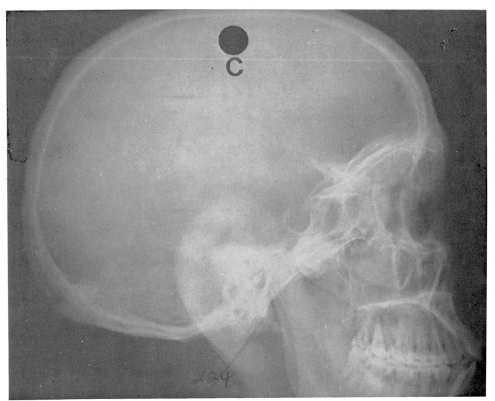

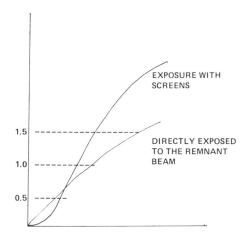

← Figure 61. To show changes in radiographic density, compare *B* using a screen exposure with *A* using a nonscreen exposure. The density in *B* was equalized in *C*, so the difference between these images shows changes in contrast caused by the screen. H & D curves are included for evaluation.

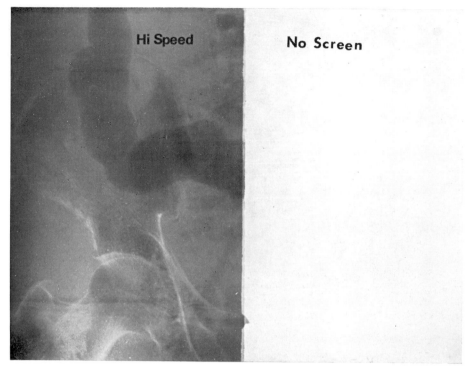

Figure 62. Obtaining quality · radiographs without the use of screens would make routine radiography impossible. Compare the radiographic effect of a high speed screen and a nonscreen exposure. A sheet of paper was used to block the screen to one-half of the film.

M.T.F. is the fundamental factor that influences screen unsharpness. In Figure 65, a beam is exposing a film, which responds accordingly and represents the body structures very faithfully. This situation will, of course, produce optimal image sharpness. In other words, there is a one to one transfer of patient information carried by the remnant beam to the film. This, unfortunately, is not the case when intensifying screens are used, because the information transferred to the film is not direct from the patient. There are three factors that affect image transfer: the manner with which the photons emit light, the thickness of the active layer, and the reflective layer.

It has already been discussed how a typical phosphor emits light in all directions, similar to a light bulb. It is obvious what might happen to a sharply defined body structure if a medium that carried the information had a spreading effect as it traveled toward the film. Also, the active layer is made up of phosphors coated to a certain thickness: unsharpness would become worse if *many* crystals carrying the body information had been

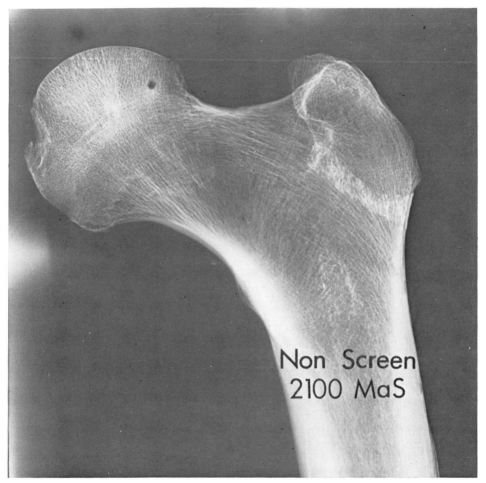

Figure 63A.

Figure 63. Sharpness decreases when changing from nonscreen to screen exposures; further decreases in sharpness result as faster screens are used.

affected by a spreading medium. The factors that influence the degree of the spreading effect are the thickness of the active layer and the efficiency of the reflective layer. Figure 66 illustrates how varying degrees of light spread affect radiographic sharpness. Although the screen's spreading effect is influenced by the thickness of the active layer and the efficiency of the reflective layer, it should be understood that the initial cause of the spreading effect is the inherent pattern of light emission from the crystal itself. A method has been devised with which screens can be tested to determine the degree of spreading ability (M.T.F.) Figure 67 shows two lead blocks precisely cut and separated by a tiny space. Three exposures were made: The nonscreen exposure faithfully duplicated the

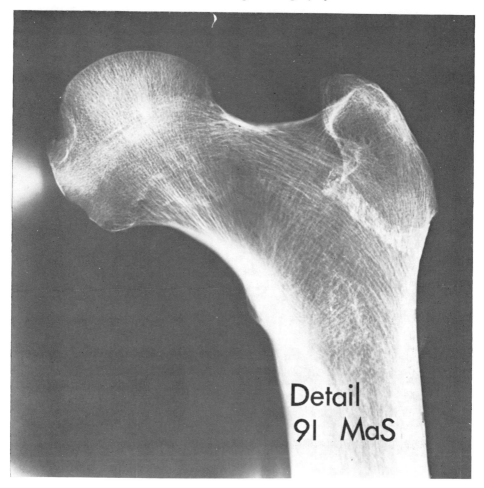

Figure 63B.

sharpness between the block; The slow speed screen gave evidence of a slight spreading function as the film has a wider line of exposure; and fast screens caused increased spreading resulting in the greatest amount of unsharpness.

In summary, modulation transfer function is the process by which the phosphors emit light to the film in a spreading fashion, which cause a loss of structure delineation. As the screen's speed increases, the spreading factor increases and sharpness decreases. It is interesting to note that the size of the crystal has no practical effect on screen sharpness because small phosphors emit light in the same diversion pattern as large ones do. A small crystal must produce the same amount of light as a large crystal to produce equal radiographic density, so a small crystal would eventually have equal spreading if a comparable density is to be maintained.

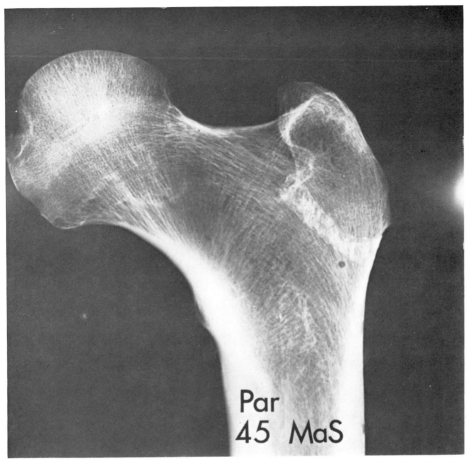

Figure 63C.

Factors Affecting Screen Speed

The prime factors that affect the intensification factor (speed) of screens are the thickness of the active layer, the efficiency of the reflective layer, and the amount of yellow dye* in the active layer. Much discussion has been made regarding the effect changes in KvP have on screen speed; however, the author has found that most conventional phosphors (calcium tungstate) used today are fairly consistent light emitters over the diagnostic KvP range.

The thicker the active layer, the faster the screen will be. The reason for this is that thicker active layers will simply contain more phosphors and thus emit a greater amount of light. In other words, more crystals

---

* Some dyes used to absorb emitted screen light are brown or pink depending on the manufacturer.

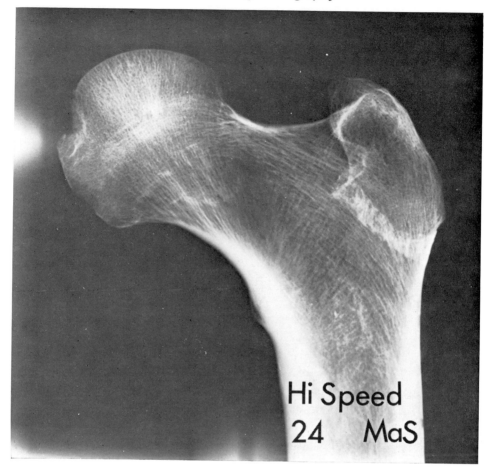

Figure 63D.

will capture or intercept more x-ray photons. This, of course, will make a darker radiograph with the same exposure. A reflective layer has been included in the screen's structure to bounce the escaping light toward the x-ray film. Thus, more screen light reaches the film per a given exposure resulting in a darker image, and, as the efficiency of the reflective layer increases, radiographic density increases. Blue light is absorbed very well by yellow substances. Some screens, usually the slow speed type, appear to be colored. Often this is the main difference between the various speeds of screens. Fast screens are usually white in appearance.

These dyes *artificially* reduce the spreading characteristic of the emitted light resulting in a positive affect on screen and image sharpness. To a lesser extent, atmospheric conditions play a role in affecting screen speed. As air temperature increases, the efficiency of a screen's ability to

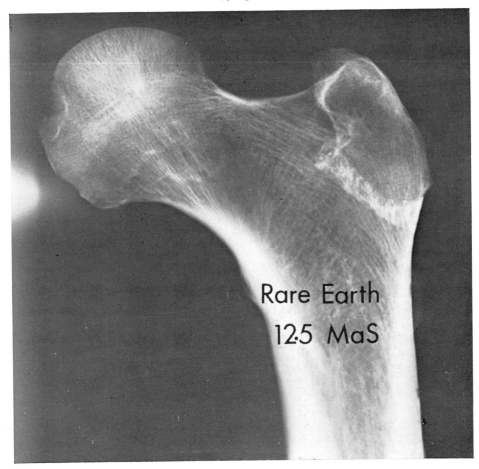

Figure 63E.

produce light decreases; however, film becomes more sensitive to exposure when used in warmer temperatures. The net effect of these opposing conditions between film and screen is that the radiographic density produced by the screen film exposure system remains fairly constant throughout most atmospheric conditions likely to be experienced in the United States.

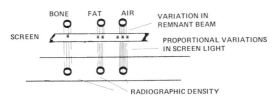

Figure 64. Screens transfer light intensities to the film, which are in turn representative of remnant beam intensities.

MODULATION TRANSFER FUNCTION

(Screen Unsharpness)

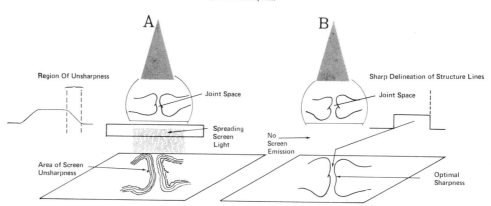

Figure 65. *B* shows that optimal sharpness has been attained because of the one-to-one transference of information carried by the remnant beam to the film. However, *A* shows a dramatic reduction in sharpness because a screen has been inserted. The screen is an intermediary which produces divergent light rays of its own which cause unsharpness.

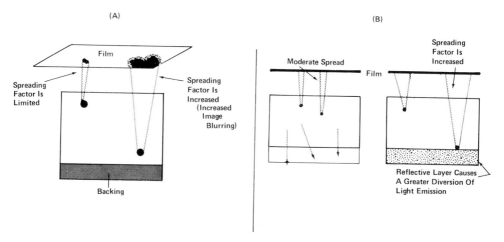

Figure 66. The normal spreading of screen light can be altered by the thickness of the active layer and the efficiency of the reflective layer. As these two factors increase, the spreading factor also increases.

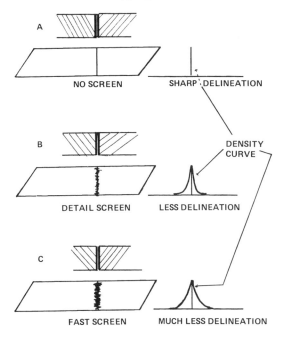

Figure 67. Schematic of modulation transfer. As faster screens are used, the image is more dependent on screen light, and sharpness decreases.

It might be noted, however, that upon freezing a pair of screens overnight and exposing them with film of room temperature, little change in density was noted, compared to the same screen allowed to warm to room temperature and exposed to another film.

Screen Artifacts

There are three fundamental ways in which screens produce artifacts on the radiographic image. One is the obstruction of blue light from the film; under these conditions of obstruction, the film will yield little or no density in that immediate area. Second, screens help to produce static electricity. These charges often expose the film and cause plus $(+)$ density artifacts. Third, poor screen contact is a more subtle artifact, nonetheless equally important as it accounts for a drastic reduction in sharpness.

With regard to the first item, if a foreign object finds its way between the film and screen surfaces and has the ability to absorb the screen's light, a minus density artifact would be caused. Examples of this are noted in Figure 68. Here we see an assortment of paper, hair, and particles of lint placed in the cassette to produce artifacts. Other minus density artifacts caused by light obstruction include stained screens and scratches.

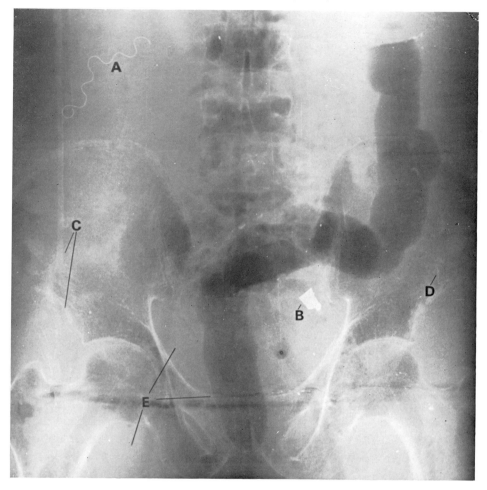

Figure 68. Foreign bodies, artifacts, absorb screen light; the film is not exposed. A—hair; B—paper; C—developer stain; D—scratched screens; E—dirt in cassette.

Screens do lose speed if used too long, but it is not because the phosphors have become less effective. Often, screens lose speed with age because the material used to suspend the phosphor crystal discolors and dries with time to a yellow tint. The blue light is easily absorbed by this discoloration of the suspension material resulting in less blue light getting to the film. There is nothing the technologist can do to correct this except to purchase new screens. Most often, speed is reduced because the abrasion layer is worn away by tiny specks of dirt and foreign material as the film is slid over the screen's surfaces while loading and unloading the cassette. When the abrasion layer is worn away, a layer of dirt forms a barrier to the screen light on route to the film and causes underexposure. From a practical point of view, this would seriously affect quality control because it is possible for the same type of screen to have different speeds if the screens were purchased at different times. The older screens would have more discoloration and artifacts and would yield less radiographic density than the newer ones.

With regard to static, if the screens are not cleaned often enough and were not coated with antistatic solution, friction will result when the film is slid into and out of the cassette. During warm, semihumid atmospheric conditions, static does not usually occur; however, during the winter months, when the air in the department has become dry through heating, friction easily causes static discharges which expose the film as noted in Figure 69. This artifact can be relieved considerably by using a reliable cleaning and antistatic application program. For smaller departments, offices, and clinics, the use of a humidifer could certainly prove beneficial in reducing static problems.

The phosphors located closer to the film surface will produce less light spreading because of their proximity to the film than those located deeper in the active layer. The more distance between the phosphors and the film, the greater the spreading factor will be (as noted in Fig. 70). Because of various problems a cassette might not be able to provide ample compression to assure that the surfaces between the film and screen will be in perfect contact. When this happens, light spreads over a wide area of the film surface causing an obvious decrease in image sharpness. Common causes for this poor contact artifact are: (1) worn felt padding, (2) loose cassette straps, (3) warped or bent cassette frames, (4) cassettes that have developed ridges with use, and (5) expanding and contracting cardboard backing.

A test for poor screen contact involves the use of a wire mesh with approximately one-fourth inch squares, the cassette under suspicion, and a light radiographic exposure: Load a film into the cassette suspected of poor contact, place the wire mesh over the front of the cassette, and make an exposure using approximately 100 Ma, 1/20 second at 50 KvP for high speed screens. When viewing the processed film the white squares of the wire mesh should be sharply defined. If poor screen contact is present, there will be localized patches of unsharpness and slightly increased density over the area. If it is found that the cassette frame is bent or warped, little can be done to improve the problem. If it is simply a matter of worn felt or loose straps, repairs can be made. If the screen itself has ridges, they must be replaced. The day light system by DuPont and the X-Omatic system by Eastman Kodak improve compression substantially. If compression is a problem, it will usually occur in large cassette sizes because it is more difficult to maintain equal compression over this large surface of film area.

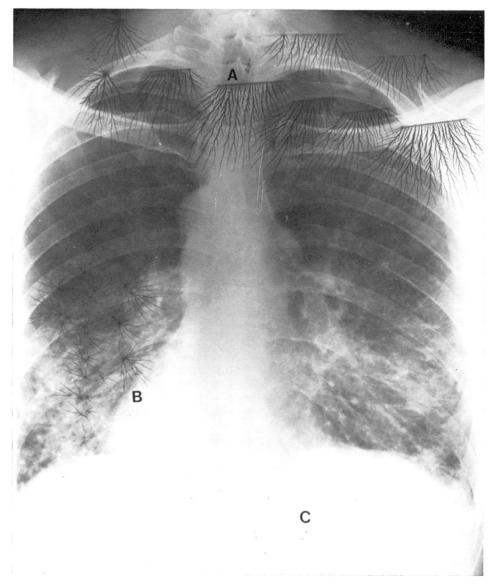

Figure 69. A—crown static; B—"tree" static; C—smudge static. Crown and "tree" static are commonly caused by friction between the film and another surface and are the most common to screens. Smudge static is usually caused by charged dirt and lint particles striking the film.

A less occurring artifact is screen lag. Screen lag is a situation in which the phosphors continue to emit light after transmission of the x-ray exposure (phosphorescence). The phosphors commonly used in making intensifying screens (calcium tungstate) do not phosphoresce. Earlier phosphors had this problem but, fortunately, with the development and

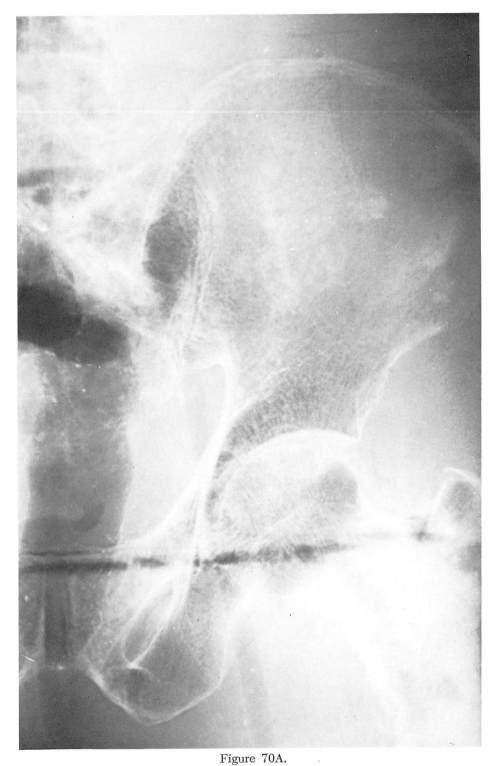

Figure 70A.

Figure 70. Poor screen contact emphasizes the inherent spreading factor of screens, causing additional unsharpness.

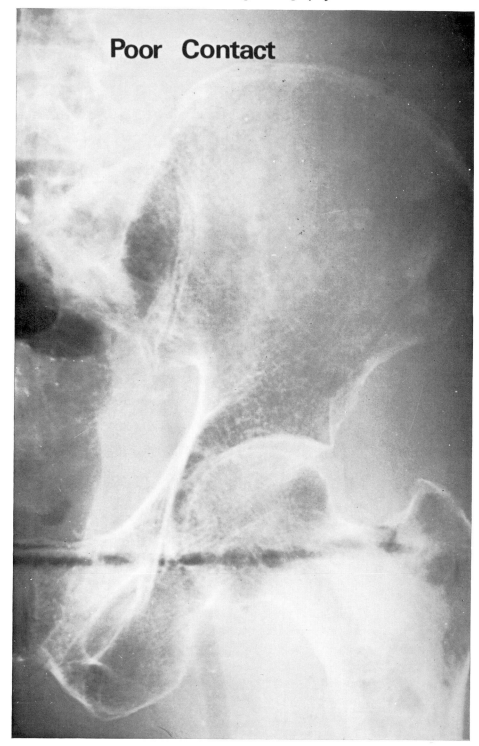

Figure 70B.

SCREEN FILM CONTACT

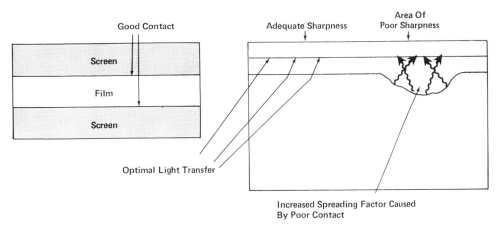

Figure 70C.

advances made with calcium tungstate, screen lag is extremely uncommon in modern radiography.

Quantum Mottle (Noise)

Quantum mottle and screen noise are one and the same. As discussed in physics, x-rays are emitted from an x-ray tube in the form of packettes or individual units of energy called *quantum* (see Fig. 71). In order to get a high quality radiographic image, it is important to have a sufficient amount of quantum to pass through the patient so that enough patient information can be transferred to the film. If not enough quantum (x-rays) are generated during an exposure, an insufficient amount of information will be transferred to the screens and film, and detail will be reduced.

To illustrate this point, Figure 73 shows what happens with quantum mottle. In A a direct exposure is made. A tremendous number of x-rays (quantum) pass through the body picking up large amounts of diagnostic information. This information is directly impressed in the film's emulsion. Because a large number of x-rays were used by the exposure, a representative amount of diagnostic information can be seen on the processed film. In B the situation is somewhat different. Because of the screen's high intensification factor, fewer x-rays are needed to produce equal radiographic density. One can easily see that the amount of useful diagnostic information transmitted to the film via the screen will decrease as the number of x-rays needed to produce a given density decrease. As faster screens are used, still fewer x-ray photons are used to produce the desired density and a greater loss of diagnostic information results. A mottled

Figure 71. This exposure was made without a collimator and with the film standing on end under the x-ray tube. One can easily imagine individual quanta exiting the tube and traveling down to expose the film while carrying patient information.

appearance is noted on the film as a result of insufficient numbers of x-ray photons, and is referred to as quantum mottle.

    Screen manufacturers were at the point where any further increases in the intensification factor would not be practical because an exposure with still fewer numbers of x-rays could not carry enough information to the screen and film to produce good diagnostic results and the mottled appearance would increase (see Fig. 72). As shown, fewer photons are needed to form a comparable image. These structures could be a part of bone or lung detail in a radiographic image. If superfast screens are used, fewer numbers of x-rays would be needed and the structure would not be seen as well. Quantum mottle appears on the film as faint blotches in the image, especially noticeable when fast screens are used with high KvP exposures.

Rare Earth Screens

It was mentioned earlier that the phosphors' ability to absorb x-ray photons and transfer that energy to visible light is vital to screen radiography. However, as just discussed, when screens become too efficient (fast) quantum mottle occurs and results in a loss of detail. Conventional screens do not absorb all the x-ray photons striking the active layer. Many x-rays are not captured by the phosphors and pass on not converted. Thus much diagnostic information is wasted or unused even though the original film density has been maintained.

Recently, however, new phosphors have been developed as a result of the aerospace program and have become known as rare earth phosphors. This term is slightly misleading as it implies the phosphors are not easily found. Actually, these phosphors are abundant; it is the process under which the phosphors must go in order to be used in intensifying screens that make them less available and thus the term rare earth. There is also a considerable difference in cost between conventional screens and rare earth screens. A conventional pair of calcium tungstate 14 × 17 inch screens lists for approximately 50 percent of the cost if compared to rare earth screens.

The advantage of these screens simply lies in their ability to capture more of the x-rays that have penetrated the body. In fact the capture ratio is approximately double. In other words, conventional fast screens use or capture only 20 percent of the remnant photons, while rare earth screens capture or absorb 40 percent, so rare earth screens can extract twice as much information from the remnant beam than had been previously possible with conventional screens. Quantum mottle is not easily seen even though rare earth screens are twice as fast as conventional high speed screens because more patient information is transferred to the film.

With the increase in capture ratio of these rare earth screens, they can produce twice the amount of light when compared to fast screens, so the exposure can be cut in half—if the original radiographic density is to be maintained. A reduction in exposure in this case would not cause quantum mottle because the screens are absorbing twice as much diagnostic information from the beam.

In summary, rare earth screens allow the technologist to reduce the exposure time by approximately one-half (if compared to fast screens) without any noticeable increase in quantum mottle. This is accomplished with the superior ability of rare earth phosphors to capture or intercept more x-ray photons carrying diagnostic information from the body.

Screen Maintenance

Even under the best of conditions, screens can be expected to last no more than five to six years; thus a good maintenance program should

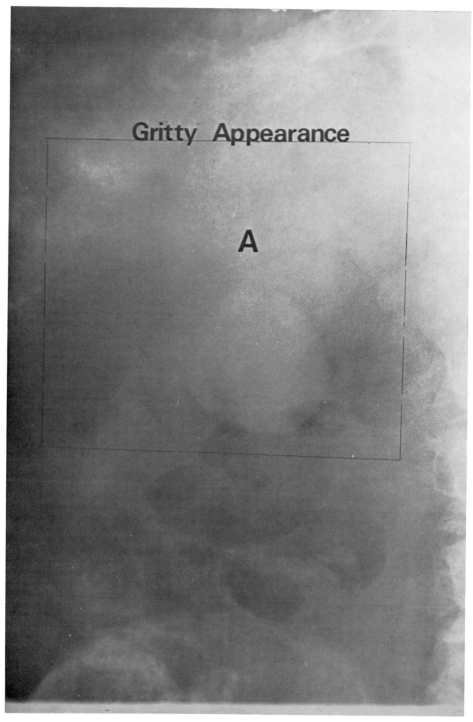

Figure 72. As faster screens are used, fewer primary x-ray photons are necessary to produce sufficient radiographic density. With the decrease in the total number of primary x-ray photons, a proportionate decrease in "patient" information reaches the film.

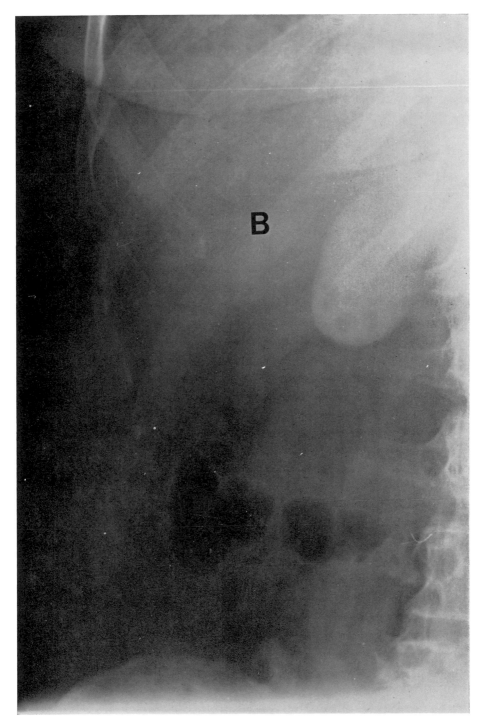

When rare earth screens are used "patient" information reaching the film is not sharply reduced, however, when compared to what occurs with fast screens. The reason for this is that rare earth crystals can capture two times more "patient" information from the remnant beam. The "gritty" appearance in A is known as quantum or screen mottle.

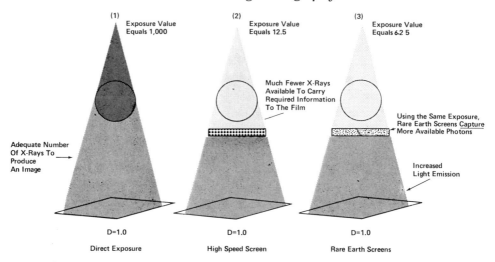

(1) Exposure Value Equals 1,000

(2) Exposure Value Equals 12.5

(3) Exposure Value Equals 6.2 5

Much Fewer X-Rays Available To Carry Required Information To The Film

Adequate Number Of X-Rays To Produce An Image

Using the Same Exposure, Rare Earth Screens Capture More Available Photons

Increased Light Emission

D=1.0　　　　　　D=1.0　　　　　　D=1.0

Direct Exposure　　High Speed Screen　　Rare Earth Screens

Figure 73. Exposure (1) requires a great many photons because it is a direct exposure; as a result, more "patient" information is carried by the remnant beam to the film. In exposure (2) there is a drastic decrease in the number of photons and less "patient" information is transmitted to the film. In exposure (3), rare earth screens are used. Rare earth screens require approximately 50 percent fewer photons than fast screens but absorb twice as much remnant radiation; thus the same amount of patient information reaches the film.

be enforced. There are many variables in screen usage, and it would be impossible to presume a standard program here. First, the number of screens in the department should be appreciated. If the particular institution has an insufficient supply, they will be used more frequently and consequently require more cleaning and damage will result more readily.

It seems reasonable that screens should be cleaned monthly or bi-monthly. A cleaning program helps to prolong the life and prevent the breakdown of the abrasion layer. If screens are not cleaned often enough, lint, dirt, and other hard foreign substances find their way into the screens, and if not removed, will cause scratches as x-ray film is slid in and out of the cassette. Eventually these particles will grind away at the abrasion layer and become embedded into the phosphor layer. Not only has the diagnostic value been reduced substantially, but the screens are costly to replace.

Cleaning instructions are well outlined in the various products designed for the job. It is important, however, not to use too much solution at any one time so that fluid will not seep into the felt padding, and to make sure the screens stand open long enough after cleaning to dry sufficiently (at least twenty minutes).

Nonscreen Versus Screen Film

Film manufacturers can make film more sensitive to certain electromagnetic energy forms; the two major types of film available for radiography purposes are screen and nonscreen film. Screen film is particularly sensitive to blue light. This type can become readily exposed by the emitted light from intensifying screens. Screen-type film can be exposed by the x-ray beam itself, but considerably more exposure must be used to yield the same degree of radiographic density. On the other hand, direct-exposure-type film, also known as nonscreen film, would yield no density if it were exposed with intensifying screens.

The emulsion thickness of screen film should not be great because the blue light from the screens cannot penetrate the silver bromide emulsion very easily. For this reason, screen film is made with emulsion on both sides of the base and is coated more thinly than nonscreen film. If nonscreen film was used in cardboard and screen film was used in cardboard, the nonscreen exposure can be reduced to one-fourth.

New advances in detail with screen radiography in recent years have encouraged more radiologists and technologists to use slow speed screens for high detail work than nonscreen techniques. There is speculation that, in the future, this trend will continue even more, to the extent that nonscreen radiography as we know it will be virtually eliminated.

CHAPTER FIVE

MILLIAMPERAGE

DEFINITION AND FUNCTION

MILLIAMPERAGE IS AN electrical term and not primarily a radiological term. Nevertheless, as you continue through this chapter you will begin to appreciate a vital relationship between the two. For our purposes, milliamperage is the quantity of electrons moving through the x-ray tube during an exposure. A term that is often used in substitution of milliamperage is tube current. The technologist has a great deal of control over the milliamperage. Although Ma indicates the quantity of electrons moving (current), we must be careful to understand that there is no mention of the speed at which the electrons may be moving. If this distinction is noted now, it will help the student to understand the more involved ideas that will be discussed later in the text. In short, the technologist adjusts the milliamperage setting to establish the desired quantity of electrons that will strike the anode of the x-ray tube during the exposure to produce various quantities of primary radiation. There is a relationship between the number of electrons striking the anode and the amount of x-rays that will be produced.

Before any attempt is made to discuss specific details, the following ideas and concepts should be firmly established. We will begin by reviewing a small section of the main x-ray circuit. Figure 74 shows a schematic of the filament circuit. The connection between the control panel and the filament circuit is drastically simplified here, but will serve well in this discussion.

You will note that, at the control panel, a knob is marked Ma selector. Through this control knob the technologist has control over the amount of current that will move in the filament circuit and eventually through the x-ray tube. The panel in Figure 74 also shows that the various Ma values are commonly broken into increments of 50, 100, 200, 300, etc. Modern panels can have Ma values as high as 1,500.

There is a relationship between the position of the milliamperage selector on the control panel and the current that is moving through the filament circuit. In fact, if the technologist moves the milliamperage

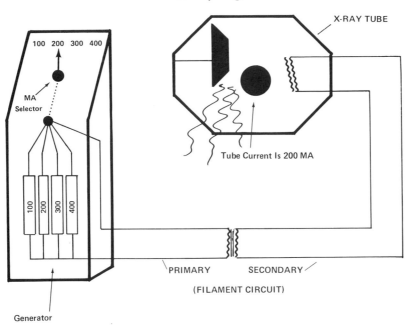

Figure 74. The filament circuit carries current to heat the x-ray tube filament so it will produce proper tube current. The x-ray tube uses the space charge provided by the filament circuit and the high voltage supplied by the high tension circuit to produce x-rays. As the filament temperature increases, more electrons are available to be used as tube current when the technologist presses the exposure switch.

control knob from the 200 Ma station to the 400 Ma station, the amount of current that will move through the filament circuit rises sufficiently to increase the radiographic density by approximately two times (see Fig. 78).

The X-Ray Tube Filament

In order to make practical radiographic use of the filament current it must be channeled by the filament circuit to something that will use these electrons for radiographic purposes. The device used is, of course, the x-ray tube. The x-ray tube will change the energy bound electrons of the filament circuit into x-ray energy (see Fig. 75).

In Figure 76, you will note that the x-ray tube has two major electrodes known as the cathode (which has a negative charge) and the anode (which has a positive charge). Another name given to the cathode is the filament, which resembles the filament of an ordinary light bulb. You will also note that the filament is actually part of the filament circuit while the anode is a part of the high tension circuit. It might be helpful to think of the x-ray tube as a bridge between two different circuits. If the

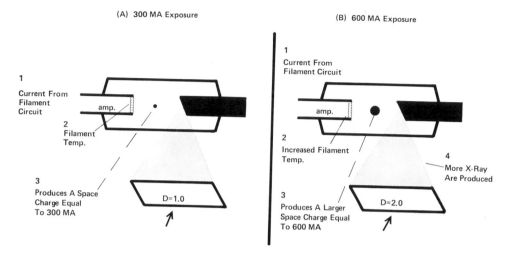

Figure 75.

technologist turns up the Ma selector at the control panel, more current will pass through the filament circuit, and this will increase the temperature of the filament in the x-ray tube. If we increase the filament temperature, the filament's atoms will become so active that it will not be able to contain all its electrons, and they will begin to "boil off" and form a small space charge just beyond the filament. This boiling-off process is often referred to as incandescence and is controlled by the temperature of the filament (see Fig. 75).

When the boiling process begins at the heated filament, the freed electrons have a tendency to remain in a relatively tight ball or cloud just beyond the cathode. The size of the cloud has an important relation-

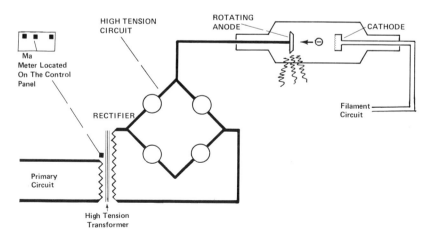

Figure 76.

ship with the temperature of the filament. As the filament temperature is increased, the size of the space charge increases, and when a larger cloud of electrons moves during the exposure, the tube current increases accordingly.

The conversion of electric energy into x-ray energy involves three factors: (1) a space charge must be generated; (2) the electron cloud must be put into motion; (3) once in motion the cloud of electrons must be abruptly stopped. Figure 76 shows that the filament (cathode) is part of the filament circuit and the anode is part of the high tension circuit. These two electrodes function as two coordinated yet separate entities in producing the desired effect of converting electrical kinetic energy into x-ray energy.

As the cloud of electrons is formed, it must be given a great deal of kinetic energy. The anode is thus given a strong positive charge by the high tension circuit. When the technologist presses the exposure button, the secondary circuit closes and the cloud of electrons (space charge) is driven toward the anode at a very high rate of speed. The *Ma meter* at the control panel then registers the quantity of current moving through the x-ray tube. This sequence of events is possible because opposite electrical charges attract each other. The final phase in the production of x-rays involves the stopping or deceleration of the electron cloud with its great amount of kinetic energy. Figure 77 shows the shape of the rotating anode. Because the anode material has a very high atomic number, it is very effectively able to stop the electrons momentarily. The kinetic energy does not simply disappear at this point but rather changes into two other energy forms: heat energy and x-ray energy (the law of conservation of energy). The number of electrons in the space charge will determine to a great extent the number of conversions that occur and the number of x-rays that will emerge as a result.

As Ma selection increases, filament current increases, filament temperature increases, space charge increases, Ma increases, and x-ray and heat production increase.

It can be seen that the technologist can exert direct control over the number of x-rays that will ultimately be produced. It should be kept in mind also that changes in Ma only alter the numbers (quantity) of primary x-rays and not the penetrating ability (see Fig. 78).

It was mentioned that milliamperage is the number of electrons moving in unison through the tube. Often students have the impression that milliamperage indicates a large quantity of electrons when actually it does not. "Milli" means one-thousandth of, so when a 500 Ma station is selected, only ½ amps are used. One must be careful not to underestimate, how-

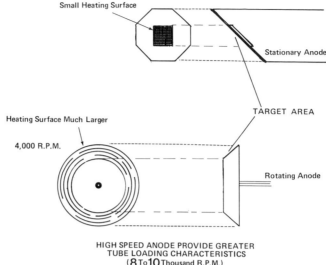

Small Heating Surface

Stationary Anode

TARGET AREA

Heating Surface Much Larger

4,000 R.P.M.

Rotating Anode

HIGH SPEED ANODE PROVIDE GREATER
TUBE LOADING CHARACTERISTICS
(8 To 10 Thousand R.P.M.)

Figure 77. The anode of the x-ray tube contains a target or focal spot which is the site of actual electron bombardment and kinetic energy transformation into primary x-ray photons.

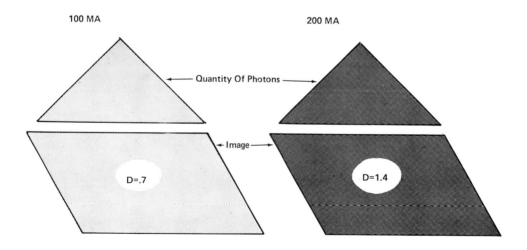

100 MA

200 MA

Quantity Of Photons

Image

D=.7

D=1.4

↟# Of X-Ray =   ↟ Ionization =   ↟ Growth Of Sensitivity Specks =   ↟ Density

Figure 78. As more photons are produced, radiographic density increases. Radiographs do not always respond as shown here because of other variables such as film speed, specific processing conditions, and whether or not screens are used. ⋏ number of x-rays = ⋏ ionization = ⋏ growth of sensitivity specks = ⋏ density.

ever, the heat and the number of photons that can be generated at the anode during such an exposure.

Ma, Heat, Focal Spot Size, and Radiographic Sharpness

High milliamperage exposures pose serious limitations on the life of x-ray tubes because of the heat generated at the anode during the exposure. The entire anode is not directly subjected to this intense heat, however. A very small area, known as the focal spot, is the primary *target* of the electron cloud. Most diagnostic x-ray tubes range in focal spot sizes from 0.5 mm to 2 mm. The focal spot represents a conflicting problem, because a small sized focal spot (0.5 mm) produces greater sharpness than does a larger focal spot size. However, the intense heat caused by the electrons striking a small area of the anode can be very damaging. Tube life becomes an important consideration in radiography on the basis of cost. The subject of focal spot size and detail will be further evaluated in Chapter 6. It was mentioned here to point out the fact that the milliamperage used for an exposure has an effect on tube life. The statement below summarizes this concept:

> As Ma increases, larger tube currents are generated; larger focal size should be used and sharpness of detail decreases.

From a theoretical point of view, it would be incorrect to say that increasing milliamperage itself causes poor sharpness. It would be quite possible to have *optimal* sharpness with a 1,000 Ma exposure, providing it was possible to use a small focal size. Each new x-ray tube delivered to an installation has a graph that gives the technologist sufficient information regarding safe exposure settings so that tube damage can be avoided. These graphs are very easy to use, but the technologist must be sure that he is using the correct graph for the tube. This can be determined by matching phasing, rectification, and the focal spot size (see Fig. 79).

Reciprocity Law

Considering the possibilities of damaging an x-ray tube because of excessive anode heat, one might question the value of using high milliamperage exposures at all. From the technologist's point of view, the primary significance of using high exposures is that they allow equally short times to be used. If an exposure is made while the subject is in motion, there exists the possibility of a blurred image. In the case of pediatric examinations, it is especially important to use the shortest exposure time possible to reduce the chances of blurring from patient motion. The relationship between exposure time and milliamperage is very nicely bal-

SINGLE RADIOGRAPHIC EXPOSURE RATINGS

SINGLE PHASE — FULL WAVE RECTIFICATION
STANDARD SPEED (60 Hertz)

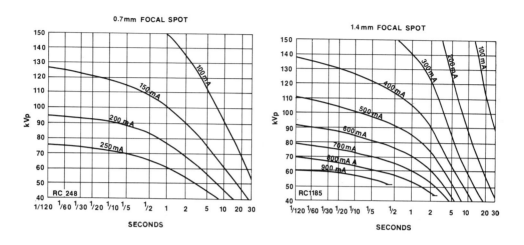

HIGH SPEED (180 Hertz)

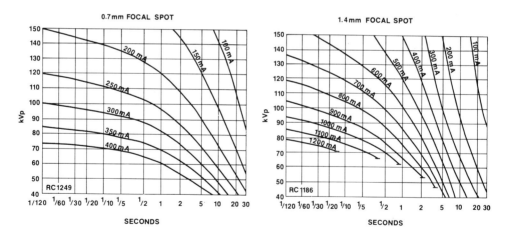

Figure 79: *Courtesy of* Picker X-Ray Corporation.

anced in the phenomenon described by the *Reciprocity Law*. This law refers to a compatibility between the accumulation of radiation on the film caused by beam intensity and exposure time. This relationship is valid, so that some manufacturers of x-ray equipment combine exposure time and Ma factors into one unit known as milliamperage-second (MaS).

MaS is simply the product of the Ma and time value used for a given purpose. Let us examine a common exposure made using 200 Ma at 1 second at 70 KvP. These settings of time and Ma will produce the same radiographic density as 200 MaS at 70 KvP.

$$200 \text{ Ma} \times 1 \text{ sec} = 200 \text{ MaS}$$

From a practical point of view, many technologists find it more convenient to think in terms of MaS when setting techniques than by calculating individual units of time and Ma. The reciprocity between time and Ma can sometimes be better understood if we consider that 200 MaS can be obtained by a wide variety of combinations as noted below (also see Table I):

$$200 \text{ Ma} \times 1 \text{ sec} = 200 \text{ MaS}$$
$$400 \text{ Ma} \times \tfrac{1}{2} \text{ sec} = 200 \text{ MaS}$$
$$800 \text{ Ma} \times \tfrac{1}{4} \text{ sec} = 200 \text{ MaS}$$
$$100 \text{ Ma} \times 2 \text{ sec} = 200 \text{ MaS}$$

Each of the above exposure settings will produce the same accumulation of x-rays striking the film resulting in equal density. Reciprocity between time and Ma exist when direct exposure films are used. There is a slight breakdown in reciprocity with intensifying screens. This reciprocity failure occurs with very short or very long exposure times: less than 1/120 of a second or greater than 5 seconds. It should be noted that the slight degree of reciprocity failure is usually not serious enough to cause practical technical problems.

MILLIAMPERAGE CALIBRATION

The technologist often has to make adjustments from what the technique charts indicate. For example, suppose an abdomen examination is to be done, and the patient requires 80 KvP, 100 Ma, at ½ second; because of the patient's inability to suspend respiration, it is necessary to decrease the exposure time but maintain overall radiographic density. Recalling the reciprocity law one may then use 400 Ma at ⅛ of a second. This would accomplish the desired result, providing the x-ray equipment is properly calibrated to produce 400 Ma accurately. In many generators, however, this is not the case. Occasionally the x-ray output of a specific Ma is out of calibration so that when 400 Ma is selected, the actual output may be 350 or sometimes even less. If a technologist is working with an uncalibrated generator, equal densities will not result at the same MaS values and the technologist will end up making repeat exposures. Figure 80 shows radiographs exposed with the same generator before and after proper adjustments were made in Ma calibrations. Since the tube

TABLE I
MaS CALCULATIONS

Pulses	Time	50	100	200	300	400	500	600	700	800	900	1000	1100	1200	1300	1400	1500
1	1/120::	0.4	0.8	1.6	2.5	3	4	5	5	6	7	8	9	10	10	11	12
2	1/60::	0.8	1.6	3.2	5.0	6	8	10	11	13	15	16	18	20	21	23	25
3	1/40::	1.2	2.5	5.0	7.5	10	12	15	17	20	22	25	27	30	32	35	37
4	1/30::	1.6	3.3	6.6	10	13	16	20	23	26	29	33	36	39	43	46	49
5	1/24::	2	4	8	12	16	20	25	29	33	37	41	45	50	54	58	62
6	1/20::	2.5	5	10	15	20	25	30	35	40	45	50	55	60	65	70	75
8	1/15::	3.3	6.6	13	20	26	33	40	46	53	60	66	73	80	86	93	100
10	1/12::	4.1	8.3	16	25	33	41	50	58	66	75	83	91	100	108	116	125
12	1/10::	5	10	20	30	40	50	60	70	80	90	100	110	120	130	140	150
15	1/8::	6.2	12.5	25	37	50	62	75	87	100	112	125	137	150	162	175	187
18	3/20::	7.5	15	30	45	60	75	90	105	120	135	150	165	180	195	210	225
20	1/6::	8	16	33	50	66	83	100	116	133	150	166	183	200	216	233	250
24	2/10::	10	20	40	60	80	100	120	140	160	180	200	220	240	260	280	300
30	1/4::	12.5	25	50	75	100	125	150	175	200	225	250	275	300	325	350	375
	3/10::	15	30	60	90	120	150	180	210	240	270	300	330	360	390	420	450
	7/20::	17	35	70	105	140	175	210	245	280	315	350	385	420	455	490	525
	4/10::	20	40	80	120	160	200	240	280	320	360	400	440	480	520	560	600
	9/20::	22	45	90	135	180	225	270	315	360	405	450	495	540	585	630	675
	1/2::	25	50	100	150	200	250	300	350	400	450	500	550	600	650	700	750
	11/20::	27	55	110	165	220	275	330	385	440	495	550	605	660	715	770	825
	6/10::	30	60	120	180	240	300	360	420	480	540	600	660	720	780	840	900
	13/20::	32	65	130	195	260	325	390	455	520	585	650	715	780	845	910	975
	7/10::	35	70	140	210	280	350	420	490	560	630	700	770	840	910	980	1050
	6/8::	37	75	150	225	300	375	450	525	600	675	750	825	900	975	1050	1125
	8/10::	40	80	160	240	320	400	480	560	640	720	800	880	960	1040	1120	1200
	9/10::	45	90	180	270	360	450	540	630	720	810	900	990	1080	1170	1260	1350
	1SEC::	50	100	200	300	400	500	600	700	800	900	1000	1100	1200	1300	1400	1500
	1 1/4::	62	125	250	375	500	625	750	875	1000	1125	1250	1375	1500	1625	1750	1875
	1 1/2::	75	150	300	450	600	750	900	1050	1200	1350	1500	1650	1800	1950	2100	2250

Milliamperage (column headers: 50–1500)

current (Ma) is so vital to the x-ray exposure, the technologist should watch the Ma meter during the exposure. If the Ma selector is placed at 300 Ma for a given exposure, the Ma meter should read 300 during the exposure and the MaS meter for very short exposures. For various reasons, problems can arise throughout the x-ray circuits that would cause the meter to read slightly higher or lower. If this variation in Ma reading is great, the x-ray generator should be recalibrated by a competent service person. In retrospect, this is frequently the reason why technique charts do not always work: the chart might call for 400 Ma, but the generator was actually producing 275 which will yield an underexposed image.

If the Ma meter makes no reading at all, there simply is no tube current and no x-ray photons are produced. This can happen as a result of the Kv or Ma selectors not being properly seated. To remedy this, the technologist should simply turn the selectors a few stops over and then back again to the original setting. The milliamperage meter can also indicate what is known as a gassy tube. In this condition, the x-ray tube for various reasons has lost its vacuum to air and during the exposure not only the desired cloud of electrons move toward the anode but a high amount of *electrons* from the air as well. This produces a tremendous volume of tube current and the Ma meter will "spike" as it monitors the additional current moving through the tube. When this occurs the x-ray unit should be turned off immediately and a supervisor notified.

PATIENT DOSE AND MILLIAMPERAGE

Although the subject of radiation safety is not within the direct scope of this text, some mention should be made that milliamperage does affect patient dose. Very simply stated, when all other things are held constant, the more numbers of individual x-ray photons (primary radiation) used during an exposure, the greater dose the patient receives. An exposure of 200 Ma for one second at 70 KvP will produce two times the radiation dose to the patient than 100 Ma used for one second at 70 KvP.

The term *exposure rate* is often used in a radiological sense to indicate the intensity or concentration of a given x-ray beam. Exposure rate is often expressed in the form of m/r. Under these conditions one could expect to see that twice the number of photons would be produced in the x-ray beam if the exposure was changed from 100 Ma to 200 Ma. The exposure rate would *not* be affected, however, if the *exposure time* were changed from one to two seconds because the intensity or concentration of photons per a given time around the measuring device would not have changed, even though the total accumulative exposure was doubled.

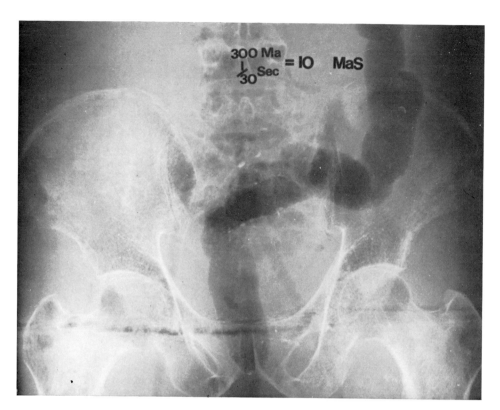

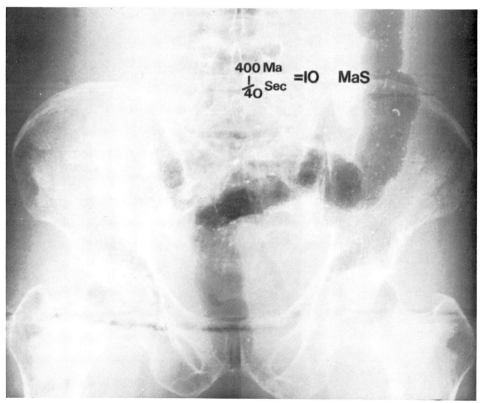

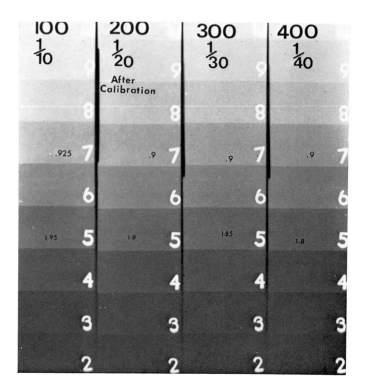

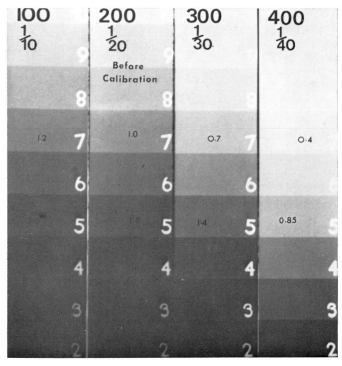

← Figure 80. Both pelvic examinations were exposed with different time and Ma values; however the MaS is equal so radiographic density should be very similar. Accompanying gray scales were made using the same MaS (10) but with different Ma stations. Note the variations on the before and after radiographs.

THE RADIOGRAPHIC EFFECT

Density and Milliamperage

In the previous pages we have discussed how the milliamperage selector is used to control the size of the space charge and how the tube current relates to the number of x-rays produced during exposure. Although this background is essential for the technologist to successfully adjust his exposure factors, he must also be able to predict how these electrical changes in the x-ray circuit will manifest changes in a radiographic image (see Fig. 81). The effect milliamperage has on the radio-

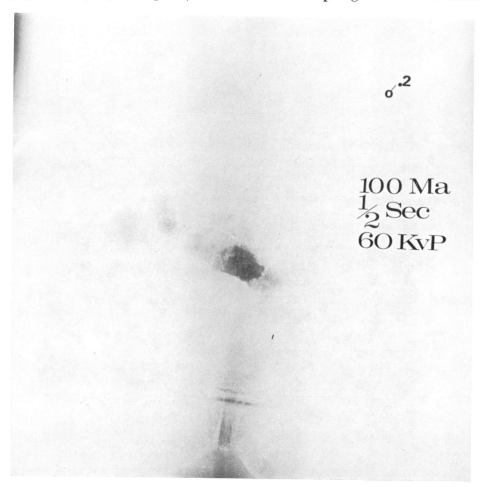

100 Ma
½ Sec
60 KvP

Figure 81A.

Figure 81. As the technologist selects higher Ma stations at the control panel, more filament heat is created in the x-ray tube and, as a result, a larger space charge is produced. This chain of events causes nearly proportional increases in film density, depending on processing and film type and whether screens are used. This series of radiographs is made with increasing Ma values.

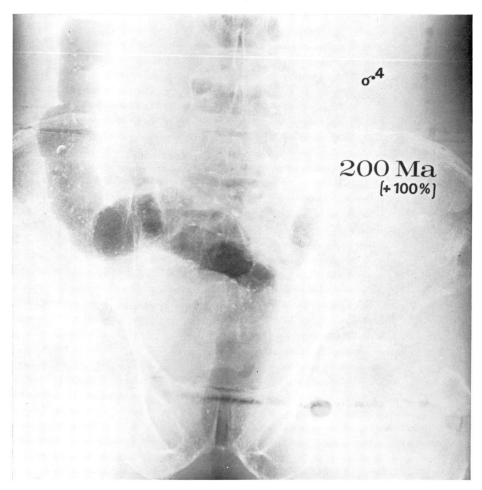

Figure 81B.

graphic image is primarily that of density. As the technologist uses higher Ma settings at the control panel, proportionately more x-rays are pro-duced, resulting in a greater accumulation of x-rays on the film's emulsion. Since radiographic density can be controlled by the amount of x-rays that have exposed the silver bromide crystals, if twice the number of crystals were ionized during an exposure, radiographic density would be doubled.

Contrast and Milliamperage

Within practical terms, milliamperage has no direct effect on radio-graphic contrast. The reason for this is that, under normal conditions, radiographic contrast is a measure of how evenly or unevenly the various body parts have been penetrated and the rate of scatter that was pro-duced by the patient during the exposure. Because milliamperage has no

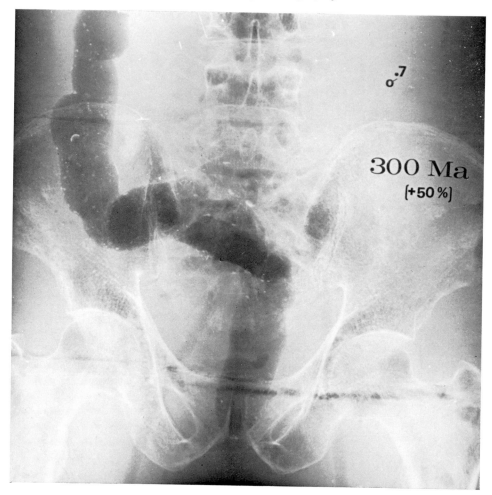

Figure 81C.

influence on these aspects of the x-ray beam, contrast is virtually unaffected by changing the Ma or the MaS. Figure 81 shows common Ma
exposure values and the resulting effect on density—as the technologist
increased the Ma, more radiation was produced and more silver bromide
crystals were ionized: As a result, the image increased in density. As
increased density occurs, radiographic contrast does not change substantially because the relationship *between* the densities remains fairly constant. As you review these radiographs, keep in mind that there is no
specific agreement, except for extreme cases, as to what level of radiographic density will produce optimal visibility of detail. In general, the
degree of radiographic density desired mostly depends on personal preference.

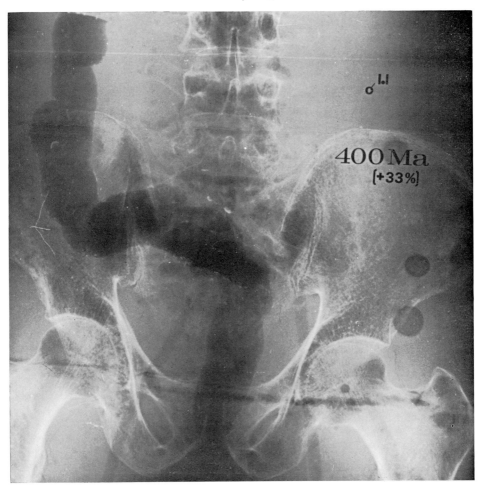

Figure 81D.

EXPOSURE TIME

The function of exposure time is to provide a duration in which the x-ray beam is allowed to irradiate the patient and the film. In general, radiography requires the shortest exposure time possible. The primary reason for this is to reduce the chance of patient motion which results in decreased sharpness. Exposure times commonly range from $\frac{1}{120}$ sec to 1½ sec depending on the examination. Exposure time can have a great effect on the life of the x-ray tube. Extremely short times or very long exposures can cause serious tube damage and thus must be taken into account when establishing technique charts and when calculating single exposures. The technologist should always have a tube rating chart at his disposal so that a given technique can be easily checked if it is suspected of being in the range of the tube limit.

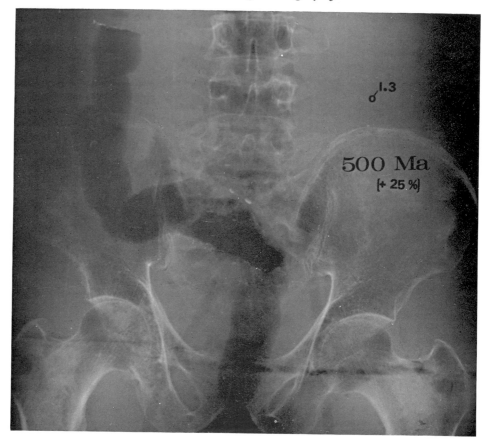

Figure 81E.

Very short times in combination with very intense exposures will cause tube damage as well. The reason for this can be understood if we consider how anode heat is generated. Each pulse of tube current that spurts from the cathode generates a tremendous amount of anode heat which must be dissipated very quickly. The heat is dissipated by the spinning anode and is carried by conduction to other parts of the tube's component parts as seen in Figure 82. If the total exposure necessary for the required image is spaced over one second for example, the total amount of heat generated by each burst of tube current can be properly spread over the entire surface of the spinning anode. If, however, an exposure time of $\frac{1}{40}$ sec was used, the intensity of the tube current must be increased approximately forty times to produce the same total amount of radiation for an equal exposure. This increased intensity of tube current causes a tremendous increase of anode heat which in a long run will decrease life expectancy of the tube or even cause serious damage immediately, depending on the actual exposure factors used and the capacity of the tube. With the $\frac{1}{40}$ sec exposure, the generated heat is not spread as uniformly over the spinning anode. If very short times are routinely used in combination with very intense exposure, uneven anode heat will result which, in turn, will cause uneven expansion of the anode. This condition if continued will eventually produce small cracks on the anode and eventually bring about the demise of the entire tube. It would be very difficult to state categorical rules regarding exposure factors and tube life because of the large number of variables in the type of x-ray tubes.

ANODE HEAT CONDUCTION

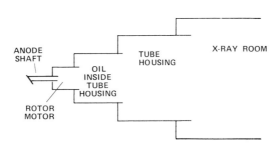

Figure 82. Without a method of anode heat conduction, the x-ray tube would become damaged after only a few exposures. In addition to what is shown, tubes can be purchased that have special oil or water circulating devices.

Some exceptions in the policy of using the short exposures to reduce motion: ribs, lateral thoracic spines, and sternum radiographs make long exposure times very beneficial. A long exposure time will permit enough motion of the surrounding body parts so that they blur, while the body part being studied remains still and more visible.

Exposure Time and the X-ray Beam

Exposure time has no influence on the composition or the characteristics of the x-ray beam. It does not influence the beam's intensity or penetrating power. It does have an obvious effect on radiographic density. As longer exposures are used (assuming all other factors are constant) the accumulation of radiation on the film increases. If exposure time is excessively long, more and more silver bromide crystals will acquire a latent image (become ionized), and radiographic density becomes excessive. Thus, exposure time has a strong influence on visibility of detail because of its effect on radiographic density. It also has some influence on sharpness of detail because of its effect on patient motion. The relationship between exposure time and radiographic density is nearly direct (as will be discussed in Chapter Eleven); however, a number of variables do influence this relationship so that once again a categorical statement regarding the exact effect exposure time has on density cannot be correctly made.

Exposure Time and Scatter Radiation

There is often a misunderstanding regarding the role time has in producing secondary scattered radiation. Scatter radiation (S/S) is somewhat proportional to changes in time and Ma. If a technique of 70 KvP, 100 Ma, at ½ sec is chosen, and the part to be radiographed is an abdomen, the primary to secondary ratio might be 1:7. This ratio will play an important role in the resulting radiographic contrast. If the exposure time is increased to 1 sec, roughly twice the total number of x-rays will pass through the body, causing an increase in radiographic density. However, since the kilovoltage was constant, the type of interactions produced (which is what determines the primary to secondary ratio) would be nearly the same although the numbers would be different, perhaps 2:14.

Automatic Timing Devices

An alternate method of accurately timing the x-ray exposure is commonly called phototiming or automatic timing. It involves the use of a device generally referred to as a photo pickup. It is composed of a small intensifying screen which is positioned in front of a small photoelectric cell. The photoelectric cell is capable of producing a small electric current after it has received a certain amount of light from the intensifying screen. The photoelectric cell is connected to the x-ray circuit in such a way that when it is stimulated by the fluorescent light to a predetermined limit, it will generate enough current to "trip" the exposure switch in the x-ray circuit thus terminating the x-ray exposure. There are many different designs and variations in manufacturing phototiming equipment. A discussion of their individual advantages and disadvantages is not as important as understanding the basic concept and how it relates to the radiographic image.

When using such a device, the technologist need only set the desired KvP (penetration and subject contrast) and Ma. The sensitivity of the photoelectric cell is usually set during the installation and except for minor adjustments left alone. It is the sensitivity of the photoelectric cell to the fluorescent light that determines radiographic density. When the photons strike the tiny fluorescent screen, the screen will give off light. As more photons reach the screen, its light output increases. This continues until it produces enough light to energize the photo cell, which then produces a small current and trips the remote exposure switch.

Automatic timing might at first appear to be a technologist's dream;

however, all the good features are balanced by a few important limitations. One of these is that phototiming often has difficulty in producing consistent radiographs when radiographing patients with a variety of diseases, i.e. excessive water accumulations. If the pickup (the fluorescent screen and photoelectric cell) happens to lie over a tumor for example, very little radiation will reach the screen, and the photocell will not become excited and produce current to signal the exposure switch. With this situation, the exposure will continue endlessly while the other areas of the body are being overexposed. The result of this would be that the film containing other body parts would have been "burned out".

A second limitation to phototiming involves positioning of the body part. With phototiming, the technologist has little positioning latitude. If the pickup is not positioned directly under the body part of interest, the probability is high that the resulting radiograph will be either too light or too dark. If the technologist will remember that the pickup is working "blind" and that it only "reads" the amount of radiation coming from the body part positioned immediately overhead and will control the entire exposure accordingly, it will be easier to understand the relationship between the equipment, positioning of the body part, and resulting radiographic density.

Reaction time of the photoelectric cell and its signal to the exposure switch is another important limitation with phototiming equipment. Reaction time is the total amount of time it takes for the fluorescent screen to emit light, plus the time it takes for the photoelectric cell to send its signal to the x-ray exposure switch, plus the time it takes the exposure switch to finally terminate the exposure. In many modern phototimed systems, regardless of the claims from manufacturers, one should not expect *consistent* automatic exposure control for times less than $\frac{1}{60}$ sec. If a chest examination is to be done on a moderate weight or thin patient using high KvP values of perhaps 140, and the Ma is set at perhaps 300, the exposure time should normally be about $\frac{1}{60}$ second depending on processing, speed of the film, and speed of the screens. However, if a phototiming device is used that has a reaction time of $\frac{1}{30}$ sec, the light from the screen may stimulate the photocell quickly enough, but the other components cannot respond to terminate the exposure before $\frac{1}{30}$ sec has passed. This would, of course, produce approximately twice the radiographic density that is needed. The only adjustment that can be made to compensate for these conditions is to reduce the Ma and/or the KvP. Figure 83 shows the major component parts as they might be arranged in a typical phototiming system. Only the specific body part being monitored for density by the phototimer is located immediately over the phototimer pickup (see Fig. 84).

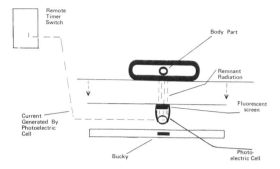

Figure 83.

One last factor is that the manual time selector will automatically override the phototimer device if it is set by error or intentionally to a lesser time value than might be needed by the phototimer. For example, an abdomen exposure is to be done using the phototimer and is automatically terminated at $\frac{4}{10}$ of a second; if the manual timer is set for $\frac{2}{10}$ of a second, it will conclude the exposure at $\frac{2}{10}$ sec. Thus, one must be careful to keep the manual timer selector at a point where it will not conflict with the automatic timing device. It is good practice, however, to keep the manual timer set close to what the automatic exposure will be in case the automatic system fails and a backup timer is needed to prevent a x-ray tube overload.

Despite the limitations pointed out above, automatic timing offers some important advantages. Considering these limitations, the only one

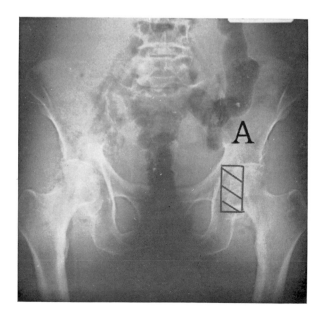

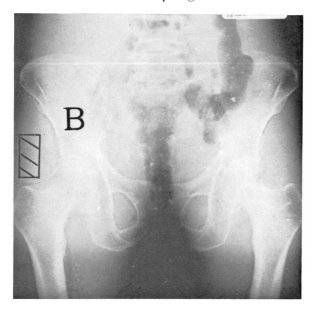

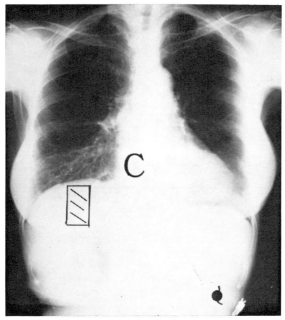

← Figure 84. If the body part is not correctly positioned over the photocell, improper radiographic density will result. In *A*, correct placement produced the desired results. In *B*, acceptable soft tissue density was produced, but the hip is too light because the photocell received enough radiation to terminate the exposure. In *C*, the rib study is too dark because the cell never received enough radiation to excite it and terminate the exposure.

that is beyond the technologist's control are those involving pathology. The technologist can soon master the problem of positioning with a reasonable amount of interest and ability. The major advantage with automatic timing devices is that they provide consistent radiographic densities from one patient to another and, as a result, over-all quality control is considerably more predictable. Because the KvP is held constant for each phototimed exposure, radiographic contrast is also relatively consistent. Along with this, one can assume a lower repeat rate as well, so patient traffic through a busy department can be managed more effectively. This will, in the long run result in a notable improvement in overall efficiency. In short, the phototiming device if properly used and calibrated cannot only improve overall departmental radiographic quality but increase efficiency as well.

CHAPTER SIX

FOCAL FILM DISTANCE

Introduction

THERE REMAINS ONE PRIME factor that has a profound effect on the radiographic image and does so in a much different way. In fact, focal film distance can complicate an exposure to such an extent it is seldom used as a variable when formulating x-ray techniques. In viewing this subject, it might be useful to once again consider a separation or independence between what has been termed the projected beam and the technical beam. If these concepts can be appreciated as two separate entities, and if it is kept in mind that the conditions affecting the projected (or geometric) beam *and* the technical beam must be optimal before a high radiographic image is realized, the study of radiography in general becomes more manageable. The technical beam consists of variation of intensity and quantity of remnant photons.

When the sun's rays shine through the atmosphere they pass many objects. The ability of the rays to pass through clouds of various thickness is primarily what can be considered its technical aspects (quantity and quality). Principally, these involve the frequency or energy of the sun's rays which are caused by the tremendous heat produced from the sun's burning gasses. The energy of the sun's light rays could be related to KvP, and the *amount* of gasses available for burning can be loosely related to the milliamperage if we were to use radiologic terms. As the sun's rays move toward the earth they carry bits of information about the contents of the atmosphere. On a cloudy day, for example, the bright sun is not seen because clouds have absorbed its light and there is some degree of darkness. If, on the other hand, the sun is shining brightly and there are only a few clouds scattered throughout the sky, we will observe shadows, which tell us something is above which the sun's rays are having difficulty penetrating. By observing these shadows, we may be able to tell something about the makeup of the clouds in terms of shape and thickness or density.

THE GEOMETRIC BEAM (OR PROJECTED BEAM)

The common sundial can help explain the projected or geometric beam. The projected beam automatically changes when there is a change in the direction or pattern of the rays emitted by the x-ray tube. Figure 85 shows a sundial and its shadow at various points as the geometry of the sun's rays change throughout the day. The shape and the direction of the shadow produced by the sundial change as the angle of the sun's rays alter in position. Imagine for a moment what might happen to the

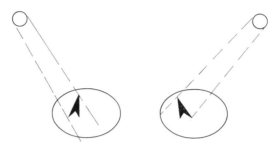

Figure 85. As the sun projects its rays toward the sundial, a geometric pattern (shadow) is projected onto its base. In radiography, similar geometric patterns are responsible for transmitting the various sharply defined structures seen in the radiograph.

radiographic image if the technologist is using the correct technical factors (MaS and KvP) but has incorrectly positioned the angle of the x-ray beam during an examination of a joint space. On reviewing Figure 86, it is easy to see that the position of the tube in A has *closed* the joint space, decreasing the diagnostic value of the examination. Thus, the projected beam affects the shape and the location of the structures seen on radiographs and can produce aberrations (distortions) of the actual body part. In general, such distortions or aberrations are not wanted because they make it very difficult for the radiologist to make a correct interpretation or assessment of the structures under examination.

FOCAL FILM DISTANCE AND THE INVERSE SQUARE LAW

As focal film distance (F.F.D.) increases, radiographic density decreases. The reason for this can be seen in Figure 87. Density readings change as the x-ray tube is moved further from the film. As a prime factor in radiography, focal film distance affects the concentration of the photons moving in the beam and as a result radiographic density is affected. It does this because as the focal film distance is changed, the configuration of the beam changes as well. In Figure 87, we see that if an x-ray tube is placed at twenty inches, with the collimator held in a fixed position, a certain field size is achieved. However, if the tube is moved to forty inches F.F.D., diverging photons allow an increased field size. The reason lies in the fact that x-rays travel from the x-ray tube in straight but diverging angles. If the technical factors were held constant,

SID = FFD (handwritten annotation)

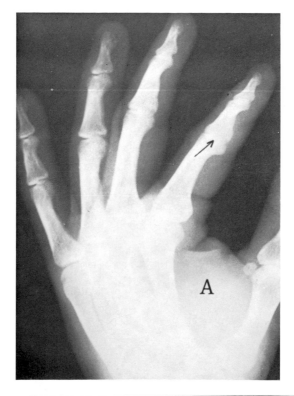

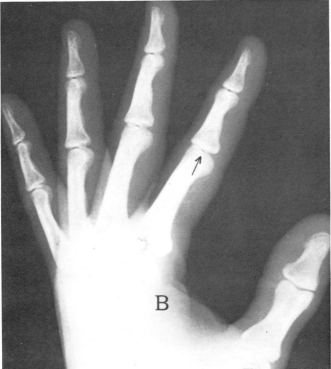

Figure 86. The geometric pattern of the beam requires the technologist to position the tube so the central ray (CR) passes through the body part properly. Angulation or decentering will "close" a joint on the radiograph.

there would be a marked reduction in the density of the image because the same number of x-rays are exposing a much larger area. The net effect of this is that x-ray photons are considerably less concentrated when they reach the film, which results in less ionization of the film's silver bromide crystals over a prescribed area of film.

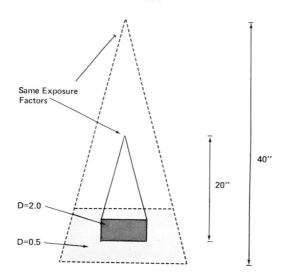

Figure 87. Radiographic density can be altered by changing the F.F.D., because as the focal distance is increased, the *same* quantity of primary radiation is spread over a larger area. The size of the area increases to the square of the distance and radiographic density changes proportionately.

This concept of the same number of x-rays covering larger or smaller areas of film as the result of changes in focal film distance can be expressed in a mathematical formula which tells precisely what the intensity of the beam will be as the focal film distance is changed:

New Beam Intensity $=$ (Original F.F.D.)^2 $\div$ (New F.F.D.)^2 $\times$ Original Intensity

Example: If the original beam intensity is 5r min and the F.F.D. is thirty inches, what would be the beam intensity if the F.F.D. was increased to sixty inches?

$$30^2 \div 60^2 \times 5 = 1.25$$

Example: If the original beam intensity is 10r min and the F.F.D. is fifty inches, what would be the beam intensity if the F.F.D. was decreased to twenty-five inches?

$$50^2 \div 25^2 \times 10 = 40$$

This indicates that the beam's intensity will decrease with increased focal film distance. The technologist must be able to determine the extent of the decrease in order to determine the effect this change will have on the radiographic image. For example, if a given set of exposure factors (MaS and KvP) is used with a twenty-inch focal film distance, the beam's intensity might be 2r per hour. If the F.F.D. is changed to forty inches, it can be determined from the formula that the beam's intensity will be ½r per hour. As a result of this decrease in beam intensity, there would be a corresponding decrease in radiographic density and the technologist must then increase his other exposure factors accordingly to regain the lost density.

The Inverse Square Law is stated as follows: The intensity of the beam is inversely proportional to the square of the distance. Very simply stated, if the focal film distance is doubled, the intensity of the beam will be reduced to one-quarter, as will be the density of the image. On the other hand, if the focal film distance is halved, the intensity of the beam as it reaches the film will be four times greater and the radiographic density will be increased approximately four times. The approximation is made here because the type of film used, as was stated in Chapter Two, affects how it will respond to changes in beam intensity. Figure 88 shows three radiographs using the same technical factors but different focal film distances. Because focal film distance has such a strong influence on radiographic density it can easily be seen why it is considered to be a prime radiographic factor.

FOCAL FILM DISTANCE AND RADIOGRAPHIC CONTRAST

It would be helpful to reflect for a moment on what has been said regarding radiographic contrast. First, radiographic contrast is *not* determined by the *amount* of x-rays produced in the x-ray tube but, rather, by their penetrating ability in combination with the different absorbing properties of the body part under examination (and the ratio of scatter to primary rays exposing the film). Considering the fact that focal film distance does nothing to the beam that would cause changes in penetrating ability, focal film distance would have no effect on radiographic contrast. It might be thought that focal film distance has an influence on the production of scatter and thus would have a strong influence on fog and radiographic contrast. Fog caused by scatter radiation is determined by the ratio of scatter to primary rays *in* the remnant beam and focal film distance has no effect on that ratio.

Choosing the Correct Focal Film Distance

Longer focal film distances will produce a radiograph with better sharpness. However, if too long a focal film distance is used, technical problems would result because other exposure factors would have to be adjusted to compensate for the loss in radiographic density. It has been pointed

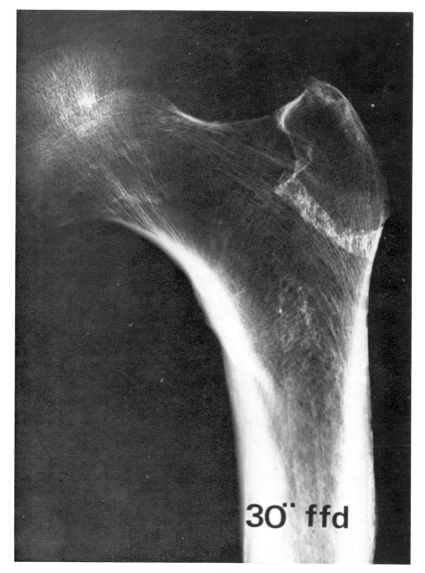

Figure 88A.

Figure 88. These radiographs show the dramatic effect that changes in F.F.D. have on radiographic density.

out that changes in focal film distance cause obvious changes in radiographic density, so a long focal film distance would require a significantly greater exposure to maintain the original density. The increase in exposure could be accomplished by one of three methods: (1) increased Kv; (2) increased Ma; and (3) increased exposure time.

If KvP is increased, there would undoubtedly be a change in radio-

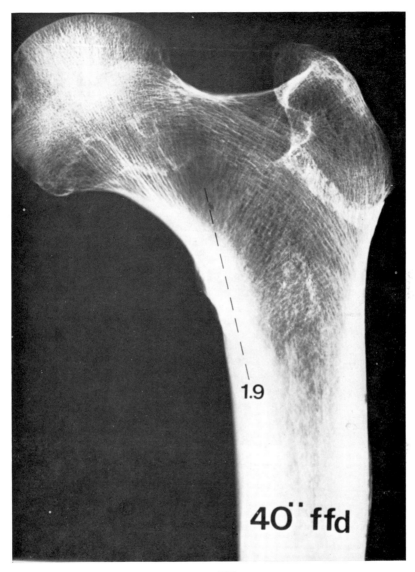

Figure 88B.

graphic contrast which might not be beneficial to the overall radiographic image. If exposure time is increased, the exposure would probably be so long that unsharpness would result from patient motion. If Ma is increased to compensate for increased focal film distance, the life of the x-ray tube might be put in jeopardy because of an overheated anode. For these reasons, focal film distance should be kept constant whenever possible. In the end, the bucky assembly has proved to be an important determining factor in establishing the appropriate focal film distance for most general

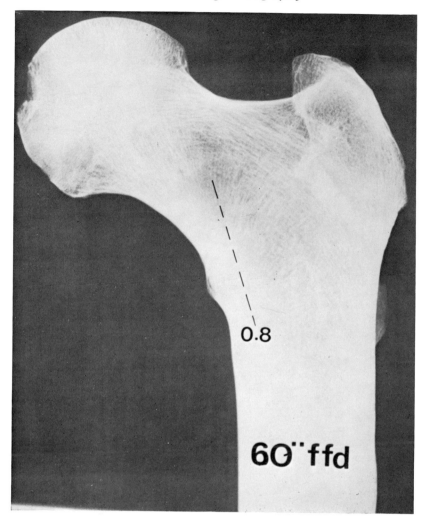

0.8

60" ffd

Figure 88C.

radiography. There is an air space of approximately 2 to 2.5 inches between the patient and the film. This space is needed for various thicknesses of cassettes and for the bucky to work properly, and a forty inch F.F.D. helps to reduce the unsharpness caused by the patient's distance from the film.

In addition to the effect focal film distance has on radiographic density, there is another factor in the geometric image which is very much affected by changes in target distance. As will be seen later, adjustments in the focal film distance has an effect on the sharpness of the various structures in the radiographic image. As focal film distance increases, sharpness or definition of detail increases; short F.F.D. exposures would result in de-creased radiographic sharpness.

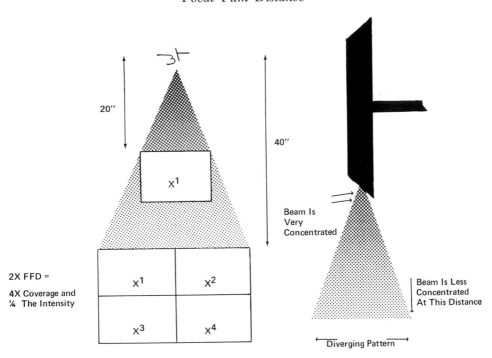

Figure 88D.

THE LINE FOCUS PRINCIPLE

In Figure 89 it can be seen that if a moderately large focal spot is used but is set at an extremely high angle from the horizontal plane, the film *sees* a smaller target area. This is known as *line focus principle* and is an inherent characteristic of all diagnostic radiographic x-ray tubes. In Figure 89, we see the differentiation between the actual focal spot and the effective focal spot. The actual focal spot is a reference to the actual area emitting x-ray photons and the effective focal spot is the area seen by the film. This inherent characteristic of the x-ray tube is basically advantageous for the technologist because it allows use of a larger target area of the x-ray tube, while a smaller focal spot is projected toward the film. When a large exposure is used there is, of course, more heat generated at the target. Unless this heat can be dispersed quickly, the anode of the x-ray tube will begin to crack and become permanently damaged. The use of larger focal spot sizes allows the tube to dissipate the heat away from the target more quickly, thereby decreasing the possibility of tube damage. Also a larger focal spot will permit the space charge to strike a greater area of the target reducing heat intensity over any one area.

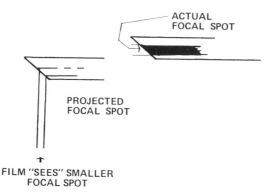

Figure 89. The line focus principle is the relationship between the size of physical focal spots and the size of the projected focal spot seen by the film.

The Point Source

An imaginary type of focal spot is known as the *point source*. A point source refers to an imaginary *point* on the anode from which x-rays might be emitted. For reasons involving tube heat loading, such a target is not possible in practical radiography, but it does serve nicely to illustrate an important concept regarding image sharpness and focal spot size. If an exposure was made with a point source, the projected beam would be slightly different when compared to one that was produced with a focal spot. Specifically, all of the x-rays would be emitted from one tiny spot, whereas with the focal spot or target x-rays would be emitted from all points along its surface (see Fig. 90). When using a focal spot, there are many points of origin for the x-ray photons, and as a result, the beam is composed of many photons criss-crossing each other as they move toward the film and through the body part under examination.

Contrasted with this is the beam produced by the point source, which is *clean* geometrically. In Figure 90, an object has been placed in the beam's path. The resulting image is known as an umbra; this is an ideal situation because there is no criss-crossing of individual photons, so a perfectly sharp image results. However, a focal area, not a point source, is used in radiography, and x-rays are emitted from many points of origin along the target. Under these conditions a notable decrease in sharpness occurs which is known as penumbra. The presence of a penumbra gives evidence that the *entire image* is unsharp. There is always some penumbra present in the radiographic image because of the many x-ray emission points from the target area; however, it can be properly controlled by using the smallest focal spot possible. A comment should be made from a practical point of view regarding this topic. It appears that the effect focal spot size has on radiographic sharpness is limited by the type of

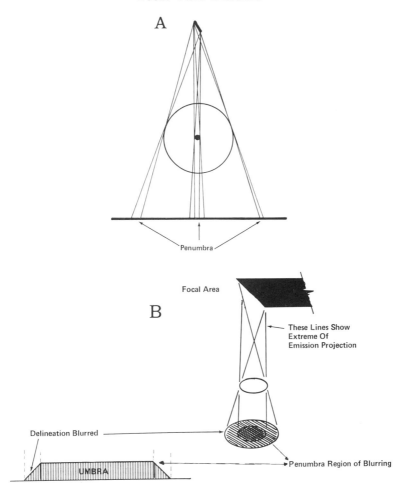

Figure 90. *A,* although a focal spot is the basic cause of penumbra, it must be used to reduce tube loading. *B* shows poor sharpness which is always present with commonly used focal spot sizes. This unsharpness is present throughout the entire image and not *just* at the penumbra.

screen used. It has been the author's experience that no clearly observable difference in sharpness occurs between large and small focal spots until detail screens are used. Thus, it seems changing focal spot sizes for the sake of showing recognizable improvements in sharpness, while using fast screens, is pointless and even counter-productive considering the increased heat load on the x-ray tube.

As noted a certain amount of object film distance, in addition to screens, causes more unsharpness. Thus, when one asks what effect focal spot size has on sharpness, the answer must be it has a limited effect depending on the type of screens used.

Focal Spot Size

In a practical sense, the focal spot size possesses an interesting problem to the technologist. Small focal spots produce sharper geometric images than larger focal spots, but larger focal spots can withstand a greater heat load which is a very important factor in tube life. Figure 91 shows the same x-ray tube's exposure capacity for various focal spot sizes. Generally, all radiographic x-ray tubes have two focal spots which are usually in the range of from 0.5 mm to 2.0 mm. Newer x-ray tubes, however, are being constantly developed so that smaller focal spots can be used without great fear of damaging the anode. The first of these developments was to change the metal used for the target itself, so it has increased resistance to heat. The x-ray tube is made of tungsten; however, it is coated with other substances that allow quicker heat dissipation than would tungsten alone.

Another advancement in x-ray tube design is the development of the high speed anode. Typically, a high speed anode turns at approximately 8,000 to 10,000 RPM, as opposed to the standard speed anode that rotates at approximately 3,000 RPM. The faster anodes have the effect of spreading the heat generated during the exposure over a greater area, thereby allowing greater heat dissipation and cooling. This is important when very high exposures are made singularly, and especially important when serial exposures are made. When heat loading becomes an especially serious problem, air or water cooled tubes can also be installed which have a definite advantage by increasing the tube loading capacity.

Three Ways to Control Penumbra

Penumbra is the area of unsharpness around the structure in the radiographic image. Figure 92 shows three ways in which penumbra can be controlled: (1) by reducing focal spot size; (2) by increasing focal film distance; (3) by reducing object film distance. Object film distance is the distance between the film surface and the object under examination. The use of a small focal spot size is important where small structures are being examined such as in mammography and various other types of special procedure work. The reason for this is that if the focal spot is larger than the object under examination, *that structure* will become less visible because of extreme penumbra (see Fig. 93).

Because of anode heat problems, manufacturers of x-ray tubes have been very concerned about making focal spots for special purposes as small as 0.3 mm. However, their use is considered by many radiologists to be absolutely essential in such special examinations as magnification techniques. Some extend this 0.3 mm prerequisite to include skull work and tomographic work as well. Here again options should be carefully con-

SINGLE RADIOGRAPHIC EXPOSURE RATINGS

SINGLE PHASE — FULL WAVE RECTIFICATION
STANDARD SPEED (60 Hertz)

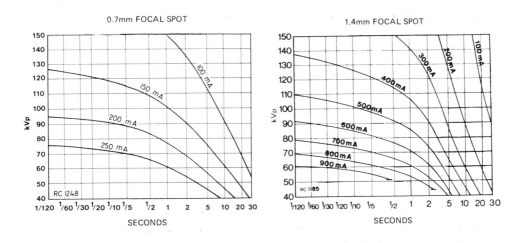

HIGH SPEED (180 Hertz)

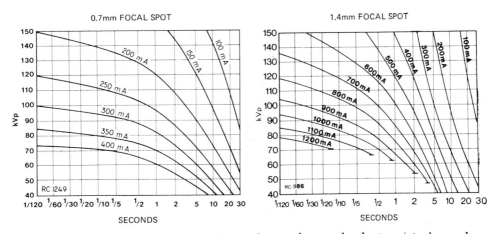

Figure 91. The technologist should know the anode speed, phasing (single or three phase), and the focal size of the tube being used before consulting a tube rating chart. *Courtesy of* Picker X-Ray Corporation.

sidered as small focal spots used with tomography could cause serious tube loading problems, especially with multidirectional tomography.

Figure 94 shows the difference in sharpness between a 2.0 mm focal spot and a 1.0 mm focal spot. In reviewing Figure 120, you will note that the effect of the focal spot size plays a more important role when the object film distance is increased.

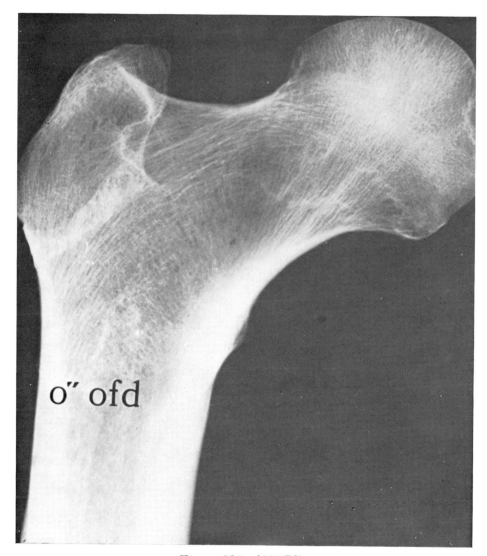

Figure 92A. (40″ ffd)

Figure 92. As the F.F.D. increases from 40 to 80 inches, sharpness improves even when using a larger focal spot.

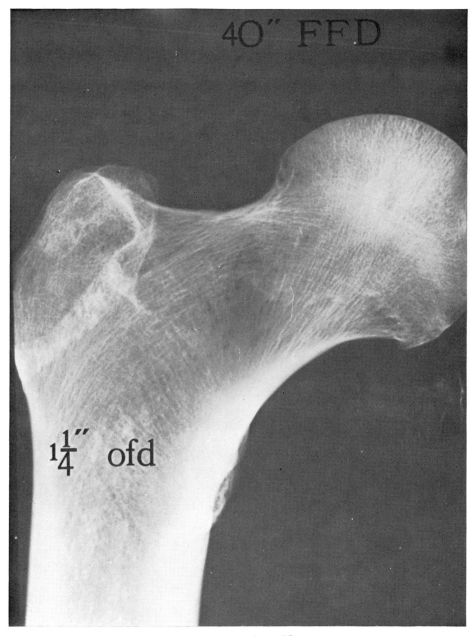

Figure 92B. (40″ ffd)

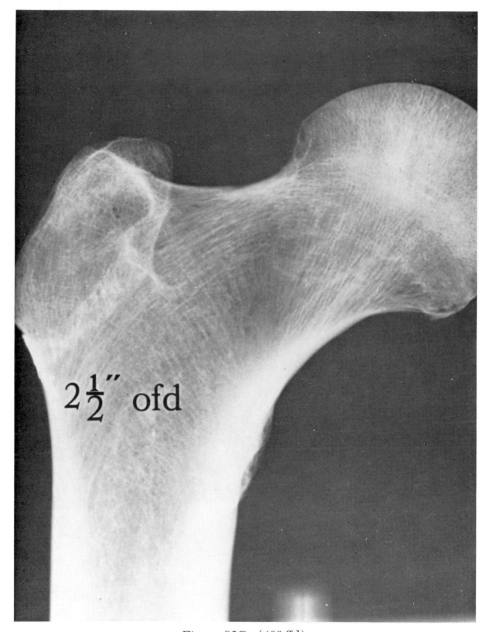

Figure 92C. (40″ ffd)

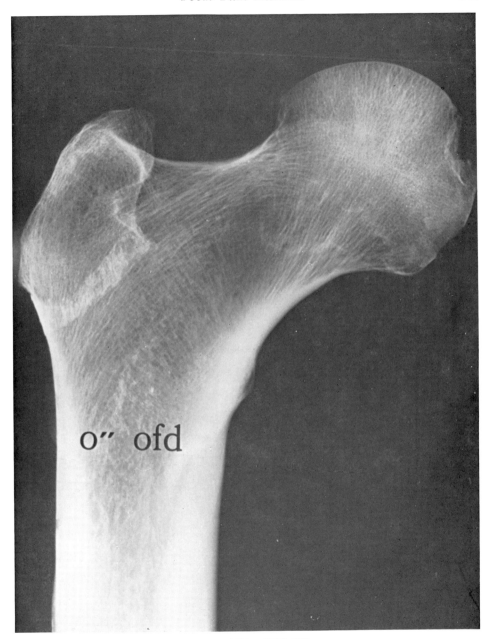

Figure 92D. (80″ ffd)

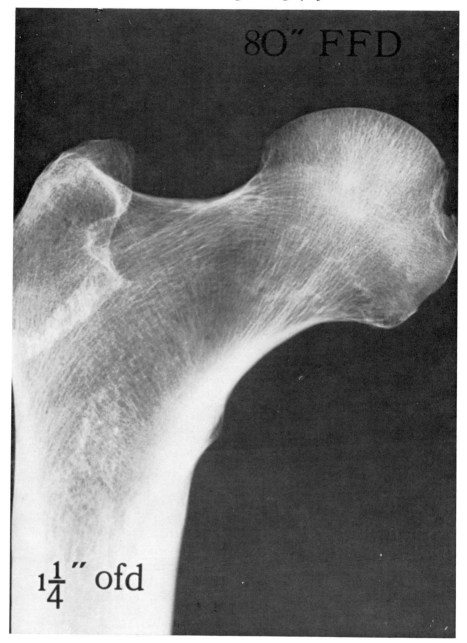

Figure 92E. (80″ ffd)

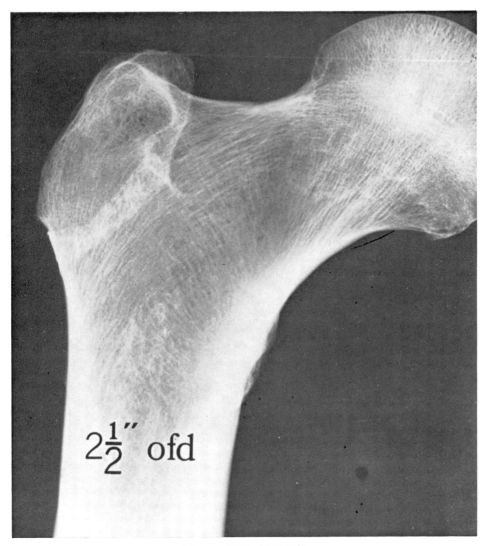

$2\frac{1}{2}''$ ofd

Figure 92F. (80″ ffd)

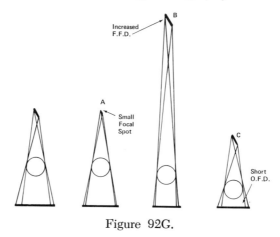

Figure 92G.

OBJECT FILM DISTANCE AND MAGNIFICATION

Object film distance (O.F.D.) must be kept to a minimum. In Figure 95 it can be seen that object film distance plays no minor role in its effect on the geometric image and the resulting radiographic sharpness. In general radiography, there is a necessity for at least a 2.5-inch O.F.D. so that the bucky assembly can work properly. As object film distance increases, penumbra increases. The reason for this is that O.F.D. causes enlargment or magnification of the image which makes the unsharpness more visible, and also the criss-crossing pattern of the beam is greater. Thus, increases in O.F.D. increase the combined effect of enlargement and beam cross-over so body structures become considerably more blurred. It is important to note also that magnification itself is not necessarily a negative aspect in radiology. It might seem so at first, because whenever magnification is produced without taking special precautions, a great deal of unsharpness and blurring results. It is not the enlargement or magnification per se that causes the unsharpness, it is the focal spot or target *area*. Magnifica-

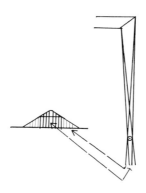

Figure 93.

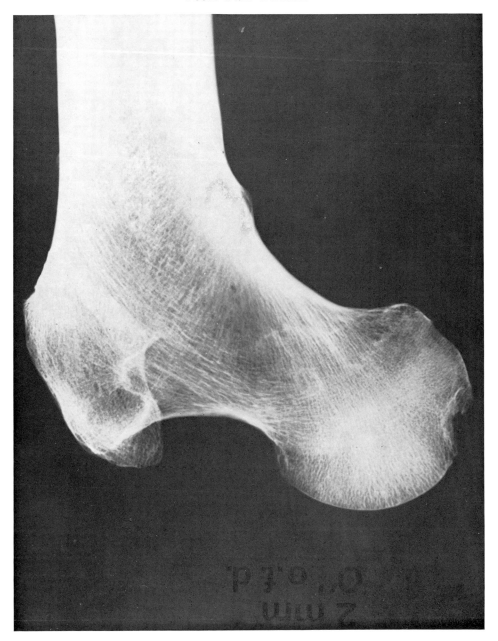

Figure 94A.

Figure 94. The effect of a small focal spot in improving sharpness is more obvious when there is at least a 2 inch object film distance. Otherwise, screen unsharpness has a more dominant role in affecting radiographic sharpness.

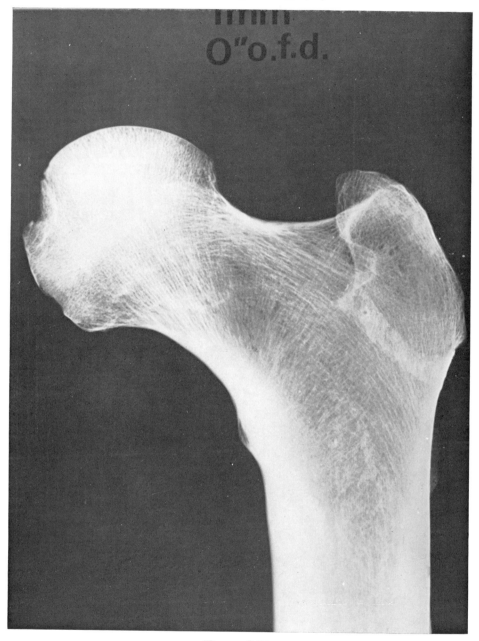

Figure 94B.

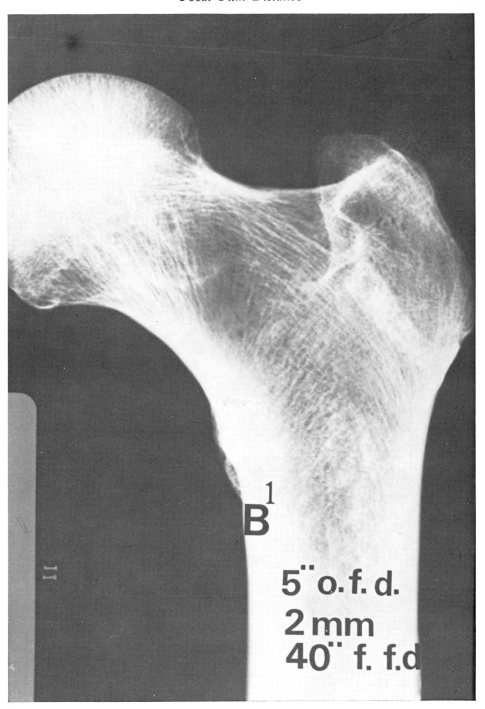

Figure 94C.

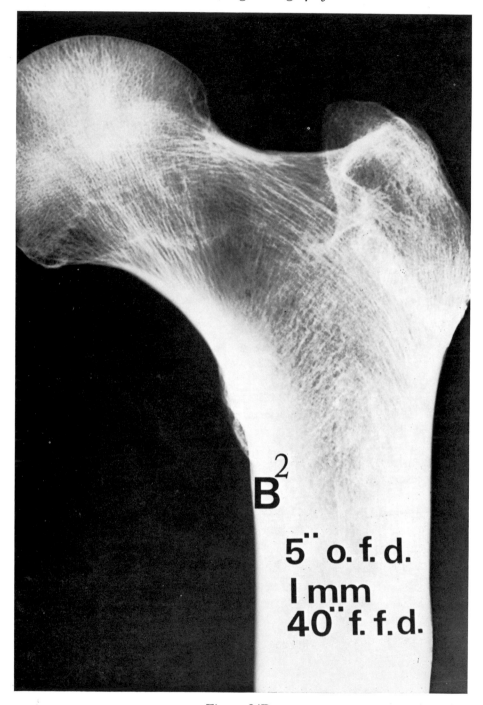

Figure 94D.

tion simply makes the existing penumbra more obvious. Magnification studies are used very often to show tiny but very important structures that might otherwise have gone unnoticed; when using magnification work, a fractional focal spot size is essential.

Common studies are performed when the location of a specific body part gives rise to a kind of inherent object film distance because of its location in the body. Thus, it is important to know that the body itself can often produce object film distance on its own. Occasionally, more than the usual 40 inch focal film distance is required to produce good radiographic sharpness of these body parts. In such an instance the body can be obliqued or rotated so that the immediate structure under examination is at the closest possible distance to the film. When the technologist comes upon a situation in which the object film distance of the body under examination cannot be changed by rotating the patient, the focal film distance should be increased. An example of this is the lateral cervical spine film. If the patient is in the lateral position and the film is touching the shoulder there is approximately a 6 inch object film distance, and if the focal film distance was not changed, magnification and decreased sharpness would result in the various structures of the cervical spine vertebral bodies. As a result of this, lateral cervical spine films are routinely done at 72 inches. Making spot films of a lateral thoracic and lumbar spine using 48 inch focal film distance would improve the radiographic sharpness of the vertebral bodies. Of course, when the focal film distance is increased in this way adjustments in the other exposure factors must be made to maintain radiographic density.

In summary, unnecessary object film distance should generally be avoided whenever possible because it usually results in decreased sharpness and in magnification which makes the unsharp structures more obvious. The technologist should increase focal film distance whenever possible to compensate for object film distance.

The Magnification Technique

Because radiologists have become accustomed to the present size of the various body parts as seen in the radiographic image and considering that one of the most important indicators leading to a roentgen diagnosis is the proper evaluation of structure size, there is some reluctance in using a magnification technique. However, the magnification technique has proven to be an important asset in roentgen diagnosis. This method is simply known as the magnification technique and it is most commonly employed while doing special studies involving vascular opacification. Cullinan and others, for some time, have advocated its use and laid important groundwork for its technical development and improvement. In Figure 96, you will see an examination of a renal arteriogram comparing conventional and magnification technique. Very little imagination is required to see the advantage this procedure offers when viewing small, complex vascular configurations.

Prerequisite for Magnification Technique

The primary prerequisite for accomplishing good magnification work lies in the use of what is known as a fractional focal spot. You will recall from the previous discussion that using a smaller focal spot increases radiographic sharpness. With magnification technique, the object film distance is varied to obtain the desired degree of enlargement. If the object under examination is located at the midpoint between the tube and the film, the resulting image will be approximately two times greater than the actual body part. The potential danger in doing magnification work is that the very small focal spot sizes required have very poor heat loading characteristics because of the tremendous heat concentrated on the relatively small target area. With this in mind the technologist must be extremely careful to choose exposure factors that will not exceed the tube limits of a single exposure or a rapid series of exposures. A second problem common to magnification technique is in centering the beam correctly. One should be able to imagine how a longer object film distance would affect the possibilities of having the image centered on the

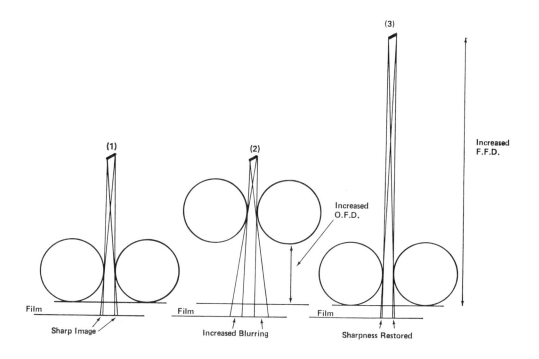

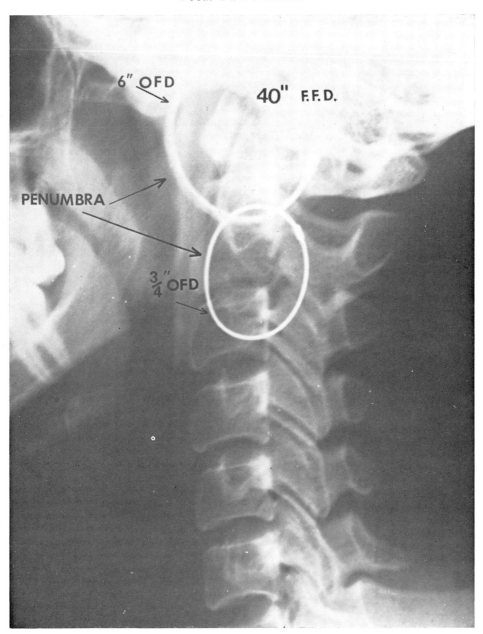

← Figure 95. Object film distance does not "cause" penumbra while focal area does. Increased O.F.D. enlarges penumbra and makes it more obvious. The root cause of this type of unsharpness is that photons are being emitted from a focal area in opposing tangent planes. The F.F.D. should be increased when possible if a large O.F.D. is present. If the cervical spine had been exposed at 72 inches, the variation in magnification and sharpness between the rings would be greatly diminished.

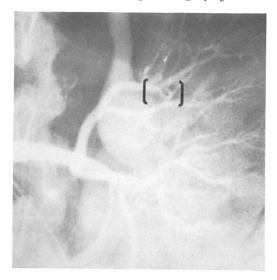

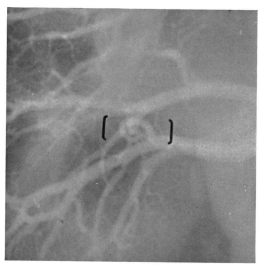

Figure 96. The magnification technique can be more easily appreciated after reviewing these radiographs.

film as the result of the diverging rays. The problem of centering is important in routine radiography, but with magnification technique it becomes crucial. If an object is not placed in the core of the beam but rather among the more divergent angled rays, the image will be shifted laterally to a certain extent. The degree of lateral projection would depend on how far away from the center ray the object is located. With magnification technique a small amount of decentering produces a disproportionate increase of lateral shifting and distortion (as noted in Fig. 97).

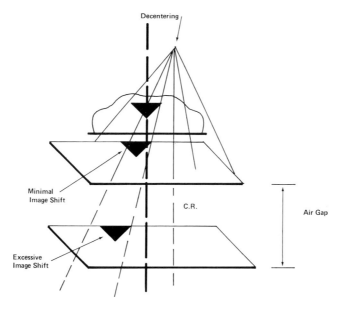

Figure 97. Decentering will be discussed later along with the air gap technique. It is imperative that the CR be centered directly over the body part.

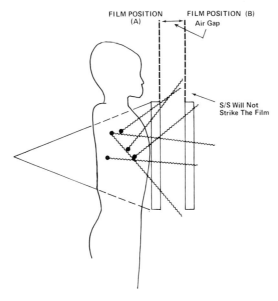

Figure 98. The air gap technique is much less effective when higher KvP levels are used, because scatter photons travel more directly toward the film.

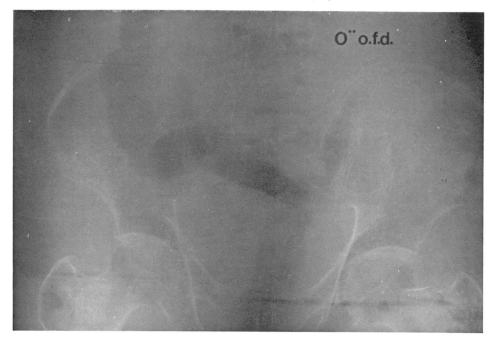

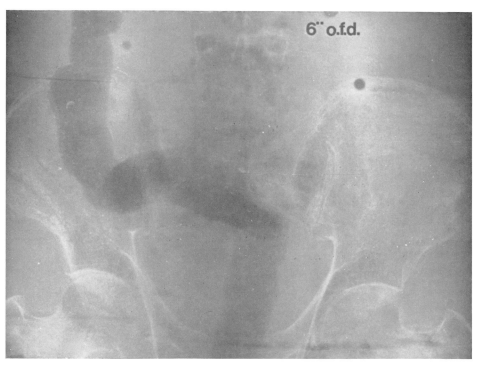

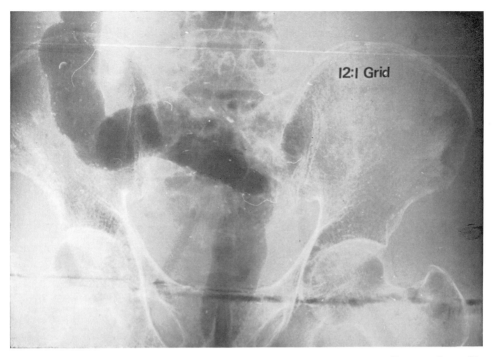

← Figure 99. Pelvis using the air gap technique to demonstrate its effect on the radiographic image. Because of the absence of scatter photons reaching the film, radiographic contrast increases and density decreases. Also shown is a comparison using a 12:1 grid.

In summary, magnification technique is an important tool to have at one's disposal and depending on the radiologist it is something that is not merely an option but essential. The major purpose is to enlarge the image and thus make more visible very tiny body structures that might otherwise go undetected.

Usually, an enlargement factor of 100 percent is used and is achieved by placing the object under examination at midpoint between the film and the x-ray tube. The primary technical problems with doing magnification work are tube loading and decentering. In reference to tube loading the technologist may consider one of the new rare earth type screens and moderately fast film. The rare earth screens fortunately help this problem by allowing a considerably shorter exposure to be used.

Object Film Distance and Scatter Radiation

Although it is important in routine radiography for the object film distance to remain as short as possible, occasionally a long object film distance can be used to the technologist's advantage. If the beam's normal diversion projection pattern is brought to mind, it can be seen that the projection patterns of secondary and scatter rays are much different (as noted in Fig. 98). Because scatter rays usually travel at a greater angle compared to that of the primary photons, if enough space is provided between the object and the film, many of these scatter photons emerging from the body may never reach the film. The primary rays, by comparison, are much more predictable and certainly more direct so we can be assured they will emerge from the body part correctly and will continue to travel in straight lines to the correct area of the film. The reason for our interest in reducing the amount of scatter is, of course, that excess scatter produces radiographic fog and decreased radiographic contrast. Thus, one can see how changes in object film distance can influence the amount of scatter radiation that will ultimately reach the film. When focal film distance is deliberately increased to produce this effect it is known as the air gap technique. The air gap technique requires a definite increase in focal film distance so that penumbra can be controlled. As with magnification technique, improper centering is also a potential problem. The author's feeling regarding air gap procedures is that, considering the problems even well trained technologists may run into with centering large patients for chest examinations, the added inconvenience associated with an object film distance of approximately 10 inches gives the procedure questionable value. Also, although many scatter rays are eliminated from reaching the film, many more could be eliminated with the use of a 10 or 12:1 ratio fine line stationary grid. Figure 99 shows the effect of an air gap with a pelvic examination.

Additional Advantages in Using Increased Object Film Distance

A more effective use of long object film distances is to improve visualization of structures that lie over one another in the body and result in superimposed structures on the film. Such superimposed body parts cause confusing densities in the radiographic image and often make the particular part under examination barely visible. In such instances, object film distance can help the radiologist and technologist. Figure 100 shows an area of the body that possesses this problem and thereby lends itself well to using an increased object film distance. There are other such areas of the body and if the technologist would consider this point for a second these may come to mind. In the meantime, the mandible is an easy and convenient example to consider. With the head in the lateral position one mandible is approximately 6 inches away from the film. With a 40-inch focal film distance, the two mandibles are superimposed making it very difficult to distinguish both structures radiographically. However, if the

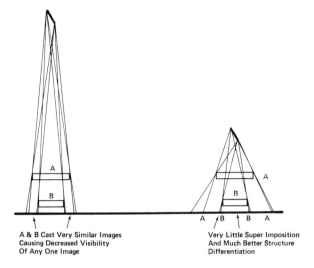

A & B Cast Very Similar Images
Causing Decreased Visibility
Of Any One Image

Very Little Super Imposition
And Much Better Structure
Differentiation

Figure 100.

focal film distance is reduced to 20 inches, the effect of the 6 inch inherent body object film distance becomes very much exaggerated and produces an extreme amount of penumbra. With a greatly increased penumbra, the sharpness of the structures begins to fade slightly as the more sharply defined structures of the body part close to the film become more visible by comparison. Occasionally technologists have reduced the focal film distance to the point where the collimator is almost touching the patient's face. This is successful only when the specific body part under examination is directly against the cassette. Very often bucky work, because of the 2.5 inch object film distance, is not as successful in producing the desired effect. In fact, the condyle away from the film is almost invisible because of the extreme penumbra produced by the combination of increased object film distance and short focal film distance. The net result of this technical manipulation of O.F.D. and F.F.D. is to reduce the visibility of the condyle which is not under examination, thus making the other condyle more visible. Once again it should be emphasized that the key to success when using this method is to keep the body part under examination not close but immediately against the film.

DISTORTION OF THE RADIOGRAPHIC IMAGE

Shape Distortion

Shape distortion occurs when the radiograph shows the object under examination to be different in form from the actual body part of the patient. Shape distortion (or true distortion as it is often called) is a condition that occurs automatically when x-rays pass through a body part at an incorrect angle or when the plane of the film is tilted (see Fig. 101). Magnification is a quite different situation than shape distortion. Often it appears that dramatic angulation of the center ray and the film often causes poor radiographic sharpness. However, the actual reason for this is not a result of the angled beam per se but, rather, a result of the increased object film distance that often occurs when the beam is angled (as noted in Fig. 102). It can be seen here that the projected image of the part under examination travels a further distance to the film as the tube is angled, thus producing increased blurring. Occasionally, angulation of the tube is necessary to improve the demonstration of certain body parts. However, if the beam is properly centered, shape distortion will be kept to a minimum. The Towne's view, for example, requires the center ray to be angled at least 30°. The resulting radiographic image gives the skull an oval shape which, of course, is shape distortion. The body parts inferior to the sella turcica seem to stretch endlessly toward the bottom of the film. Because the beam is properly centered, distortion of the body part immediately under examination is kept to a minimum. Tube angulation can also be used to reduce the superimposition of various body parts (as noted in Fig. 103). Here a sigmoid view of the colon made with the tube angled 35° causes an "artificial" straightening of the sigmoid. Many other possibilities can be mentioned here, however, it is sufficient at this point to simply present the idea of angulation of the beam to achieve certain positive results, and alert the reader to be watchful of the negative effect that improper angulation can have with regards to shape distortion.

Size Distortion

Size distortion is frequently used as a synonym for magnification. Size distortion is caused by the fact that x-rays are emitted from the target in increasing divergent angles as they pass through the body and reach the film. Object film distance and focal film distance determine the extent to which magnification or size distortion occurs. It might be helpful to use some examples so the relationship between focal film distance, object film distance, and magnification can be more easily analyzed. It should be understood that there is no standard object film distance or focal film distance that can produce optimal results. Magnification is controlled by the relationship between focal film distance and object film distance. For example, a good radiographic image can be obtained with a 2 inch object film distance and a 40 inch focal film distance, but an equally good image would be obtained if a 4 inch object film distance were used, providing an 80 inch F.F.D. is used. This is an important point to understand because

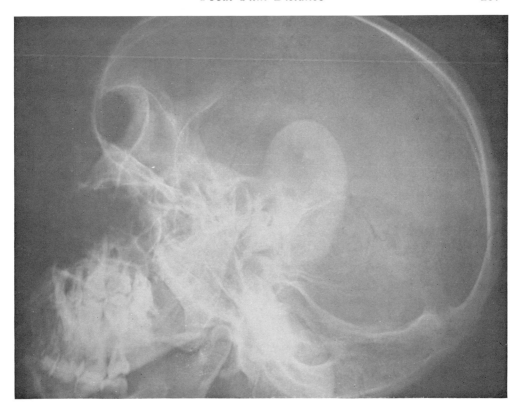

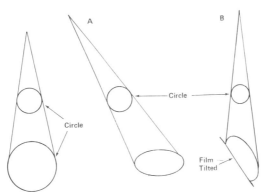

Figure 101. This properly positioned lateral skull is distorted because of faulty tube alignment.

when doing unusual or difficult cases, such as cross-table, operating room, or portable work, placing the body part directly against the film is very often impossible. However, if the technologist keeps in mind that the critical factor for good sharpness and minimal size distortion is to produce a proper relationship between object film distance and focal film distance, a sharp and nondistorted radiograph of the body part can be produced. As noted earlier, when the technologist comes upon a situation where the object film distance may not be or can not be decreased, the focal film distance should automatically be increased. Figure 104 summarizes these concepts.

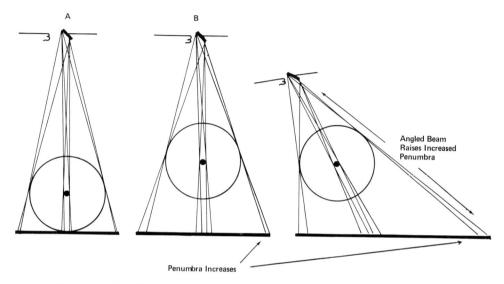

Figure 102. The O.F.D. increases automatically as the tube is angled.

Stereotactic Procedures

There are some occasions where proper centering is required well beyond what is normally achieved in routine radiography. A surgical procedure known as stereotactic demands unusually precise centering and no distortion whatever. The procedure itself involves the placement of a small electrode into the pituitary gland of the brain. This is done by advancing the electrode through a bore hole in the patient's skull and inserting the probe through a very narrow space between the two hemispheres of the brain. It is impossible for a surgeon to know exactly where the electrode is during this procedure without x-ray control. During the procedure, a series of radiographs are thus made to pinpoint the progress and location of the electrode as it is moved through the brain to the pituitary gland. Needless to say, anything less than absolute precise calculation is not acceptable as permanent brain damage could result. Once the electrode is in place, a small electrical charge is placed to the gland with the intention of permanently deadening its activity. Precise centering with zero magnification and shape distortion are characteristics of the stereotactic x-ray procedure.

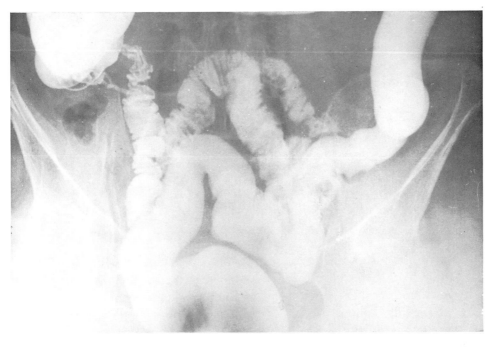

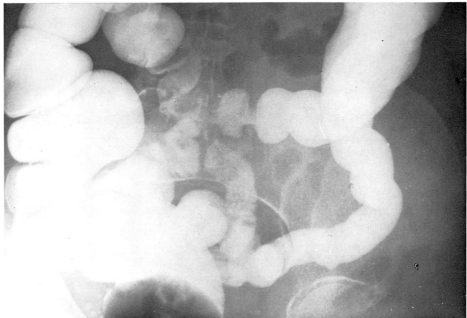

Figure 103. The angled CR makes an area of many superimposed structures more visible.

MAJOR GEOMETRIC FACTORS		Radiographic contrast	Radiographic density	Size Distortion	Shape Distortion	Geometric Unsharpness
F.S.S.	Focal Spot Size					X
F.F.D.	Focal Film Distance		X	X		X
O.F.D.	Object Film Distance	X*	X*	X		X
A.C.R.	Angle of Central Ray				X	

* Only when increased o.f.d. allows scatter radiation to escape and not expose the film.

Figure 104. How the various geometric factors affect the radiographic image.

A pelvimetry examination is also very important geometrically. The purpose of this procedure is to supply the gynecologist with information pertaining to the size of the baby's head relative to the dimensions and position of the mother's birth canal so that a decision can be intelligently made regarding delivery technique. Accurate calculations regarding the degree of magnification are very important here as well. The relative sizes of the baby's head and the mother's birth canal are very important measurements and thus must be radiographed with a minimal amount of distortion using proper centering and positioning of the center ray.

The procedure for radium implant localization is another example of how much the projected image influences the radiographic image. Here, again, measurement calculations are necessary and distortion must be kept to a minimum by using proper centering.

Although the various items discussed in this chapter may at first seem confusing or complex, actually they are not. The illustrations shown will help you understand how each of the geometric factors discussed affects the projected image and ultimately the radiographic image. There are a number of variations in which these factors relate to each other and produce various effects, however, if the basic concepts of focal film distance, object film distance, focal spot size, and tube angulation are kept in mind these variations can be worked out with some individual thought.

In time, a confidence will begin to emerge that will considerably benefit the technologist when he is confronted with a difficult emergency radiographic situation. The sense of accomplishment as a result of performing one of these exams is well worth the initial effort.

The Anode Heel Effect

A chapter covering the projected or geometric image would be incomplete without at least a brief discussion of the anode heel effect. This

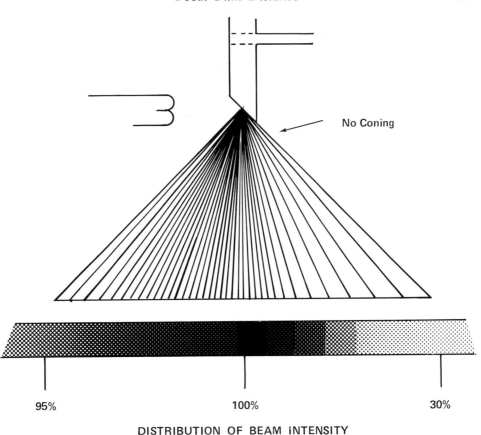

No Coning

95%   100%   30%

DISTRIBUTION OF BEAM INTENSITY

Figure 105. The variation of beam intensity is primarily at the peripheral edges, not in its core.

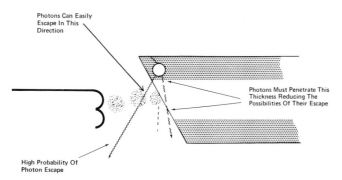

Photons Can Easily Escape In This Direction

Photons Must Penetrate This Thickness Reducing The Possibilities Of Their Escape

High Probability Of Photon Escape

Figure 106. As the newly "born" x-ray photons pass through the thicker portion of the target, fewer are able to escape causing an inconsistency of beam intensity.

area of radiography is so often confused and misused that some clarification needs to be made with regard to its practical application in radiography. This phenomenon of the heel effect is caused by the angle (bevel) of the target. Figure 105 illustrates schematically the anode heel effect as it relates to the variation in intensity along the long axis of the x-ray tube. It should be pointed out that the anode heel effect has little practical influence on the radiographic image. As was pointed out earlier in this chapter, the x-ray beam travels from the target in a divergent pattern with increasing angles from the core of the beam to its perimeters. The extreme changes in beam intensity at the perimeters is the result of partial absorption of the x-ray photon in the anode material itself (see Fig. 106).

It is true that such intensity in beam variations exists; however, you will note that this variation occurs primarily at the outer peripheral rays— during routine radiography an assortment of beam-limiting devices are always used, thereby (see Fig. 107) eliminating these peripheral rays from affecting the radiographic image. If two chest exposures are made, one with the patient's diaphragm under the cathode portion of the tube and the other with the patient's diaphragm under the anode end of the tube, there will be very little difference between them.

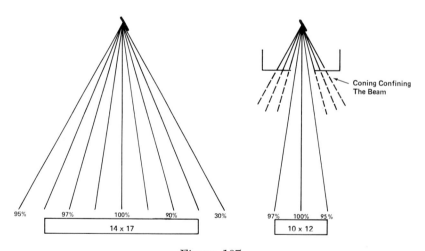

Figure 107.

In summary, the anode heel effect is caused by partial absorption of the x-ray photon by the anode; it should be understood that there are three factors that affect the extent to which this occurs radiographically. One, of course, is the use of beam-limiting devices. As the collimator is closed more tightly, the possibility of an x-ray film experiencing different intensities of the beam is decreased substantially. Second, as the angle of the anode increases from a horizontal plane, the heel effect becomes more exaggerated. The third major factor that influences the anode heel effect is the focal film distance, as shown in Figure 108. As focal film dis-

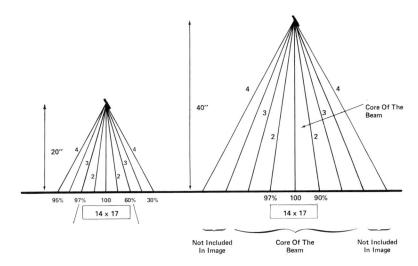

Figure 108.

tance increases, the possibilities of the anode heel effect being seen in the image become considerably decreased because more of the center portion or core of the beam is utilized for the exposure.

CHAPTER SEVEN

KILOVOLTAGE

CHAPTER FIVE described the method by which electrons are produced to form the space charge. The space charge alone would have no value to the technologist if it did not also possess an additional form of energy that could later be converted to x-ray energy. Frequently technologists feel that it is the electron itself that undergoes the conversion process into x-radiation. Instead, it would be more correct to say that the electrons making up the space charge are simply vehicles upon which kinetic energy rides to the anode. The electrons collide with the anode but then pass through the anode into the high tension circuit and past the Ma meter. We will discuss in this chapter why kilovoltage, an electrical term, is so important to the technologist in producing a radiographic image.

DEFINITION AND FUNCTION

Kilovoltage is electrical pressure caused by an abundant supply of electrons at one side of the circuit. It is the nature of this imbalance to "want" to become balanced. This "need" manifests itself in a situation where the electrons at the oversupplied (negative) side of the circuit will begin to move toward the deficient side to effect the balance (see Fig. 109). When one electrode with a very strong negative (−) charge is placed near a second electrode with a positive (+) charge, there is a natural tendency for these charges to attract each other and eventually equalize or balance. The electrical pressure that was caused by the inequity between the two electrodes causes electrons to flow, and in so doing, current is registered. The level of kilovoltage in the circuit is determined by the degree of imbalance between these charges, and as the electrical imbalance of the charges increases, electrical pressure would likewise increase. This would result in greater voltage (faster moving electrons through the circuit). The voltage will continue as long as there is an electrical imbalance, but as soon as the imbalance of charges is equalized, the voltage will drop to zero and the electrons, of course, will stop moving. This concept is relevant to radiography because the technologist uses

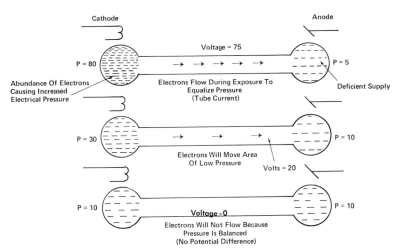

Figure 109. The speed of electrons moving from cathode to anode is dependent on the imbalance of electrons in the x-ray tube.

KvP to set up various levels of electrical pressures within the x-ray tube. This in turn has a very strong effect on the quantity and quality of x-ray photons that will ultimately be produced. It is a common misconception that Ma is the only factor that influences the total quantity of primary radiation, however, it has been well documented that KvP plays an important role in determining the quantity of primary radiation in a given x-ray exposure as well.

Kilovoltage, Tube Current, and the X-Ray Tube

As you read through this chapter, many important factors will be discussed. However, they will all be totally dependent upon the influence Kv imposes on tube current. In order to produce an x-ray beam, three criteria must be met: (1) A supply of electrons is needed; (2) they must be given a high level of kinetic energy; and (3) the electrons must be suddenly stopped thus converting their kinetic energy into radiation.

The supply of electrons, of course, forms at the filament of the x-ray tube as a result of filament heating (see Fig. 110). These electrons (in the tube) act as vehicles upon which kinetic energy rides. Earlier, it was said that uneven or imbalanced charges seek to balance themselves. The x-ray tube has two electrodes known as the cathode and the anode. Under normal conditions, the cathode has the negative (−) charge, which means it has an abundant supply of electrons, and the anode has a positive (+) charge, which means that it has a deficient amount of electrons; under these conditions, electrons will begin to flow from the cathode toward the anode. X-rays are produced when the electrons of the tube current strike and interact with the atoms of the anode.

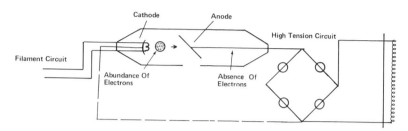

Figure 110.

To explain the process of how kinetic energy is converted into x-radiation, an analogy of an automobile collision has some application. If we observe an automobile moving at the speed of ten miles per hour and it collides with a stone wall, a transformation or conversion of energy will result. The kinetic energy built up by the motion of the car suddenly becomes nonexistent when the car reaches the wall and is stopped. Under the law of conversion of energy, something must happen to that kinetic energy, because "energy cannot be created nor destroyed." What actually occurs is that the kinetic energy transforms into heat and sound waves. If we observe the same type of car moving with a higher level of kinetic energy of 40 miles per hour, the higher level of kinetic energy built up in the increased speed would be transformed into a higher level of heat and sound.

A similar kind of energy conversion takes place in the x-ray tube as the kilovoltage is raised and lowered to control the speed of tube current. It should be pointed out that the car functioned as tube current does in carrying kinetic energy. However, the kinetic energy in the x-ray tube is converted into heat and x-radiation. The x-ray tube then is simply a device used to convert kinetic energy into x-ray energy, but it is very inefficient because up to 99.8 percent of the kinetic energy of the tube current becomes heat and only 0.2 percent is eventually converted to x-radiation. The x-radiation emitted by the anode in this conversion process is known as primary radiation (see Fig. 111). It should be noted that the *actual percentages* will increase or decrease as different KvP levels are used; however, it has been generally held that 0.2 percent is a fairly accurate figure to use for discussion purposes.

In summary, this discussion can be outlined as follows: X-rays are formed when a high level of kinetic energy is achieved by the space charge in traveling from the cathode to the anode. This radiant energy is of a high frequency (short wavelength) and the specific level of the kinetic energy of the tube current is, of course, dependent upon the kilo-

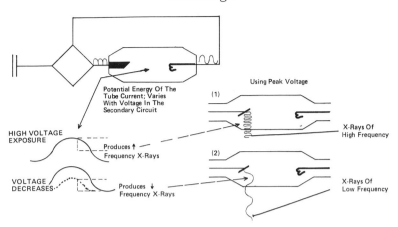

Figure 111. Whether high or low energy x-rays are produced, interactions between the anode and the space charge cause primary radiation. As tube voltage increases, more efficient conversion takes place and a higher primary beam is produced.

voltage set by the technologist before the exposure is made. The higher the kilovoltage used, the greater the electrical pressure will be between the cathode and anode, and the faster the electrons will move toward the anode. This will cause greater quantities of primary x-rays and greater x-ray energy (penetrating power).

KvP = ⋏ Kinetic energy = ⋏ Conversion Numbers = ⋏ Energy of x-ray photons

Kilovoltage and the X-Ray Circuit

It is important that we draw some concepts regarding how the electrical pressure is generated in the x-ray tube. Figure 112 shows a very simplified drawing of an x-ray circuit. We will begin at the primary circuit where the incoming current is taken from a main 220 volt supply. The autotransformer is responsible for adjusting this incoming voltage to achieve higher or lower kilovoltages. It does this by using a series of "taps" that are directly connected to the KvP selector located on the x-ray control panel. The selected voltage passes through the primary circuit to the high tension transformer where it is boosted approximately 500 times its previous value. If the autotransformer produces a voltage of 100 on the primary side of the circuit, it will become 50,000 volts after it reaches the secondary of the high tension transformer.

Figure 112 shows the relationship between the high tension transformer, the cathode, and the anode. It can be seen that, when a technologist chooses a high KvP setting, an electrical pressure is set up accordingly in the tube. It can also be seen from the illustration how the x-ray tube actually bridges the connections between the filament circuit and the high tension circuit. During the exposure, the filament circuit

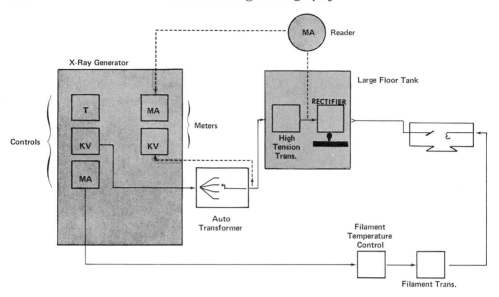

Figure 112.

produces the space charge which is then driven toward the anode by the high voltage.

BEAM QUANTITY AND QUALITY

If one considers *primary radiation* as possessing two separate aspects or characteristics, it will be easier to understand the concepts of beam quantity and quality. Further, if the technologist can gain an understanding of how these characteristics can be adjusted and balanced in each exposure, he will have gone a long way toward becoming a competent radiographic technologist. One can correctly say that, under normal working conditions, a radiographic image is very dependent on the balance between quantity and quality of radiation produced and, as was pointed out earlier, the KvP chosen by the technologist for a given exposure has a great deal of influence on both these beam characteristics. As was noted earlier, KvP was traditionally not thought of as playing a significant role in affecting the quantity of x-ray photons produced at the anode; however, this myth should be discarded, but this certainly does not deny the very strong effect MaS has on beam quantity. However, MaS does not *exclusively* control beam quantity. This can be summarized in the statement below:

$$\text{KvP} = \frac{\text{Beam Quality}}{\text{Beam Quantity}} = \frac{\text{Radiographic Contrast}}{\text{Radiographic Density}}$$
$$\text{MaS} = \text{Beam Quantity} = \text{Radiographic Density}$$

Kilovoltage and Beam Quality

An x-ray photon can have various penetrating abilities depending on the KvP. A low KvP exposure will produce x-ray photons with relatively low penetrating abilities, and a high KvP exposure will produce a beam with higher energy and penetrating abilities. If all other factors are held constant, as KvP increases, more photons will penetrate the patient and make a darker image. Penetrating ability, beam quality, and beam energy are terms that are synonymous. The terms wavelength and frequency can be confusing. Figure 113 shows two x-ray beams: As the frequency increases, wavelength decreases.

Wavelength Distribution

The x-ray beam is made up of rays known as photons or quanta. In a normal diagnostic exposure, these individual *packets of radiant energy* (quanta) vary from one another in their individual frequency (see Fig. 71). For example, if the technologist chooses 80 KvP at the control panel, a wavelength of perhaps 0.3 angstroms will be predominant, but the beam will also contain photons ranging in wavelength from several angstroms to a minimum of 0.155 angstroms. Each diagnostic exposure has some variation in wavelength, but they do make up a "common" identifiable beam. If an x-ray beam contained photons that were very much different from each other with regard to frequency, the term heterogenic, poly-

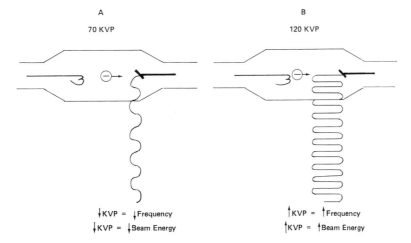

Figure 113. Increases in KvP produce a beam with overall greater frequency (energy) x-ray photons even though not all of these primary photons are equal in frequency. These gray scales demonstrate the effect KvP has on contrast (the same objects were used in all cases shown). As the KvP was increased, the objects' parts experienced different absorption rates with respect to each other, altering subject contrast and ultimately radiographic contrast.

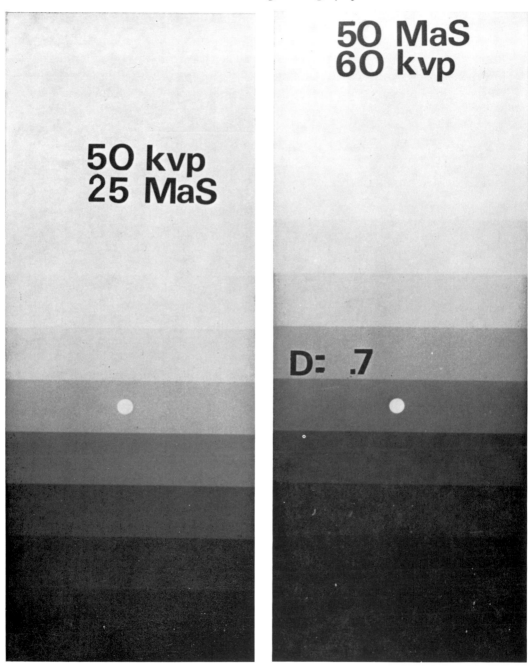

Figure 113A.   Figure 113B.

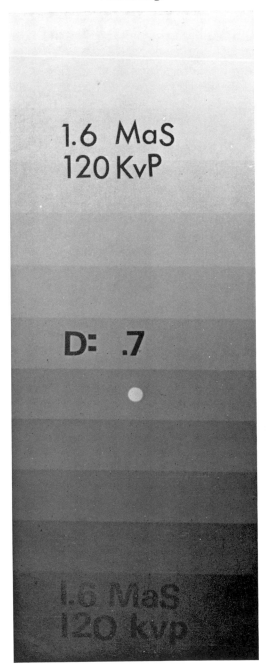

Figure 113C.

chromatic, or polyenergetic would apply. However, if another beam were emitted from an x-ray tube that contained photons with the same frequencies (equal penetrating power) the term homogenic or monoenergetic beam would apply. This is an extremely important concept because the predominance of high or low energy photons in a beam almost totally influences subject contrast and, ultimately, radiographic contrast. The thickness of filtration also influences the mixture of photons in a given beam.

KILOVOLTAGE AND SUBJECT CONTRAST

Controlling subject contrast is absolutely vital to good radiography, and the main factor used to control subject is kilovoltage. Subject contrast exists when two or more body parts absorb x-ray photons differently. Figure 114 illustrates the radiographic effect of subject contrast. As was mentioned earlier, the body has a complex arrangement of absorbing tissues and structures. Indeed, the body has characteristic "regions" of subject contrast, such as the chest, abdomen, extremities, and the skull.

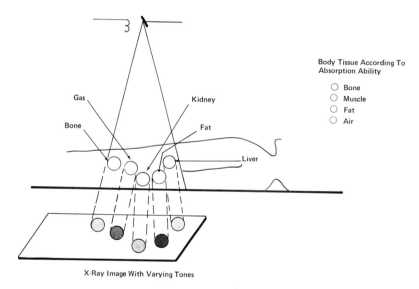

Figure 114. Some of the absorption differences in the body.

The chest exhibits the ability to absorb x-ray photons differently and, accordingly, it is often referred to as a high subject contrast region. The abdominal region, however, is something quite different because its individual body parts are composed in such a way that they absorb the beam much more uniformly and, understandably, this region is known as a low subject contrast area. As will be discussed later in the text, one of the

most important steps the technologist must take in producing a high quality radiograph is to know how to adjust KvP so the beam exiting the body
(remnant radiation) will produce the best possible image. The remnant
beam will be given special attention in Chapter Eleven, but a brief discussion is required here to show its relationship between subject contrast
and the radiographic image.

In order to explain subject contrast, we will use the example of a
stained glass window made with pieces of variously sized and colored
glass. These variously colored pieces of glass *individually* allow only so
much outside light to pass through. The observer is actually experiencing
how these opposing *dark-* and *light*-colored sections are absorbing the
sun's light to cause different degrees of light intensity. The observer's eyes
sense these wide *variations* in *light intensities.*

In Figure 115, we can imagine that one glass section is made with
similar light-absorbing abilities, such as light green, light blue, and beige;
these *similar* colors will absorb the sunlight uniformly, causing low subject contrast. In a section of window with an arrangement of sharply con-
trasting colors of glass, the sunlight is absorbed by the individual pieces
of glass with greater variance, and we perceive this as having a high subject contrast.

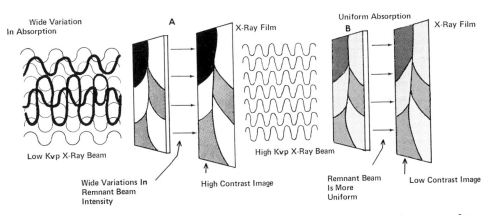

Figure 115. The primary x-ray photons vary in their absorption, resulting in subject
contrast. Subject contrast is affected by the characteristics of the body and the energy
of photons attempting to pass through.

The various body parts also allow x-ray photons to penetrate and be
absorbed. If a body region contains structures that absorb the beam at
widely varying degrees, the remnant photons emerging from the body
part will have widely varying intensities and, as a result, will cause corresponding densities on the x-ray film producing a radiograph with high
contrast. A lower subject contrast would result if individual structures

produced a remnant beam with somewhat uniform intensities (resulting in a more uniform density pattern on the film's emulsion).

Later, we will discover more about why different body parts have different absorbing abilities; however, for now it is important that we understand that the beam's ability to pass through the body is controlled by the technologist as he manipulates KvP. These two factors determine whether or not the intensities of the remnant beam will be uniform. Thus, the degree of variation or uniformity in the intensities of the remnant beam will ultimately affect the degree of radiographic contrast. This important concept is outlined in the two statements below:

(A) High KvP exposures result in photons of higher energy, yielding uniform penetration and producing films with low subject contrast and low radiographic contrast.

(B) Low KvP exposures result in photons of lower energy, yielding a variance in penetration of body tissues and producing films with high subject contrast and high radiographic contrast.

EXPOSURE LATITUDE AND KILOVOLTAGE

The word latitude, in general, is a term that tells the degree of acceptance or tolerance. When exposure latitude is used in a radiologic sense, it tells to what extent an exposure can be changed before the diagnostic quality of the image is threatened. In other words, exposure latitude tells how much an exposure can be altered before a noticeable change results in diagnostic value. Kilovoltage has a unique effect over exposure latitude, as you will see in the following example. If an exposure was made at 60 KvP, an increase to 69 KvP should approximately double the radiographic density over what was obtained with 60 KvP. If, however, the *initial* KvP was 110, an increase of 16 would be necessary to achieve the same radiographic effect. Because of this, it is said that KvP influences exposure latitude (see Fig. 116). As KvP increases, exposure latitude widens or increases as noted in this example. With this in mind, it can be seen that there are some advantages to using high KvP exposures. In general when high KvP techniques are used, technique adjustments become less critical. However, there are two side effects associated with high KvP exposures (100 KvP and over) that are not necessarily advantageous.

1. As KvP increases, additional scattered photons reaching the film can cause fogging.
2. Often the body parts under examination have a very low inherent subject contrast; when a beam is used that penetrates that part too uniformly, subject contrast becomes too low, producing low radiographic contrast.

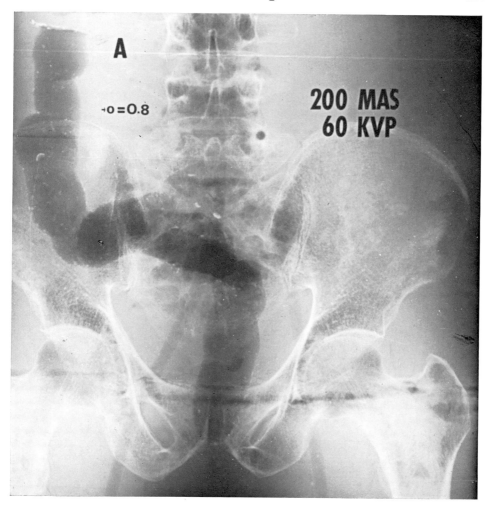

Figure 116A.

Figure 116. Similar density changes have resulted between *A* and *B* and between *C* and *D*: A 9 KvP increase in the 60 KvP range produced approximately the same density changes (*A* and *B*) as a 16 KvP increase in the 110 KvP range (*C* and *D*).

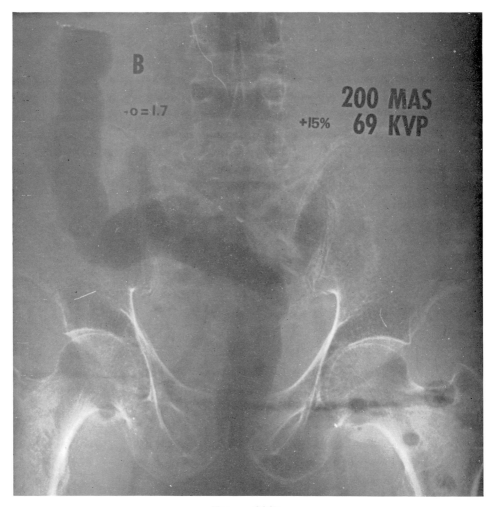

Figure 116B.

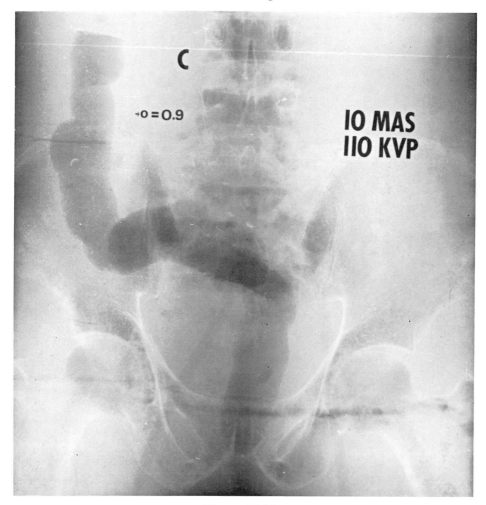

Figure 116C.

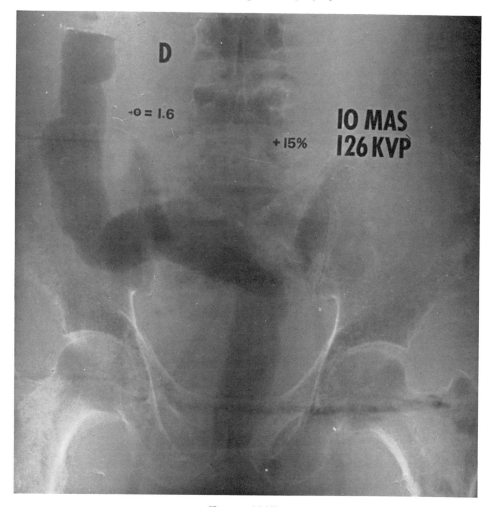

Figure 116D.

Kilovoltage, Patient Dose, and Beam Efficiency

Not only does KvP have a strong effect on the radiographic image, it affects the patient dose as well. As we know, the body has great variation in structure, which causes the beam to have more or less difficulty in penetrating the body while en route to the film. As an x-ray photon passes through the patient, it will undergo one of the following: (a) total penetration; (b) total absorption; (c) partial absorption. (See Fig. 117.) The type and number of interactions produced in the body during an exposure has an effect on the quality of the image as well as patient dose. The level of KvP used for each exposure determines, to a large extent, which of the resulting interactions will be dominant. A high KvP setting, for example, will produce a higher level energy conversion at the anode

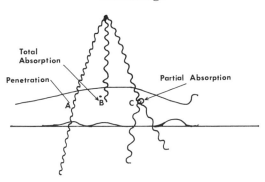

Figure 117: The *type* of photon-tissue interactions that result during an x-ray exposure is greatly affected by the energies of the individual photons—which are controlled by KvP.

and will, in turn, produce a more penetrating beam. With such an exposure, fewer numbers of photons are needed to obtain the same density and result in an overall lower patient dose. A lower KvP exposure, on the other hand, will produce a lower energy conversion at the anode and the beam's penetrating ability will be greatly diminished as more absorption type interactions are likely to occur. If a beam of low penetrating energy were to be used, a greater quantity of radiation would have to be produced to assure sufficient radiographic density. Thus, beam efficiency is an important concept for the technologist to keep in mind when formulating x-ray exposure factors: As KvP increases, beam efficiency increases and patient dose decreases (Fig. 118 illustrates this point). The first condition noted in this figure indicates comparable radiographic density is achieved with less patient dose and less tube current. Keep in mind that when less tube current is used, as seen here, the anode can operate at a lower heat

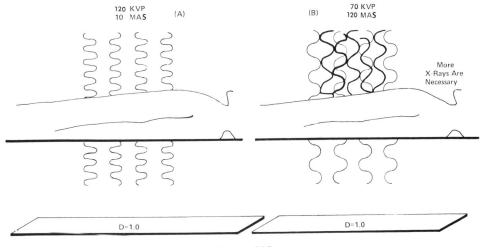

Figure 118.

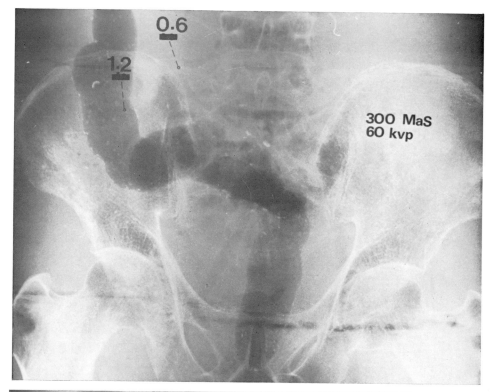

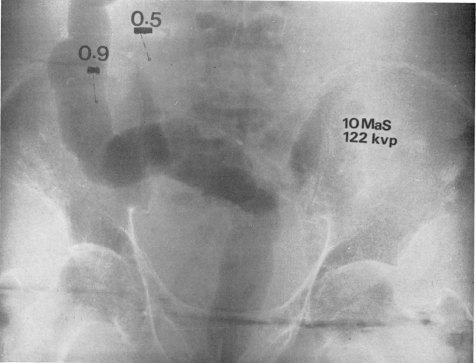

Figure 119. The effect KvP has on subject and radiographic contrast is indicated by the actual densities shown. Observe the MaS values necessary to produce similar densities when the KvP is changed. Also the densities shown indicate changes in contrast.

level which enhances its life expectancy. With this, one can correctly say that kilovoltage levels have an influence on patient dose by virtue of their effect on beam efficiency. If higher KvPs are used, fewer x-ray photons are needed to produce the same radiographic density. Figure 119 shows radiographs that have been exposed with high KvP and low MaS compared with low KvP and high MaS exposures.

KILOVOLTAGE AND SCATTER PRODUCTION

Diagnostic exposures cause secondary radiation. However, for practical purposes, the energy of secondary photons produced in a typical diagnostic exposure have no effect on the radiographic image. On the other hand, scatter production is very influential over the appearance and diagnostic value of the radiographic image. The total amount of scatter production is dependent on a number of factors. Although they will not be discussed completely until Chapters Eight and Nine, it is appropriate that they be listed here. As long as the amount of filtration is held constant, the five items below will control the total amount of scatter radiation produced in a given diagnostic exposure.

1. Tissue density
2. Thickness of tissue
3. Atomic number
4. KvP level
5. Field size

Between the two extreme cases of total penetration and total x-ray absorption, the intermediate case of partial absorption (or scatter production) exists. One may correctly say that scatter rays are a type of "byproduct" that if not controlled will cause radiographic fog. When one realizes that up to 90 percent of the x-rays reaching the film during an exposure of an abdomen are scattered rays, one can more realistically appreciate how important it is to choose the proper level of KvP in order to prevent an excessive amount of scatter production. Although there is scatter production present in all x-ray exposures, it must be kept in mind that radiographic contrast does not necessarily deteriorate due to large quantities of scatter, but rather because of an imbalance in the ratio of scatter to the number of primary photons that strike the film. The proper ratio of scattered to primary photons in the remnant beam must be maintained for good radiographic quality to result.

When the ratio of scatter to primary photons is out of balance, radiographic fog results. Figure 120 illustrates this important concept. The term scatter ratio, of course, refers to the relative amounts of scatter and primary photons that are present in the remnant beam. The most im-

portant type of interaction producing large amounts of scatter in the diagnostic exposure is known as the Compton interaction: As kilovoltage increases beyond 70 the relative numbers of Compton interactions produced
in a given exposure increase substantially. In other words, as the KvP increases over 70 scattering type interactions are more common when compared to the total absorption type, known as photoelectric interactions,
and will result in a condition known as scatter fog (which greatly reduces radiographic contrast and visibility of detail by producing an additional veil of density over the radiographic image).

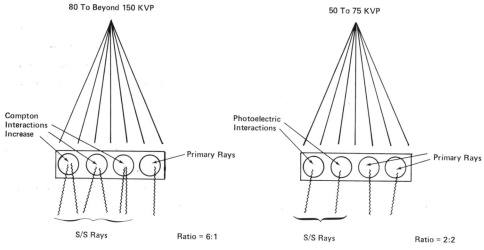

Figure 120. As the quantity of scattered photons increases relative to the quantity of
primary radiation, the possibility of fog increases dramatically. The figures used are
hypothetical.

There is another phenomenon that occurs with high KvP exposures
which increases the likelihood of scattered rays reaching the film. With
lower KvP exposures, the scattered rays travel in greatly divergent angles
as compared to the primary photons. Because the scattered photons are
so divergent in direction many do not reach the film. However, as higher
KvP values are used for a given exposure, the scattered rays produced
tend to move in a more direct line toward the x-ray film. The result of
this, of course, is that the film experiences a greater exposure by scattered
rays and the probability of fog increases. Also, these more direct scattered photons have greater penetrating ability, causing even more scattered photons to reach the film with predictable results on the radiographic
image.

Thus, kilovoltage has a tremendous effect on radiographic contrast
because of its control on scatter production and its effect on subject con-

trast. Lower kilovoltages produce more of the total absorption type of interactions in the body (commonly known as the photoelectric effect). This type of interaction is much more sensitive to differences in tissue composition, i.e. tissues with varying atomic numbers and densities. It should be pointed out further that this type of interaction, although it may produce a radiographic image with higher contrast, also produces an increase in total patient dose. Thus, once again, a kind of trade-off must be made between the type of radiographic contrast desired and patient dose when establishing a technique chart.

After reading through the material in this chapter, one might have the impression that it is generally a good practice to lower the kilovoltage to control the relative amount of scatter and increase subject contrast. Before any such changes are made one must carefully consider that kilovoltage is the only prime factor we have available that controls the penetrating ability of the beam, and if it is lowered too much, the body part will not be penetrated adequately (see Fig. 121). KvP levels should be set to primarily establish proper contrast and adequate penetration of the part being examined, and then be left alone using MaS to regulate radiographic density. The method used to control scatter should be left to the use of proper coning and proper selection of grid ratio.

Base KvP is a term that is used to describe the least amount of KvP that will provide adequate penetration for a particular body part. Although the base KvP does not apply perfectly for every radiographic condition, it is generally an adequate guide and will serve the technologist well in arriving at a valid starting point for establishing new exposure factors.

Base KvP = body part thickness in cm × 2 + 30 KvP

Since the average KvP applied across the x-ray tube is greater with three phase equipment, lower KvP settings will be required. To find base KvP for three-phase equipment, it would be more correct to add 20 KvP instead of 30.

Kilovoltage, Radiographic Density, and Contrast

No mention has been made, up to this point, specifically regarding KvP and radiographic density. We already know that, if the milliamperage is raised, the quantity of primary photons being emitted by the anode is raised by roughly the same proportions and radiographic density increases. If KvP is increased from 70 to 80, the second exposure will produce approximately twice the density. Thus kilovoltage does have an effect on radiographic density, caused by the fact that KvP produces more energetic photons which will more easily pass through the body. Also as KvP increases, the number of primary photons increase as well because more

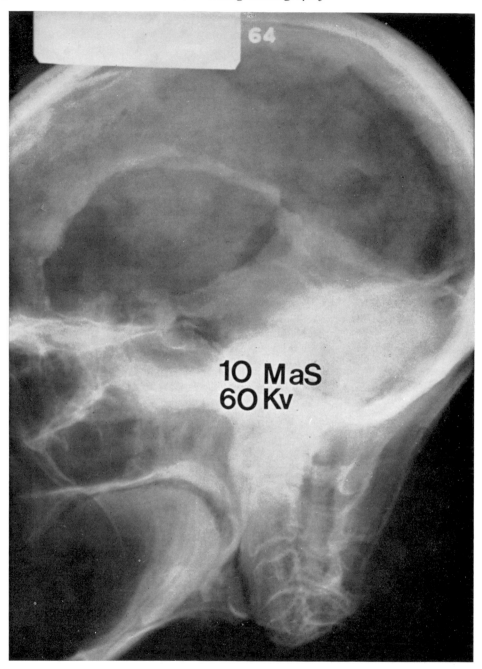

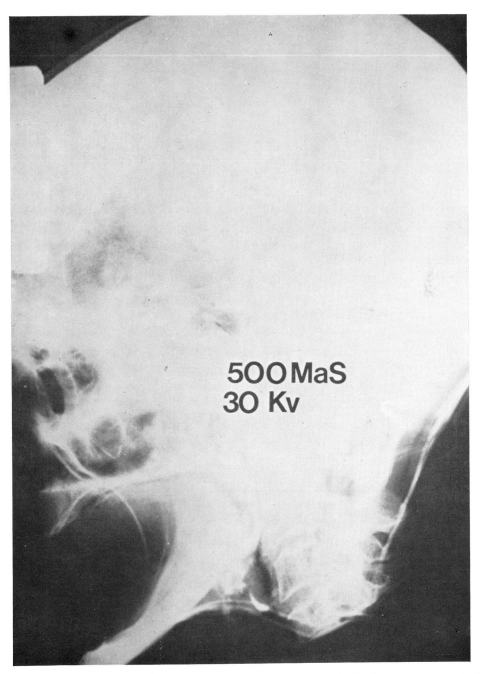

← Figure 121. If the KvP is lowered too much for a particular body part, no practical increase in MaS will produce a diagnostic image.

efficient conversion occurs from the tube current's kinetic energy into x-rays.

Radiographic contrast is the difference between two tones or densities. For example, if density "A" is 2.0 and density "B" is 1.0, the contrast between the two is 1. If density "C" is 0.5 and "D" is 0.75, there is a 33 percent change in density if we measure "C" against "D". However, if we wanted to measure "D" against "C", the value would be 50 percent.

As was pointed out above, KvP influences radiographic contrast as a result of its effect on the subject. Recall the description of the stained glass window: The eyes responded to the glass pieces that absorbed the sun's light as radiographic film responds to the varying intensities of the x-rays passing through the body. In one area there was uniform light penetration and, as a result, variation in light intensity was low, yielding a low contrast image. In the other section of the stained glass, the light penetrated nonuniformly. In this instance, a high contrast situation existed which was transferred to the film. In a sense, the x-ray film "sees" and records these various intensities of the remnant beam (see Fig. 115).

An IVP examination is illustrated in Figure 122. The area of the opacified kidney and those structures immediately adjacent to it demonstrate a high subject contrast. Note how this changes as the kilovoltage is increased. Lower KvP exposures produce a predominant number of total absorption type interactions (photoelectric) that are very sensitive to body parts with respect to their varying atomic numbers and tissue densities. This can be seen in the statements below:

(1) Decreased KvP produces increased numbers of total absorption interactions, yielding increased subject contrast and increased radiographic contrast (also less scatter).

(2) Increased KvP produces increased partial absorption interactions, yielding more uniform penetration of body parts and decreased subject contrast (also increased scatter production and decreased radiographic contrast).

From the statements above and by reviewing Figure 122, we can begin to see how KvP levels should be chosen for a particular body part. When radiographing a body part with a high inherent subject contrast, low KvP chest exposures probably would yield excessive radiographic contrast. Chest examinations lend themselves well to high KvP exposures. Also, because the chest region contains a large amount of air along with low density lung tissue, scatter production can be controlled very readily by using a high ratio grid. When radiographing an abdomen, it would be wise to use a lower KvP; this would produce more total absorption versus total penetration type interactions, and greater variation in tissue absorption would result. Subject contrast would increase and scatter production would be kept to a minimum as well. The net result of this would be higher radiographic contrast with better visibility.

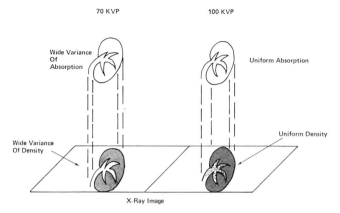

Figure 122. Since many absorption interactions are present in a low KvP exposure, the contrast medium in the kidneys allows fewer x-rays to reach the film, producing a higher radiographic contrast.

In summary, it can be said that the body is composed of very complex structures with many different absorption properties and it behoves the technologist to appreciate how the various beam characteristics (frequency and type of interaction) affect subject contrast and ultimately radiographic contrast (see Fig. 123). With this knowledge the correct choice of KvP can be made.

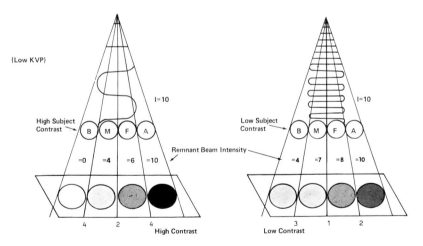

Figure 123. Radiographic image is dependent on subject contrast (differential absorption).

Fixed versus Variable KvP Techniques

With the mass of information discussed thus far, only a few final observations need to be made regarding KvP. One of these is how a technique chart can be written; there are two basic systems commonly used today: (a) optimum or fixed KvP and (b) variable KvP. A fixed KvP technique chart is one in which the KvP level is held constant for all thicknesses of a given body part. Abdomens, for example, might be exposed at a 75 KvP regardless of the measurement. Of course, some compensation is necessary for various thicknesses and MaS is used to serve the purpose for adjusting radiographic density. Most often, the exposure time is adjusted to compensate for patient thickness. In general, the KvP initially chosen for each body part must meet some prerequisites to insure against underpenetration. Also, it is important to choose a KvP that will provide optimal radiographic contrast, as noted earlier. To help establish the proper level of KvP the five criteria listed below should be examined:

1. Ability of the body part to absorb the primary beam.
   a. tissue density
   b. atomic number of the tissue
   c. tissue thickness
2. Inherent subject contrast versus the type of radiographic contrast scale desired.
3. The ability of the body to produce S/S.
4. Grid ratio available.
5. Special examinations where subject contrast must be enhanced or possibly de-emphasized, i.e. chest work, injected contrast media, mammography.

A fixed KvP technique chart does have some important advantages. In short, the major advantage in maintaining consistent KvP for each body thickness is that radiographic contrast tends to be more consistent. A disadvantage of optimal or fixed KvP is, of course, exposure time is increased with body thickness resulting in the possibility of increased patient motion. The principle of fixed KvP technique is used in all manufacturers' phototimed equipment. When a phototiming system is used, the technologist selects the proper KvP and the automatic phototimer changes the exposure time according to the patient. Because of the serious limitations that high milliamperage exposures have on tube loading, it is generally considered that varying the Ma to compensate for body thickness and radiographic density is a very poor practice. If a sufficient number of Ma stations were made available, this disadvantage could be eliminated or greatly reduced and the problem of excessively long exposure times would be eliminated.

With variable KvP techniques, a KvP is chosen for an average thickness of a body part. This is often a subjective choice made in accordance with the criteria listed earlier. The kilovoltage is then increased or decreased according to body thickness as the milliamperage and time factors are held constant. Figure 124 shows an illustration of both the variable and fixed Kv type methods. Note that with a variable KvP chart, the

FIXED KvP   VARIABLE KvP

ABDOMEN A.P.	CM	Kv	Time	Ma	Grid	F.F.D.	Scn		CM	Kv	Time	Ma	Grid	F.F.D.	Scn	
	17-18	80	2/10	300	12:1	40″	Fast		17-18	80	2/10	300	12:1	40″	Fast	
	19-20	80	3/10	300	″	″	″		19-20	84	2/10	300	″	″	″	
	21-22	80	4/10	300	″	″	″		21-22	90	2/10	300	″	″	″	
	23-24	80	1/2	300	″	″	″		23-24	96	2/10	300	″	″	″	
	25-26	80	7/10	300	″	″	″		25-26	102	2/10	300	″	″	″	
	27-28	80	9/10	300	″	″	″		27-28	108	2/10	300	″	″	″	

Figure 124. The relative techniques (for an abdomen, fast screens, slow film) needed for variable and fixed KvP methods.

KvP is increased by increments of two for every cm of patient thickness. As was pointed out earlier, time and Ma do not change, so relatively short exposures can be used regardless of part thickness. However, because of the effect that Kv has on exposure latitude, thicker body parts require higher KvPs and so greater increases than two Kv are required per cm thickness. For example, if a knee measuring 12 cm requires 70 KvP, a second knee measuring 13 cm would require 72 KvP. If the KvP level were 95 for the 12 cm knee, a knee measuring 13 cm would require 100 KvP. The major problem with variable KvP charts is that radiographic contrast is not consistent. As higher KvPs are used, subject contrast decreases and the degree of S/S fog often increases. A fixed KvP technique chart using high KvP exposures is accepted in modern radiography and would have the advantage of decreased patient dose as well as shorter exposure time.

CHAPTER EIGHT

THE HUMAN BODY AS AN EMITTER AND BEAM MODIFIER

THE MANY CONCEPTS presented thus far are extremely important to good radiographic quality. Yet one important item has not been considered: the human body. Its complexity and function are matched only by its variation in form and absorption characteristics, and it is these variations that can cause technical havoc for the technologist. When we think of a tall, lean individual, a certain body form comes to mind; this type of body form (habitus) can produce a great number of different x-ray absorption rates that, if understood, will make it much easier for the technologist to choose correct exposure factors. Such body variations can come in the form of an abdominal obstruction, cardiac enlargement, evacuation of air from the lungs, atrophy of bone and soft tissues, and many other possibilities. Figure 125 demonstrates an assortment of these pathologic problems; the reader should notice how they vary from normal body composition.

Recently, the use of automatic timing devices has helped considerably in making automated exposure adjustments for radiographing such maladies, and this trend will probably continue. Even with phototiming, the technologist must have a sound working knowledge of how the body modifies the primary beam into a series of intensities that ultimately expose the film's emulsion. With this introduction to the "patient problem," it is certainly no understatement to say the human body is by far the single greatest variable in radiography. Often technologists can develop a type of intuition that will serve them well in making the final decision in favor of one specific set of exposure factors as opposed to another. With some practical understanding of how body habitus relates to radiography, the technologist will eventually be able to *predict* with confidence how he can alter the primary beam via exposure factors to produce the desired radiographic result. A seasoned technologist can almost "see" how the beam will pass through the body part under examination and imagine the beam as it is absorbed differently by the internal structures, *before* the exposure is even made. In short, no matter how accurate or complete a

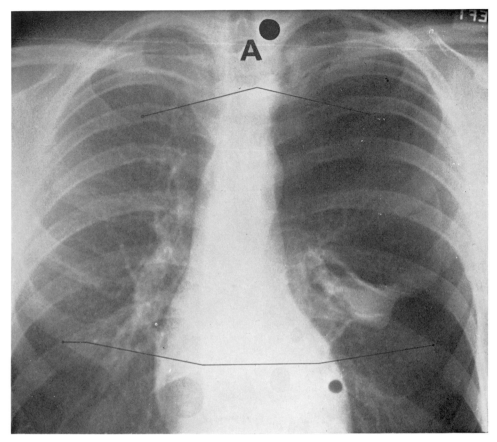

Figure 125A. A dramatic change in appearance between the right and left side of the chest due to a collapsed left lung.

technique chart may be, the technologist is responsible for making the final adjustments to achieve optimal diagnostic results.

MAJOR ABSORBERS OF THE BODY

Fortunately, humans do have some basic similarities in composition. The major substances that account for most of the x-ray absorption are fat, fluid, muscle, and bone. Equally sized fat globules from one patient will be equal in absorption to that from another person, and the other four body substances are also equal from one to another person. The *distribution* and *quantity* of these are usually responsible for the confusion when choosing the correct set of exposure factors for a given patient.

Another aspect that can weigh heavily in formulating a technique is the presence of air in the body. As with the other major body components, its presence alone is often not as important as how much there is. In short,

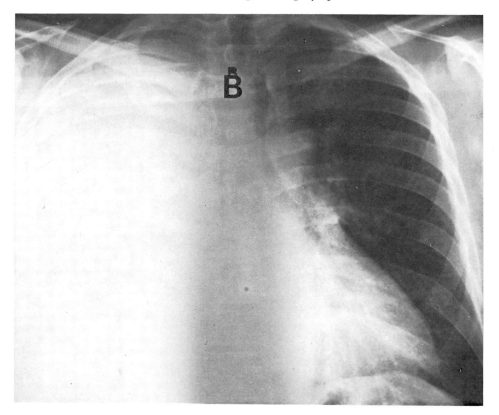

Figure 125B. The entire right side of the chest is filled with fluid; it is difficult to penetrate the chest with this type of pathology.

the technologist can more effectively manage the situation at hand by "technically pigeonholing" the patient as to (1) the quantity and distribution of the major tissues being radiographed, (2) age of the patient, (3) the pathology that might be present within the patient, and (4) how the primary beam may be altered (what interactions will take place) by these various characteristics. In the remainder of this chapter we will attempt to point out how all these body aspects both hinder and help the technologist and offer some suggestions for making the problem more manageable.

Some General Facts Regarding Body Habitus

The body is made up of approximately 62 percent water, 15 percent fat, and 23 percent bone; approximately 40 percent of the total body weight is muscle (at age seventeen). Of the soft tissue substances, muscle has the greatest water content and this has an important bearing on the radiographic image. Fat contains less water as compared to muscle tissue.

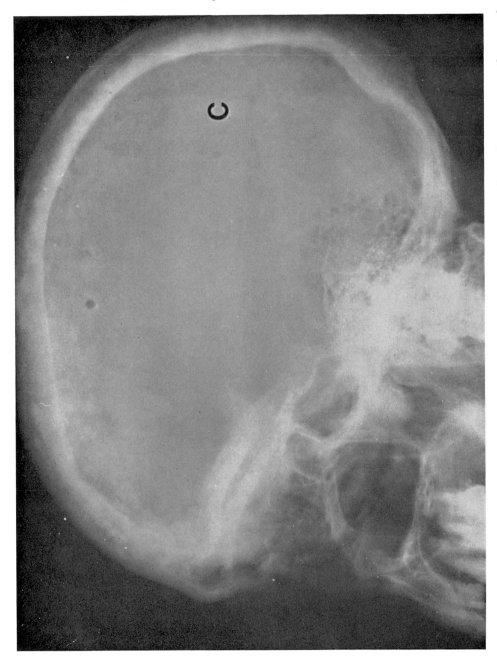

Figure 125C. A skull revealing a condition known as Paget's disease where the bone becomes quite dense and difficult to penetrate.

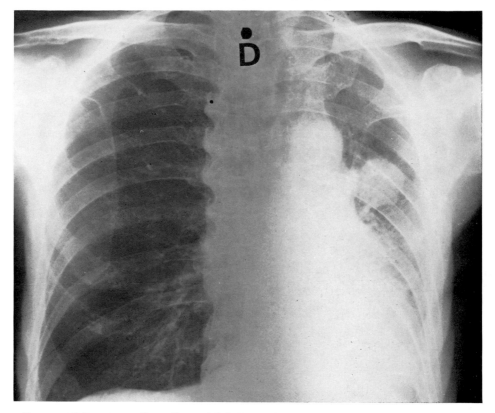

Figure 125D. A totally collapsed left lower lobe, secondary to bronchogenic Ca.

Body fluid itself is more absorbent to x-rays than muscle tissue. Healthy bone absorbs more x-ray photons than body fluids, fat, or muscle; however, the greatest absorbing substance in the human body is tooth enamel.

Fat Content

As noted above, when fat is compared to muscle it has a lower water content and slightly different atomic number. It should be kept in mind that water is extremely efficient in producing scatter radiation. As noted earlier, the water content alone will have an important effect on the radiographic image. Large amounts of muscle tissue would produce more scatter than an equal amount of fat, and thus muscle is likely to cause a considerable decrease in radiographic contrast. Radiographic density will also decrease because muscle absorbs more photons than fat. There is no doubt about the deleterious effect S/S fog has on radiographic contrast. It is interesting to note that most technologists expect to see a decrease in radiographic contrast resulting from radiographing fat people. A typical image that results when radiographing a fat patient is increased

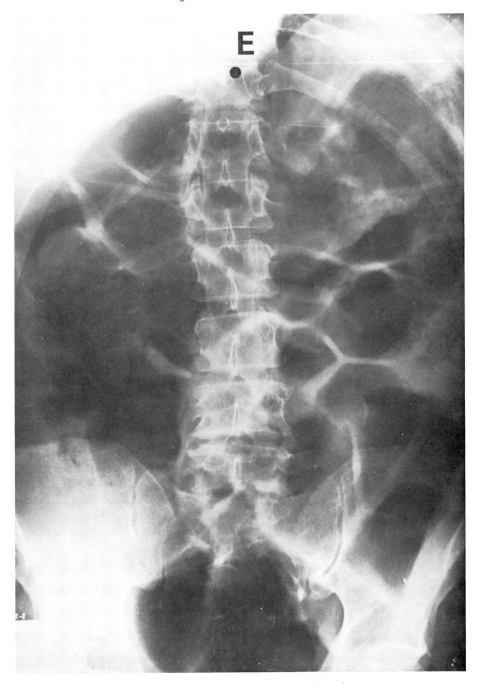

Figure 125E. An abdomen distended with air.

density with a gray, flat appearance. The explanation for this lies in the fact that when obese patients are radiographed, the organs themselves are encapsulated amidst thick fatty layers which generally reduce the relative differences in x-ray absorption between the organs and the tissues immediately surrounding. Thus, obese patients simply have much less subject contrast and Figure 126 schematically demonstrates this concept.

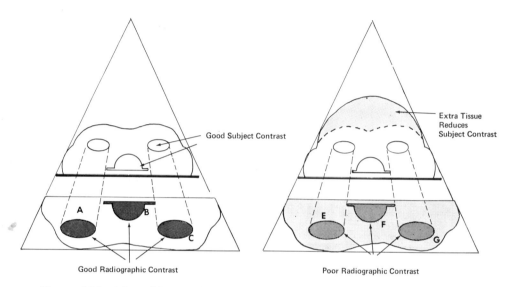

Figure 126. The additional fatty layer causes a reduction in subject contrast.

Also, if carefully analyzed, a radiograph of an obese patient may be poor in quality simply because the film has been over-exposed. Fat is not very absorbent to x-rays, i.e. if two large patients measuring approximately the same were radiographed using the same exposure factors, the fatter patient could produce a darker image. With this in mind, one should consider how technique factors might be chosen for obese patients. Many times the technologist, because of patient size, becomes slightly intimidated and thus has a tendency to simply overexpose the radiograph. Often the technologist would obtain a better radiographic image of such a patient if the technique were actually reduced slightly, preferably with KvP; this would cause a moderate to obvious increase in subject contrast. The author has seen many obese patients, measuring as much as 28 cm for an A-P abdomen, done very successfully with 70 KvP. Figure 127 shows two radiographs of the same patient. Figure 128 shows two different patients, one with a high fat (A) content and the other with a low fat (B) content.

Muscle Content

Patients who have excellent muscle tone and are very muscular pose a slightly more difficult problem in obtaining high quality images. The principal problem here is that the KvP should be increased due to the higher absorption rate of muscle. This adjustment compounds the radiographic contrast problem because, as was already mentioned, muscle produces large amounts of scatter by virtue of its high water content and increasing the KvP contributes additional quantities of scatter reaching the film. It has been the author's experience that it is very difficult to obtain a radiograph of high contrast in large, very muscular patients. In older patients, this muscle tissue will begin to slowly break down and will eventually become dehydrated to some extent, allowing considerably more x-rays to penetrate the body.

Water Content

Water content is a key indicator in detecting a healthy body, and too much or too little water has a deleterious effect on radiographic quality. Figure 125B shows a patient with excessive water in one side of the chest. Usually, it is unnecessary to penetrate the excess water; however, when it is necessary to do so, a more than moderate increase in KvP (approximately 15 to 30 depending on the initial KvP level and the quantity of water present) is usually indicated. Any additional increases to improve radiographic density should be made by increasing MaS and, as a last resort, a shorter focal film distance. The *proper content* of water in a healthy person helps to establish subject contrast of the body. With elderly patients, for example, dehydration and poor muscle tone require the technologist to reduce KvP, sometimes by as much as 10 to 20 in extreme cases, to help build subject contrast. As a result, the MaS should be increased to maintain proper radiographic density. Decreasing KvP under these conditions produces more differential absorption with less Compton interactions. At age 60, body water content is reduced to approximately 52 percent.

In general, radiography of older patients usually requires an overall reduction in exposure because of their poor absorption characteristics. The author's first rule of thumb is to reduce the KvP used to help increase subject contrast. In addition to this, a MaS reduction of 25 percent is often required for people between the ages of sixty-five and seventy-five, and for patients beyond eighty years of age a MaS reduction of 50 percent is often required.

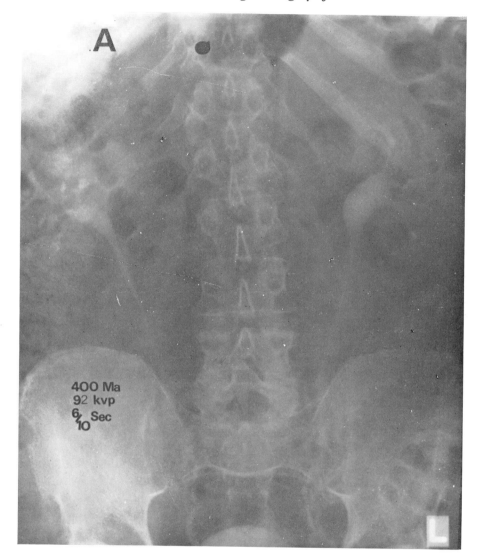

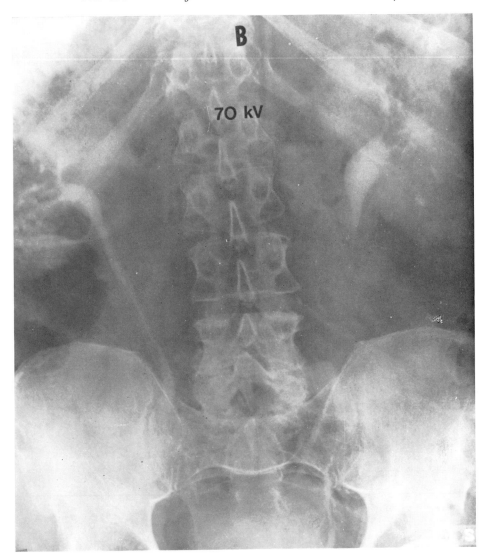

← Figure 127. Low KvP settings can improve subject contrast. Contrast is vastly improved in *B* as compared to *A*—the psoas muscles are more visible and the pelvic bone reveals considerably more detail. This was a large patient who measured 28 cm when lying supine.

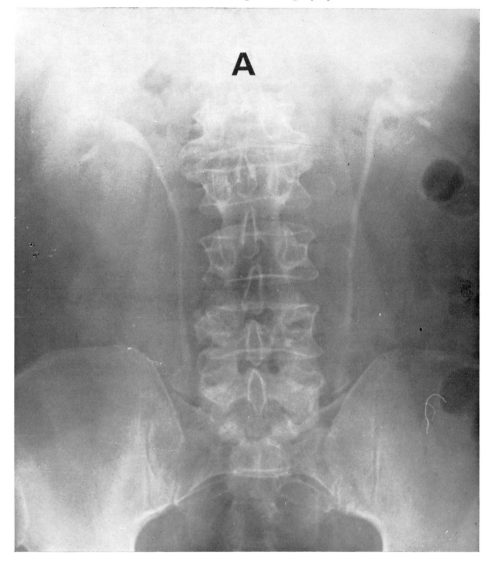

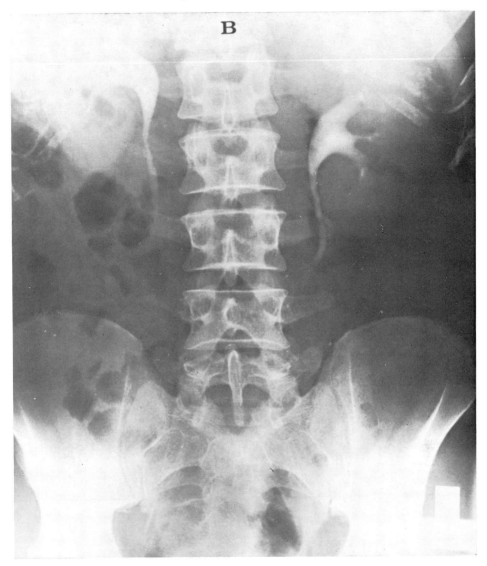

← Figure 128. Patients with different body compositions present vast differences for radiography. *A* was a middle-aged patient with a low subject contrast. *B* was a young patient with high subject contrast.

Bone Content

Bone content is the most consistent major body substance in terms of volume per patient (approximately 23 percent of body weight). It has a very low water content. It is interesting that bone has one of the most variable absorption qualities in radiography. The principle absorbing element in bone is calcium; the calcium content, however, changes widely with age. Figure 129 shows two radiographs demonstrating how the appearance of the bones of elderly people can be improved by reducing KvP. With elderly patients bony structures actually appear to be darker than the surrounding soft tissue.

Young children and infants also have low subject contrast between

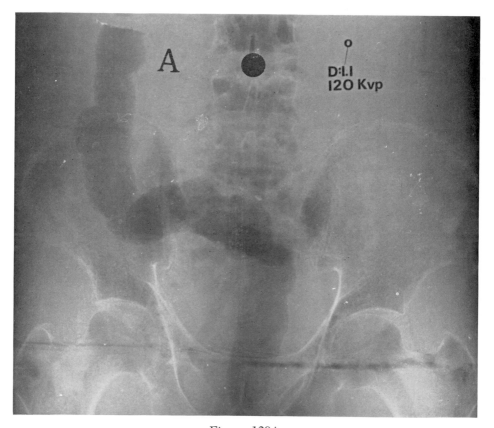

Figure 129A.

Figure 129. Greater differential absorptions can be achieved by reducing KvP and producing more photoelectric-type interactions; overall radiographic density has been maintained. Different KvP levels have been noted to demonstrate their effect when making such deliberate contrast adjustments. H & D curves are also shown. (A 12:1 grid was used throughout.)

bone and soft tissue. In either case, radiographic contrast between bone and soft tissue diminishes and makes diagnosis a very tricky proposition, especially when a fracture is small and the cortex remains in good alignment. The choice of technique for good bone radiography should be adjusted for age and possible pathology. As age increases, bone demineralizes and the beam's ability to penetrate quickly increases, thus producing an overall darker radiographic image. Due to this lack of absorption, it is felt by many that elderly patients should be exposed using a high MaS, low KvP type of technique. The diagnostic quality of the radiographs noted in Figure 129 is certainly an interesting comparison. The value of a less energetic beam with elderly patients becomes more understandable.

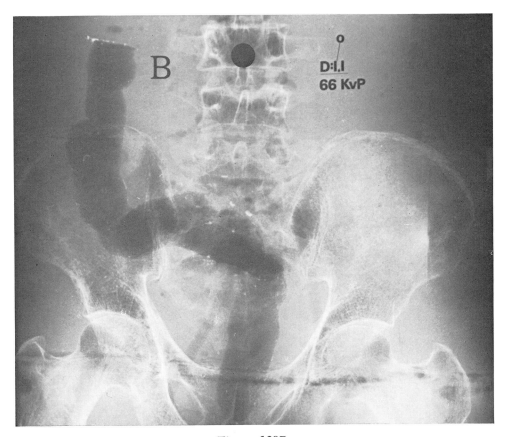

Figure 129B.

Pathology is another important item that contributes to a variation in bone absorption. A listing of common diseases that affect radiation absorption is shown in Figure 130.

From the above discussion, the technologist should keep in mind that

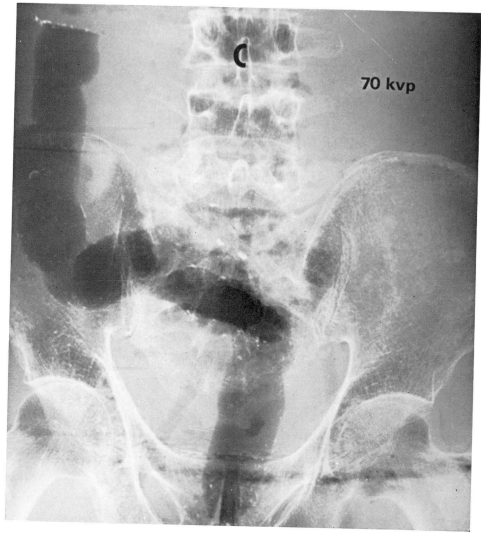

70 kvp

Figure 129C.

the mention of a specific disease in the patient's history might well be noted because possible adjustments in technique may be necessary.

Evaluating the Patient

It is important for the technologist to be able to "size up" the patient correctly before the first exposure is made. During the first few minutes with the patient as much attention as possible should be paid to such things as overall muscle tone, age, fat, muscle content, and possible pathology, and in reviewing the patient's history. While helping the patient on the table, one might be noting how firm the muscular structure is and

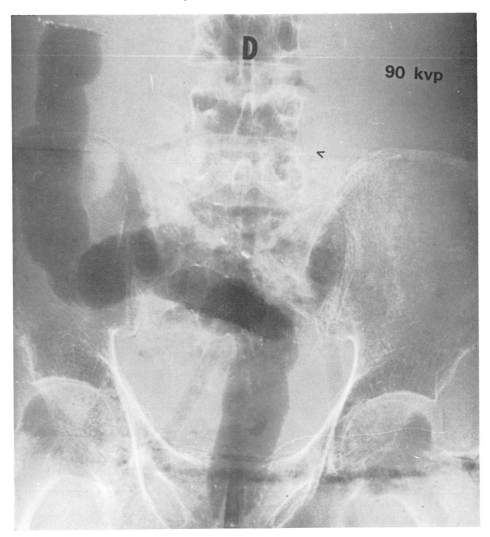

Figure 129D.

the overall mass of the patient. Considering these aspects along with age and pathology, the technologist is better able to have the body habitus "pigeonholed" or "sized up" pretty well. Remember that no one factor thus far mentioned is sufficient, singularly, as a determining factor for establishing a technique. Through experience and careful observation, a competent technologist develops an intuitive awareness as to how the primary beam will be modified by the patient's body. In Chapters Eleven and Twelve suggestions will be presented regarding how one can juggle exposure factors to arrive at a technique that would be most likely to produce an optimal radiographic image. The manner with which adjusting exposure factors is accomplished, from the author's point of view, makes the differentiation between a "button pusher" and a skilled professional radiographic technologist.

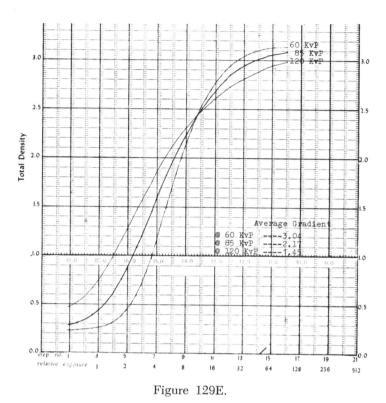

Figure 129E.

Figure 130. Common diseases that often affect the radiographic image. The technologist should refer to the patient's history for clues in determining the final exposure factors.

COMMON PATHOLOGICAL FACTORS

Increased Tissue Density

Condition	*Commonly Seen In*
Osteoblastic Metastases	Bones
Paget's Disease	Bones
Pleural Effusion (Accumulation Of Fluid)	Chest
Consolidated Lungs (Pneumonia)	Chest
Intra-abdominal Fluid (Ascites)	Abdomen

Decreased Tissue Density

Condition	*Commonly Seen In*
Osteoporosis	Bones
Emphysema	Chest
Bowel Obstruction (Excessive Abdominal Air)	Abdomen

IMPORTANT CHARACTERISTICS OF MAJOR BODY REGIONS

The body can be separated into four major regions: (1) the abdomen, (2) the chest, (3) the extremities, and (4) the skull. Each of these requires the use of a different combination of exposure factors. A brief discussion of the differences and similarities of these regions will prove beneficial in understanding how various body regions influence the overall radiographic image. Also, some insight might be gained regarding the different philosophies held by various radiologists and technologists as to which set of exposure factors may produce the best results.

The Chest Region

Earlier, the concept of subject contrast was presented. To review quickly, subject contrast is a comparison of the rates at which various body structures absorb the primary beam. If an area is constructed in such a way that it produces obvious or abrupt differences in absorption rates, it is said to have high subject contrast. If, on the other hand, the inherent absorption ability of the various body tissues in the region under examination are similar, we identify it as having a low subject contrast which would produce low radiographic contrast.

The chest region has a relatively high subject contrast. A brief analysis of its components will explain why. First, there is normally a considerable amount of air in the chest, and air is commonly considered an effective contrast media. In a manner of speaking, most of the structures in the chest are bathed in or surrounded by air, which causes great variations in absorption rate. The effect of this is shown in Figure 131. You can see the object under examination is less obvious in "A" than it is in "B" with the presence of air being the only variable. Second, the structures of the chest are not as compact as those of the abdomen and thus are generally easier to distinguish. The third major factor that contributes to high subject contrast is the density of the structures present in the chest. For example, the structures of the mediastinum are extremely dense when compared to lung tissue. The major vessels contain fluid and so they are much more absorbent to x-rays than the air filled lungs that surround them. In short, wide variations between tissue density, a moderate degree of separation between structures, the presence of large amounts of air and, of course, the various mixture of major body tissues all contribute to the high subject contrast of the chest region.

The Abdomen

The abdomen presents an altogether different problem to the technologist. Here, the structures and organs involved have a much more

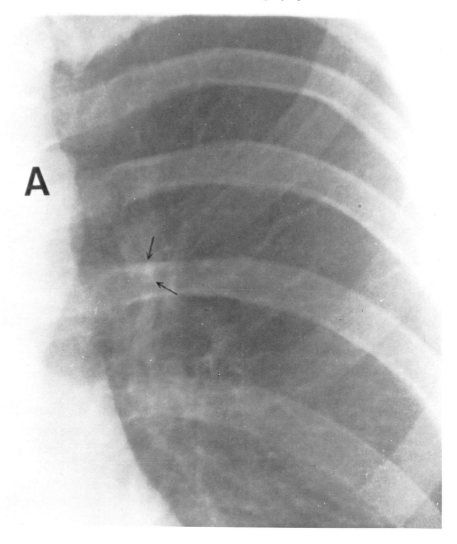

subtle subject contrast; thus, the abdominal area is known to have a
rather *low subject* contrast. The relative water content of the abdomen
area is high, and tissue density of the structures in the abdomen are very
uniform as compared to those of the chest. Another important factor is
that the various organs in the abdomen have similar absorption characteristics and are often partly superimposed on each other. This, of course,
makes it even more difficult for the beam to distinguish between the
absorption rate of one tissue as opposed to that of another. The end result
is a remnant beam that is very uniform as it reaches the film. The presence
of fat, however, tends to silhouette some organs in the radiographic image
providing a slight increase in the visibility of the structures. In fact, the

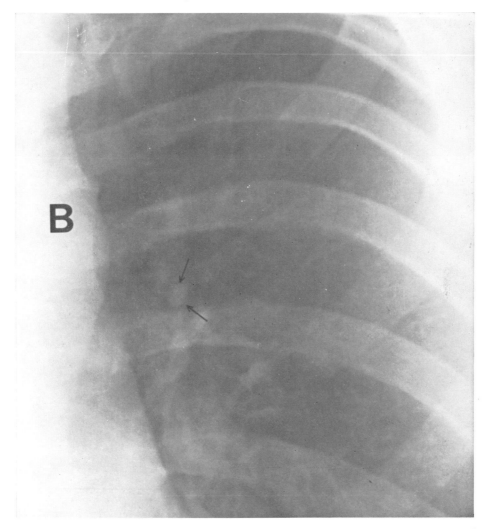

← Figure 131. The small node (arrow) in *A* is difficult to see when compared to the same node in *B*, shifted to a slightly different location by respiration.

absence of these fatty layers around various organs is sometimes indicative of disease. The kidney is a good example of this, as noted in Figure 132. Also, the thickness of each organ plays an important role in the ultimate subject contrast, and of course the absence of air makes the overall absorption rates even more uniform.

The Extremities

The extremities have the greatest subject contrast in its strictest sense. Each extremity has only two major tissues that absorb the x-ray beam:

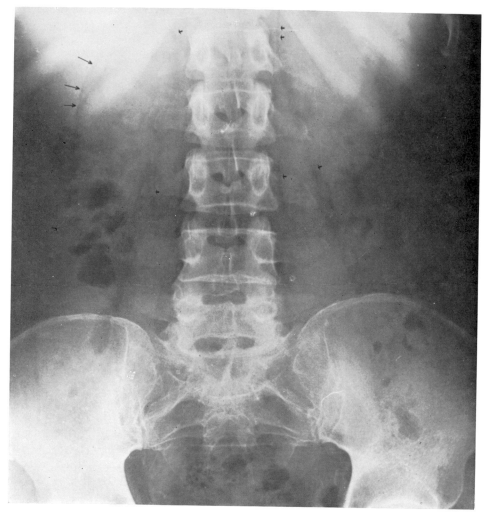

Figure 132. A number of "fatty" lines are indicated, others are not indicated.

the bone and the surrounding muscle. Thus, with extremities the two major absorption rates are bone versus soft tissue. With this arrangement of subject contrast, the differences in absorption rates are great.

The Skull

The skull offers a problem still different from those already discussed. The skull, of course, involves pure bone radiography. There is very little soft tissue to diagnose or reproduce radiographically, so the subject contrast of this area is primarily caused by various thicknesses of the bone. The petrous bone offers a definite change in absorption relative to the external wall of the skull. The areas of the face have some cavities in which air can commonly be seen. As you already know, varying thickness of any similar tissue including soft tissue can affect subject contrast.

A LOOK AT PHOTON-TISSUE INTERACTIONS

It should be kept clearly in mind that no diagnostic exposure possesses only a single type of interaction, but rather a mixture of interactions. There is, however, a dominant type of interaction for each exposure and that interaction is largely responsible for the resultant radiographic effect (see Fig. 133).

In Chapter Seven much time was given to describe how the various types of interactions affect radiographic contrast as controlled by the KvP level. We should now approach this topic with slightly more detail so the technologist's scope of understanding will be appropriately widened. Before going on, we should recall the three possible situations the primary beam may experience: complete penetration of the body parts, an interaction involving partial absorption produced by the Compton effect, and the interaction involving total absorption caused by photoelectric interactions. There are other interactions that can occur in the body as a result of irradiation; however, for the kilovoltages used in diagnostic radiography, the Compton and photoelectric interactions are primarily responsible for producing the radiographic image, and our discussion will be limited to these.

It should be very apparent by now that the interaction between body tissues and the beam of x-ray photons is the central issue in radiography. It has been the author's opinion for some time that, whether radiography is viewed primarily as a science or as an art, technologists must have a clear understanding as to how x-rays interact with body tissues to produce the remnant beam, because it is, after all, the remnant beam that will ultimately determine radiographic quality. The scope of this discussion on x-ray interactions with tissue will be brief and oriented to its very practical applications.

Photoelectric Interaction

The photoelectric interaction is interesting and very important to radiography. It involves the production of two types of secondary radiation, while the original primary photon disappears altogether. It should be pointed out and kept in mind throughout this discussion that the KvP level used for any exposure primarily determines the type of interaction that eventually results. The photoelectric effect, for example, is more likely to occur in the 35 to 70 KvP range.

A schematic of the photoelectric interaction can be seen in Figure 134. The incident photon strikes an electron located in one of the inner orbits;

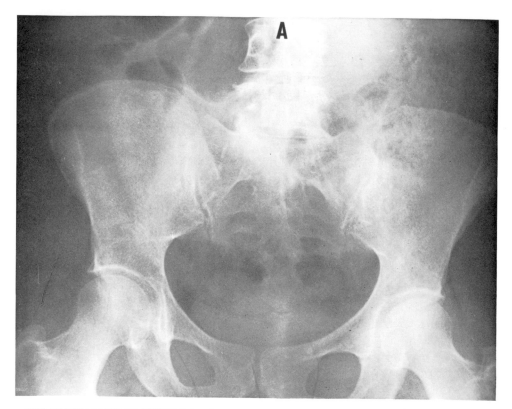

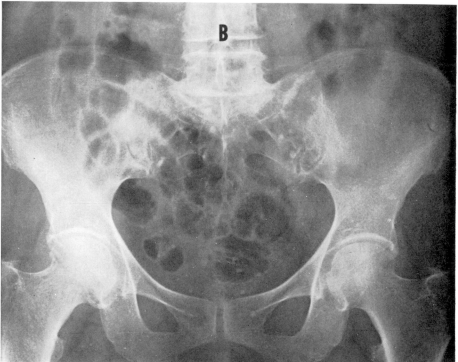

Figure 133. The KvP is increased in *A* as compared to *B*. The resulting increase in Compton interactions in *A* produced a more homogenic density pattern on the film. (These patients had a very similar body habitus.)

usually only the K or L shells are involved. The first thing that becomes apparent is that the entering photon disappears entirely. It is totally ab-sorbed by the atom of the tissue, which now suddenly has more energy than it knows what to do with. By referring to the law of conservation of energy, we know that a form of energy (such as an x-ray photon) cannot simply disappear without any other effect. Immediately, the excited atom ejects an electron from one of its innermost shells in an attempt to balance its overall energy state. The ejected electron is called, as it begins its voyage from its home shell, a photoelectron. The photoelectron has less energy than the original primary x-ray photon that struck the atom. By further inspection of Figure 134, you will see that an electron from an

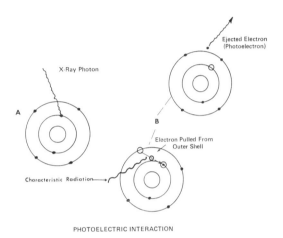

PHOTOELECTRIC INTERACTION

Figure 134. An entering primary x-ray photon strikes the L shell and is totally con-sumed. The atom cannot contain this increased energy and discards it by ejecting an electron. The emitted photoelectron will probably not go through more than a few centimeters of body tissue before being absorbed. An electron is moved inward to fill the vacancy; because it is going to an orbit with a lower energy level, it discards its excess energy, known as characteristic radiation. The characteristic radiation travels an even shorter distance in the body before it is absorbed. The photoelectric effect is thus a total absorption interaction.

outer shell moves to fill the vacancy in the inner orbital shell, but to be "comfortable" in its new location, it must adjust its own energy to the energy level of its new home. This is accomplished by emitting what is known as characteristic radiation. The term *characteristic* is used here because it tells something about the element or orbit from which the radiation came. There is a cascading series of new electrons that serves to satisfy the other orbital vacancies, and each produces its appropriate level of characteristic radiation. This process will continue until all vacan-cies of the atom are filled and the atom is once again stable. The outer

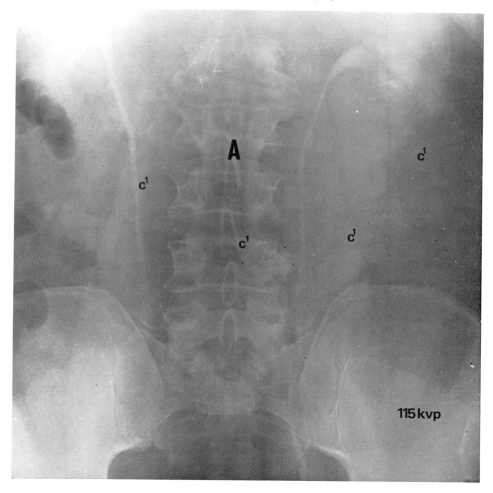

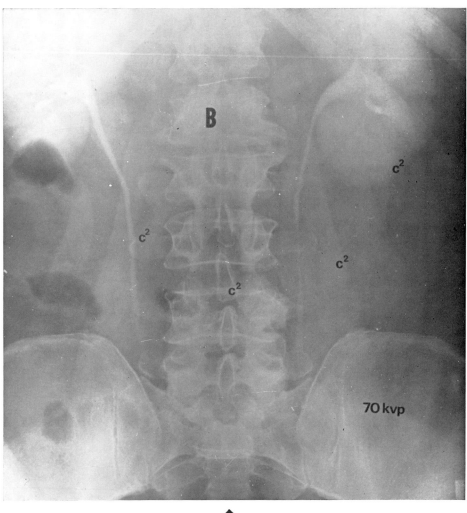

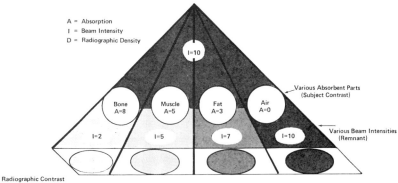

← Figure 135. Differences in body part absorptions are always present to some extent in a diagnostic exposure. The degree of difference is dependent on the type of interaction, which is determined by the beam's energy level. The "C" numbers tell the variance between the radiographic densities in adjacent body structures. This comparison indicates that a higher radiographic contrast is achieved when a lower KvP is used because it increases differential absorption.

shell will fill its vacancy by pulling in a free floating electron which may be moving in the general vicinity of the vacancy.

The importance of this interaction to the radiographic image is that it is extremely sensitive to tissues having different atomic numbers and absorption rates. This means that with a particular exposure, body tissues having different atomic numbers, i.e. fat, bone, muscle, and air, and different tissue densities would have corresponding absorption differences. These differences in body absorption will cause corresponding differences in the intensities of remnant radiation which will, in turn, expose the film to varying degrees. Figure 135 shows this concept schematically. It should be kept in mind that it was the kilovoltage level that made the photoelectric effect dominant in this particular exposure. If another kilovoltage was chosen (above 70 KvP) a different type of interaction would more likely result and the effect of this would be the production of a radiographic image with a different overall appearance.

The photoelectric interaction can be used by the technologist to gain relatively high subject contrast. There are certain examinations that demand high absorption variations (subject contrast) in the body. With IVP examinations, for example, it is important that not only the kidney be seen easily but the opacified calyces be optimally visible as well. There is obviously a difference in atomic number between the kidney and the contrast material and, by using the photoelectric effect, one can take advantage of this and produce the best technical image possible. Most iodine-based contrast material gives optimal opacification when irradiated by Kv levels from 60 to 70. Figure 135 shows two radiographs using high and low KvP levels. While reviewing these radiographs, it should be brought to mind that the type of interaction that was dominant during these exposures has a great influence on the contrast of the radiographic image and that the type of interaction can be effectively controlled by the choice of kilovoltage.

One last example might be helpful to cement this concept in the reader's mind. Rib examinations are often difficult to radiograph; in fact, it is the author's opinion that they are one of the most technically difficult studies of all. There is, however, an extreme difference in the atomic numbers of the internal thoracic structures and the bony thorax. If one keeps this in mind and uses the principles described above regarding the photoelectric effect and low KvP, one has a considerably better chance of producing a good diagnostic study. One should not let the physical size of the patient intimidate one's thinking. For a chest measuring 28 cm A.P., a surprisingly low KvP (50 to 60) is adequate for penetration. At these Kv levels, the photoelectric interaction will go to work on the technologist's behalf, taking advantage of the relatively large differences in atomic number between the lungs, bone, air, and vessels to produce optimal contrast. Of course, for chest radiography, high KvP exposures would more likely produce optimal radiographs.

Compton Interaction (Modified Scattering)

The Compton interaction or Compton effect is shown in Figure 136, where an incident photon is striking a loosely bound outer electron. The angle at which the primary photon makes "contact" with the outer orbiting electron will determine how much energy the photon loses and what direction it will take afterward. If a high energy photon strikes the outer electron at a glancing (indirect) angle, it will lose less of its original photon energy to the atom than if it were traveling on a more direct course. An analogy of how billiard balls react when they strike each other at different angles and with varying speeds can be useful here. The incident primary photon, upon contact with the orbiting electron, leaves the interaction site with a different angle as a cue ball would. The primary photon after this collision with the electron is known as a scattered photon. During this interaction the outer electron will be dislodged from its orbit and is then called a recoil or Compton electron. The vacancy is filled by a free, floating electron. There is some characteristic radiation given off as the vacancy is filled by the free electron, but it is so low in energy it has no effect on the radiographic image as with the photoelectron.

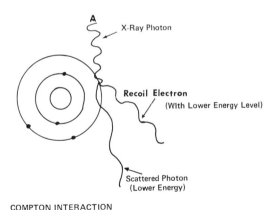

COMPTON INTERACTION

Figure 136. As a result of the collision of a higher energy primary x-ray photon with another orbiting electron, a recoil electron (called a Compton electron) is ejected from the atom. Simultaneously, a scattered photon is produced and exposes the film.

In general, as the KvP is increased, the role of the Compton interaction in producing the ultimate radiographic image is increased. There are two reasons for this. First, as the KvP is increased, the relative number of

Compton interactions as compared to photoelectric interactions increases.
Second, with increased KvP the scattered photons are more penetrating
and are directed more toward the film. The increased penetrability of the
scattered radiation compiled with a greater tendency to travel toward the
film results in a definite increase in radiographic fog. The methods used
to control radiographic fog produced by excessive scatter will be discussed
thoroughly in Chapter Nine. Figure 137 shows how scatter photons cause
this unnecessary and unwanted radiographic density known as fog. After
rambling around inside the patient, the *scattered* photons are either absorbed by the patient or go on to expose the film. Overall, considerably
more of these photons expose the film as opposed to those that are absorbed by the body. Those that strike the film do so with indifference to
the body's inherent subject contrast. In other words, the scattered photons
are likely to expose an area of the film that is supposed to be light in
density. The net effect of this is that crucial differences in radiographic
densities are greatly diminished, and the diagnostic value of the image is
seriously impaired.

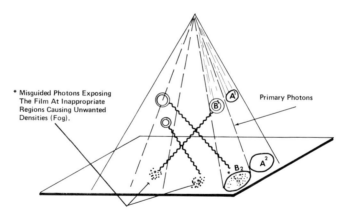

Figure 137. Object A has totally absorbed the primary x-ray and is recorded by the
film with a light density. Object B produces a scatter photon as a result of partial primary beam absorption; scatter photons have invaded the area of the film noted as
B^2, causing that area to be fogged. In reality, this invasion of scatter photons is evenly
spread across the entire film, producing an overall gray flat radiographic image.

Between the photoelectric and Compton interaction, the probability
of a photon emitted by the x-ray tube passing through the patient without undergoing some type of interaction is relatively low. It has been
estimated that less than 10 percent of the beam reaches the film in its
original form. With this fact, one can see more easily how important these
interactions are in radiography.

Summary

In all, the body causes the final adjustment or modification of the x-ray beam before it reaches the film. Differences in absorption (subject contrast) among various body tissue types will determine, to a great extent, the ultimate radiographic quality. When one considers that up to 90 percent of the radiation exiting the patient is not directly produced by the target (not primary radiation) of the x-ray tube, but rather by the body itself, we can more realistically appreciate the body as an emitter of x-rays: This knowledge demands our strict attention as technologists.

In general the patient must be evaluated in total. A patient's age or weight simply does not supply one with enough important information about the patient's absorbing characteristics. Similarly, measurements alone should not be used as a single source of data for establishing a set of exposure factors. The patient's abdomen which measures 21 cm A.P. actually tells you very little. The patient could be 90 years old; he could be athletic with excellent muscle tone; or he could be a middle-aged individual with a large degree of fat content. Each one of these conditions should signal the technologist that an adjustment might be necessary beyond the technique chart. In short, there is no real substitute for experience in making the final judgment in technique. One should be mindful, however, that mere seniority is not an accurate measure of experience. There are unfortunately a number of technologists who have been working for a number of years, but due to a lack of professional interest and self-discipline in relating theory to practical application, their "real" experience is quite limited.

FILTRATION AND RADIOGRAPHY

There is one remaining factor that must be discussed. Although important, from a practical point of view, it is a factor over which the technologist has little or no control. All *routine* diagnostic exposures are made with a filter fixed between the x-ray tube and the collimator. A filter is used to reduce the number of unnecessary photons reaching the body. The main concern here is obviously one of radiation safety and patient dose. It is well known that many low energy x-rays are emitted by the tube during each diagnostic exposure. Their effect radiographically is of no significance; however, they do produce unnecessary photon interactions (ionizations) in the body which, of course, increase the patient's dose. To prevent this, a thin aluminum filter is placed between the portal of the x-ray tube and the collimator aperature. There is always some inherent filtration provided by the glass envelope and surrounding oil which is usually equal to from 0.5 to 1.0 mm aluminum. To this, equipment manufacturers generally add at least 2.0 mm Al of "additional filtration." The inherent filtration (in the glass envelope) plus the added filtration produces a total filtration equal to no less than 2.5 mm to meet government regulations. When low KvP values are required, perhaps for special soft tissue techniques, less total filtration can be used since government standards for filtration are calculated according to the "probable" KvP level that is most likely to be used by the equipment.

The Radiographic Effect of Filtration

One can easily see by viewing Figure 138 that only moderate radiographic differences are produced as a result of commonly used filter thicknesses, as long as one does not exceed 3.0 mm aluminum. Filtration produces a harder beam because the filter absorbs the many soft rays that would cause the overall primary beam to be softer. As filtration increases to a certain point beyond normal diagnostic thicknesses, the overall beam becomes excessively harder. The effect of adding filtration to the beam has some similarities to increasing the KvP.

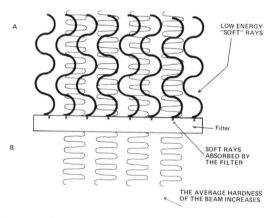

Figure 138. As filtration thickness increases, fewer and fewer x-rays can penetrate, resulting in decreased radiographic density. Fewer lower energy photons pass through the filter as thickness increases, and higher energy photons reach the film. The average energy of the beam is higher and more uniform.

The radiographic effect caused by filters is that of decreased contrast and decreased density. It was pointed out earlier that the objective of filtration is to remove lower energy, diagnostically useless x-rays. With light filtration this is true; however, as thicker filters (greater than 3.0) are used, there is some absorption of the higher energy rays. Contrast is decreased with increasing filtration thicknesses because the beam's penetrating capability is more uniform (as shown in Fig. 139). This will cause a more even penetration of the various body parts giving rise to less subject contrast. In short, the advantage of filtration is basically to prevent unnecessary exposure to the patient by absorbing those soft rays that have no radiographic effect. It should be noted from a radiation safety point of view that, if a technologist notices or even suspects that a filter is not being used properly, he should report his suspicions to his supervisor without delay. Further, one should not attempt to repair or make any adjustments in filtration, but leave this specialized work to a trained service person.

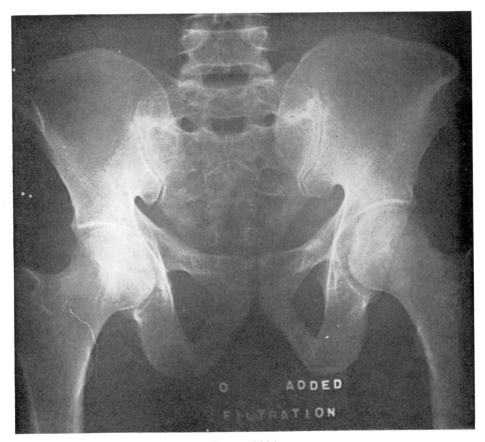

Figure 139A.

Figure 139. As filtration increases, density and contrast decrease.

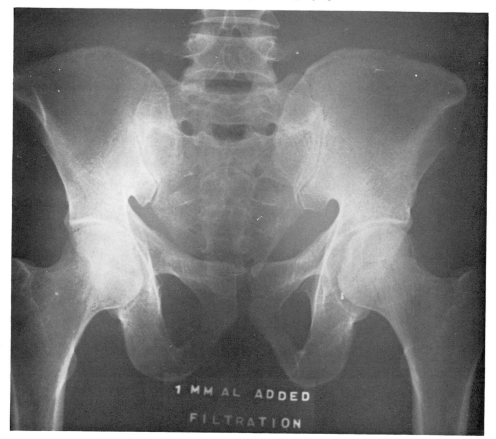

Figure 139B.

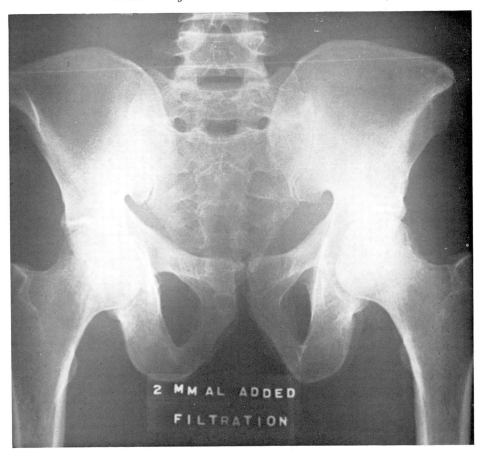

Figure 139C.

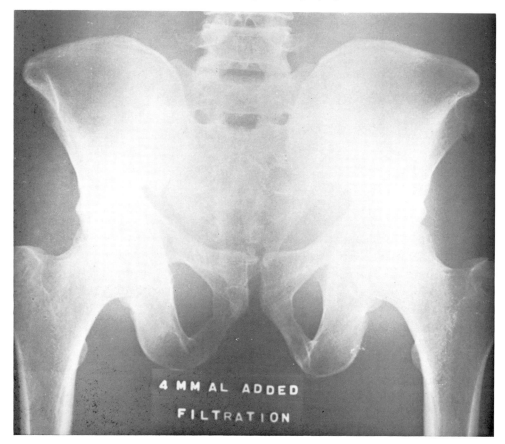

Figure 139D.

CHAPTER NINE

CONTROLLING THE REMNANT BEAM

RECOGNIZING THE INFLUENCE the remnant beam has on the radiographic image, this chapter will discuss the accessory equipment used by the technologist to control its effects on the image. Primarily, scatter radiation will be of major concern here.

THE CONCEPT OF CONING

We have established thus far that the presence of *excess* scatter radiation in a remnant beam will produce a fogged image. It should be emphasized here that excessive scatter radiation is determined by the ratio of primary photons to scatter photons making up the remnant beam. One way for the technologist to control this unwanted proportion of scatter radiation is to limit the numbers of interactions that produce scatter within the patient.

Reducing the number of interactions within the patient at first sounds like a complicated process. The use of coning, however, is extremely helpful in this regard and can be used without difficulty. Coning or beam-limiting devices control the amount of S/S interactions by simply restricting field size. Researchers realize that because S/S rays travel toward the film in such varying directions it seems reasonable to presume a scattered photon could be produced in one area of the body and strike an entirely different area of the film's emulsion. Figure 140 shows this effect. It can be understood that when a technologist is doing an examination of the gallbladder, for example, S/S radiation could strike an area of the film located in a region of the lumbar spine, kidney, or pelvis. These "transplanted" rays can be easily controlled with the use of beam limiting devices and with the use of radiographic grids. We have seen how S/S radiation can indeed stray from the point of origin to another location on the film surface. This causes what would normally be light areas of the film to have an additional density and decreased radiographic contrast results. The major contributor to radiographic fog is an excess amount of scatter rays in the remnant beam.

Types of Beam-Limiting Devices

There are four types of beam-limiting or restricting devices: (1) general purpose cones; (2) cylinder cones; (3) diaphragms; and (4) collima-

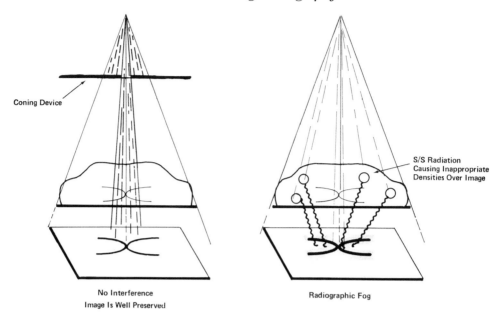

Figure 140. Unwanted densities are produced by scattered photons causing a general decrease in contrast of the object under examination.

tors (see Fig. 141). Collimators have become the most important of these because of the advantages they offer in patient safety and in convenience to the technologist. Also, government legislation aimed at reducing the patient exposure to unnecessary radiation has encouraged the use of collimators. This legislation prescribes that all newly purchased collimators must be electrically connected to the bucky tray so that it will automatically cone down to the film size used. The technologist now has considerably fewer options over coning as compared to recent years. Until about 1960, mostly cylinder and general purpose type cones were used. Because they were not adjustable, hurried technologists often did not change cone sizes with each size of film used. As a result, cones were not always properly used and the patient did occasionally receive additional unnecessary doses, while films had decreased radiographic quality. Manufacturers were urged to produce a device that would be less time-consuming for technologists to use and would provide better radiation protection for patients; modern collimators provide the answer. With this device, quick adjustments in field size can be easily accomplished.

Basic Construction of a Collimator

Collimators, regardless of the manufacturer, have basically the same components, which include a light source used for centering the central ray over the body part, a mirror to project the light beam to the patient, and a series of pulleys to move a pair of lead plates. The important value of having a light source is to show the technologist what area of the body will be irradiated so that proper positioning can more easily be accomplished.

The mirror must be held firmly in a precise position inside the collimator or the projected light coming out of the collimator will not give a true guide for centering (as noted in Fig. 141). Complaints that centering locks on the tube crane have gone out of adjustment or that the table is off center often reveal that the mirror had slipped out of adjustment and that the locks themselves and the table are in good working order.

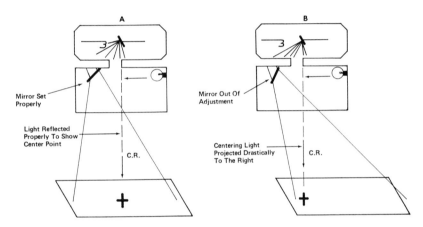

Figure 141. The mirror is responsible for projecting the collimator's light properly. The repair work should be done by a service person, not a technologist.

A well-designed collimator will have built into it a double set of adjustable lead leafs. A double pair as opposed to a single pair of lead leafs provides increased coning at the edges.

Besides the mirror slipping out of adjustment, another problem that will occur with collimators is that the shutters themselves could go out of adjustment. This is often a result of small cables slipping and getting caught in the nylon pulleys. It should be clearly understood that any corrections involving a beam-limiting device should be left strictly to a trained service person because any adjustments in field size or alignment possess a potential radiation safety hazard to the patient and to the technologist.

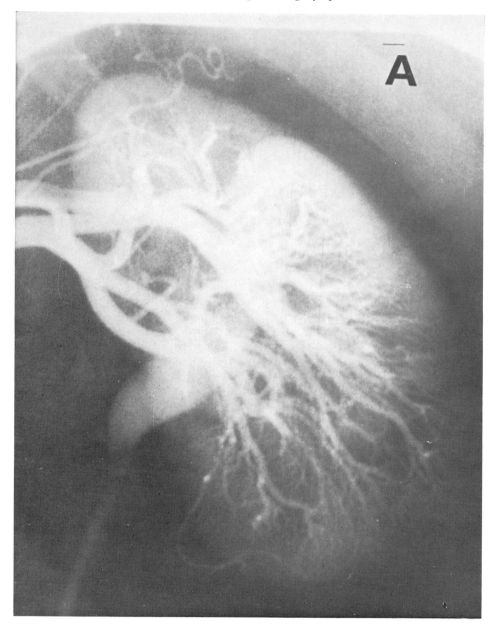

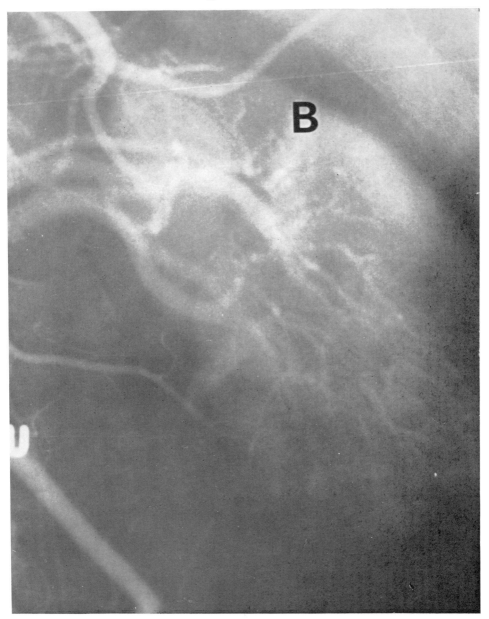

← Figure 142. Very "tight" coning in this examination has a positive effect. Increased contrast has improved visibility of detail greatly in *A*.

Uses for Conventional Cones

Of the conventional type cones that have been used before the collimator, cylinders are still favored for many special radiographic procedures such as tightly coned down selective renal arteriograms or cerebral arteriograms. Figure 142 shows a selective renal arteriogram using a cylinder cone and is compared to an open field exposure.

THE RADIOGRAPHIC EFFECT OF BEAM-LIMITING DEVICES

The purpose of using a beam-limiting device is to reduce unnecessary exposure to the patient and to improve radiographic contrast by reducing the number of scatter interactions in the patient. With this, since S/S rays are photons with penetrating qualities that help produce radiographic density, the technologist should be mindful that when their numbers are reduced radiographic density is moderately to significantly decreased. In

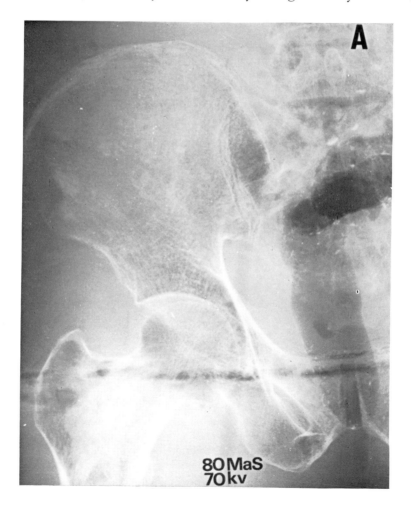

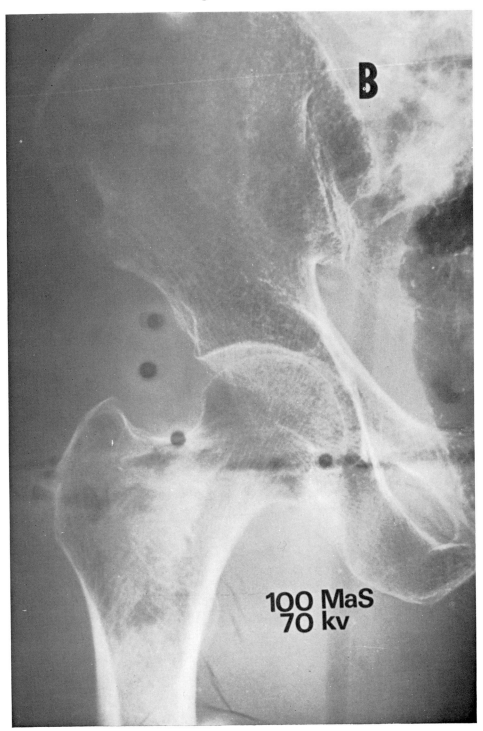

← Figure 143. The MaS has been increased by 25 percent to compensate for coning.

addition to the density issue, increased coning causes a moderate to substantial increase in radiographic contrast as well. It is generally understood that when moderately tighter coning is used where it was not previously at least a 5 Kv increase is necessary to regain the lost density. From the author's point of view, however, it would be more appropriate to increase the MaS by 25 percent to 35 percent (depending on the total amount of S/S produced). Reasons for this are, first, increments of 5 KvP do not always produce the same radiographic effect at all Kv levels because of the Kv influence on exposure latitude, and, second, increasing KvP is likely to produce more scatter interactions which will tend to defeat the purpose of the coning. The effect on the radiographic image is determined by the thickness of the body part, to what extent coning has been used, the level of KvP being used, the density of the body, and the atomic number (type of tissue) being irradiated. In Figure 143 we can see the degree of compensation used when tight coning was employed as compared to open field.

Influences on Patient Dose

As was pointed out earlier, field size has a strong influence on the total number of interactions within the patient. In reality, it is the interaction that accounts for the danger in irradiating patients. At normal settings, an A-P lumbar spine exam exposed with and without coning will result in differing exposure doses to the patient. To say the least, the technologist must be especially mindful of coning procedures when radiographing children and young adults of child-bearing years.

Summary

The important concept for the technologist to remember regarding beam limiting is that excess amounts of S/S radiation (disproportionate quantities of secondary to primary) are strongly influenced by exposing too large an area of the patient's body part. The resulting overabundance of scatter rays causes high volumes of radiographic fog which in turn produces poor radiographic contrast. With the use of beam limiting devices, smaller field sizes can be obtained, causing fewer numbers of interactions to form (less stray or transplanted rays). This sequence of events usually improves contrast and visibility of detail.

↟ Field size leads to ↟ S/S interaction resulting in S/S rays striking the film producing ↟ Density and ↡ Contrast.

Although contrast improves with progressively tighter coning, the most dramatic radiographic effect occurs when going from no coning to a cone of the correct size for the body part under examination and when the smallest cone cylinders are used for spot filming. It should not be forgotten that different body parts produce varying quantities of scatter, and that proper coning is very important for those areas that yield large quantities of scatter. With this in mind, it should be clear that when adjusting the technique for coning situations, one should be mindful of the total amount of scatter radiation produced by that particular body part. For example, tight coning will have much less effect on density and contrast when radiographing an elbow than it would if the same coning changes were made when radiographing an abdomen or a lateral lumbar spine.

RADIOGRAPHIC GRIDS

The Concept

Although coning is effective in improving radiographic contrast by reducing field size, it must be kept in mind that, even if proper coning is used, the body part under examination provides sufficient quantities of S/S radiation to cause S/S fog. Thus, a device must be placed in the remnant beam that will further reduce the amount of S/S rays.

You have already learned how adjustments in KvP will alter the number and type of interactions produced by a body part, but this alone does not sufficiently improve radiographic quality and, in addition, increases the possibility of underpenetration.

The grid is actually a kind of "filter" that absorbs radiation attempting to pass through its lead strips *at too great a divergent angle* as compared with the primary beam. Figure 144 shows that, with all the negative characteristics associated with scatter radiation, it is fortunate that they travel

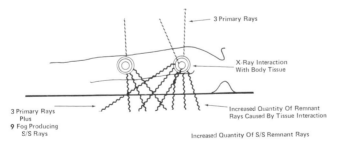

Figure 144. Many more scatter photons than primary photons reach the film. Scatter photons travel in very different patterns as compared to primary photons, especially when lower KvP exposures are used. The numbers used do not indicate actual values.

at a more divergent angle than the primary photons. In fact, this one characteristic of S/S makes the use of grids practical. The total S/S filtering effect, known as *cleanup*, is mainly regulated by grid ratio. Grid ratio is the relationship between the height of the lead strip to the distance between them.

Construction

Radiographic grids are available in a number of types and styles, yet all grids are composed of nothing more than a series of alternating lead strips (cut very thin) and radiolucent material. The type of radiolucent substance most used for this interspace material is aluminum, although in some cases, especially in the older grids, cardboard is used. Cardboard interspacing is not as popular now because it is not as strong as the other material and also has some sensitivity to humidity changes causing expansion and contraction. With this expansion and contraction over time, the very thin lead strips become warped and cause radiographic defects.

Aluminum interspacing material, as might be expected, has a slight advantage because it usually has a somewhat higher absorption rate to scatter rays, resulting in slightly better cleanup.

Grid Ratio

One of the most important characteristics manufactured into a grid is its ratio. Grid ratio, by definition, is the relationship between the height of the lead strip and the spaces between. Common grid ratios are 5:1, 6:1, 8:1, 10:1, 12:1. The use of 16:1 grids has very little advantage in cleanup and relatively high disadvantage in narrow positioning latitude which will be discussed later; as a result, 16:1 grids are becoming less popular. A 5:1 grid, for example, will contain lead strips measuring five times higher than the actual distance between them (see Fig. 145). Grid

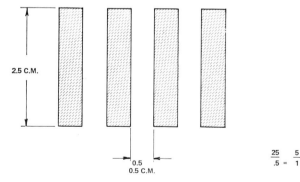

Figure 145. The relationship between the height of the lead and the interspace will greatly affect the likelihood of the grid trapping more or fewer scatter photons.

ratio principally determines what percent of scatter it can remove from the remnant beam. The percentage of scatter removed from the beam as earlier noted is called cleanup or grid efficiency. As more scatter rays are taken out of the remnant beam by the grid, the grid is said to have better cleanup and greater efficiency.

Although it will be no surprise to hear that cleanup increases as grid ratio increases, it is important to know that the increase in cleanup is *not* proportionate with increases in grid ratio. For example, the cleanup for a 5:1 grid is approximately 88 percent (removes 88% of S/S in the remnant beam); but if a 10:1 grid ratio is used in place of a 5:1, the cleanup will be approximately 91.2 percent, which represents an additional cleanup of only approximately 4 percent. As still higher grid ratios are used, additional cleanup becomes even less dramatic. When going from a 12:1 grid to a 16:1 grid, the additional cleanup in S/S is only 3 percent and the effect on the radiographic image is very slight, indeed. The radiographic effect of using a grid is more obvious when higher KvPs are used with thick, dense body parts (see Fig. 146). In general terms, it can be correctly stated that as grid ratio increases grid efficiency and cleanup increase as well. Figure 147 shows the relative increases in cleanup using various grid ratios. There is, of course, a decrease in density because (1) less S/S is reaching the film and (2) grids do absorb some primary photons.

	60 KvP		100 KvP	
GRID RATIO	% OF SCATTER ABSORPTION	% OF PRIMARY ABSORPTION	% OF SCATTER ABSORPTION	% OF PRIMARY ABSORPTION
5:1	88%	approx 30%	70%	24.2%
6:1	89.4%	- - -	81%	- - - - -
8:1	93%	31.4%	88.8%	27%
12:1	95.8%	34.3%	92.5%	29.1%
16:1	97.2%	35.6%	95.4%	31.6%
5:1 cross	97.6%	53.8%	92.6%	39%
8:1 cross	99.1%	50.8%	97%	46.3%

Figure 146.

In quick review, as grid ratio increases more S/S radiation is removed from the remnant beam thereby providing a better balance of scatter to primary in the remnant beam; less S/S radiation reaches the film, and the radiographic image will have slightly greater contrast thus providing the technologist and radiologist with much better visibility of detail. Because the amount of cleanup is mainly determined by ratio, progressively higher grid ratios will progressively produce higher radiographic contrasts. In

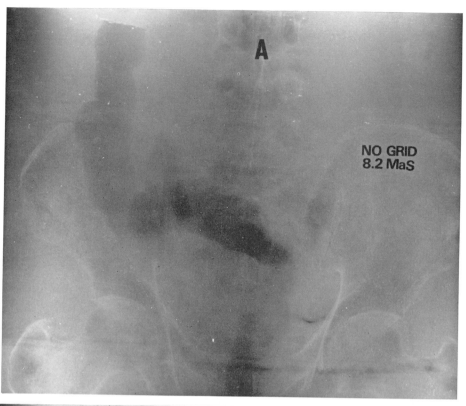

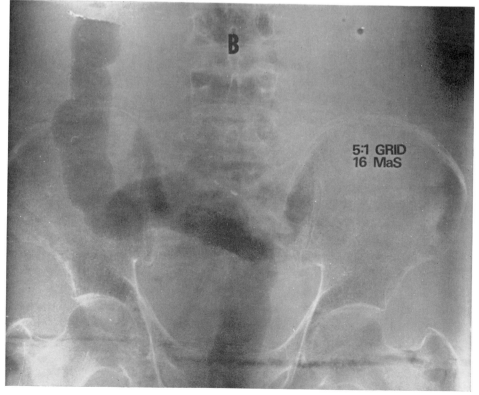

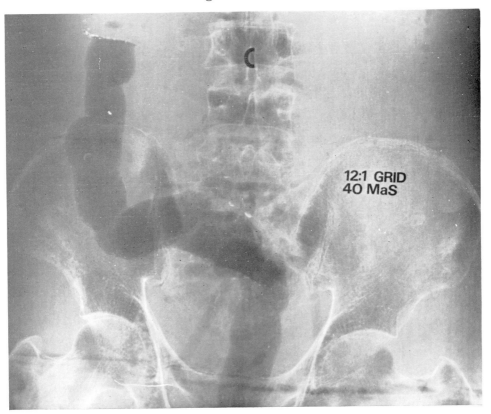

← Figure 147. The contrast seen in these films indicates the effect grids have and the importance of proper grid selection. *A* has virtually no visibility of detail. These radiographs were exposed at 80 KvP.

further summary, grids function on the principle that scatter radiation travels at a more dramatic (divergent) angle as compared to that of the primary photons. Thus, grids allow primary photons to pass through its lead strips easily while "capturing" the more tangent scatter photons and eliminating them from the exposure. Figure 148 illustrates how the lead strips take advantage of this to produce a higher quality image.

Grid Cutoff

With all the absorption of excess scatter radiation by the grid, many of these still manage to get through and reach the film. The primary rays, however, must get to the film's emulsion or a condition known as grid cutoff will result. Grid cutoff is when *too many* primary rays are absorbed by the grid causing areas of decreased density. Figure 149 shows the

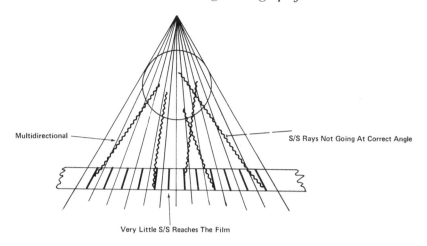

Multidirectional

S/S Rays Not Going At Correct Angle

Very Little S/S Reaches The Film

Figure 148. Scattered photons are trapped by the lead strips.

radiographic effect of grid cutoff. The technologist should be mindful that the degree or amount of lost radiographic density depends on the amount of primary radiation that has been absorbed by the grid.

Grid cutoff can result from a number of conditions which will be discussed later in this chapter; however, it might be beneficial to review Figure 150, which illustrates the two major causes of primary cutoff. Figure 151 shows an improperly positioned focused type grid.

Stereo Radiography

Stereo radiography presents a problem to the technologist with regards to grid cutoff. Figure 152 shows there is no problem in angling the beam along the length of the lead strips; however, it might sometimes be necessary during stereo examinations to angle the beam across the lead strips. The technologist should remember when he shifts across the lead strips that the highest ratio he can successfully use is a 6:1 grid. The reason for this is that if higher ratios are used for *cross grid* shifting, the two stereo radiographs will have unequal densities, and high ratio grids (10:1 and 12:1) would allow no primary photons whatever to reach the film.

High grid ratios do improve radiographic quality and in fact are considered absolutely essential for most x-ray examinations. Because grids are so effective in reducing *excess* amounts of scatter radiation from the remnant beam, they permit the technologist to use higher Kv values. The importance of high KvP exposures in providing decreased patient dose and allowing shorter x-ray exposures to be used should not be taken lightly.

High Ratio Grids and Cutoff

The major disadvantage in using high ratio grids can be seen in Figure 153. From this illustration, you can see that as grid ratio increases, more of the lesser angled rays are eliminated from the remnant beam. It can be imagined, after studying this illustration, that if the ratio is too great not only will the angled S/S rays be removed but too many primary rays will be eliminated from the exposure as well. For this reason, the maximum ratio manufactured for practical use is a 16:1 grid and, as briefly mentioned earlier, these grids have questionable value.

In short, the technologist must choose an appropriate grid ratio which will meet the criteria of (a) providing maximum cleanup and (b) permitting sufficient primary radiation to pass through its lead strips so that cutoff does not occur. A term known as positioning latitude is thus used by technologists to refer to the extent to which a grid can be mispositioned (misaligned with respect to the primary beam) before enough grid cutoff will occur to result in decreased radiographic quality. Grid ratio then controls the amount of positioning latitude (positioning error) a grid will allow. We have been discussing cutoff in terms of high ratio grids; it should be made clear, however, that grid cutoff can also result when using a low grid ratio as well; however, positioning latitude is much greater and repeat exposures are less likely to occur as a result of malalignment between the grid and the primary rays.

Major Types of Grids

There are other important considerations in choosing the correct grid for a particular exposure condition. Grids can be arranged into two *family* groups: liner and cross grids. A linear grid is any grid that has lead strips running in the same direction. A cross grid's lead strips run in opposing directions, usually at 90 degrees to each other. This type of grid forms a "criss-cross pattern" on the radiographic image if inspected closely. Figure 154 shows the two basic designs.

Nonfocused and Focused Grids

The earliest type of grid was the linear grid. For some time this grid seemed more than sufficient; however, as various ideas for new techniques in radiographic procedures were considered and explored, the original linear grid was in need of improvement. The major problem with the linear grid was that it did not allow the remnant beam to pass between its lead strips located at the outer edges of the film. This is because the primary photons coming from the tube are more divergent at the outer beam perimeter as compared to those moving down through the center

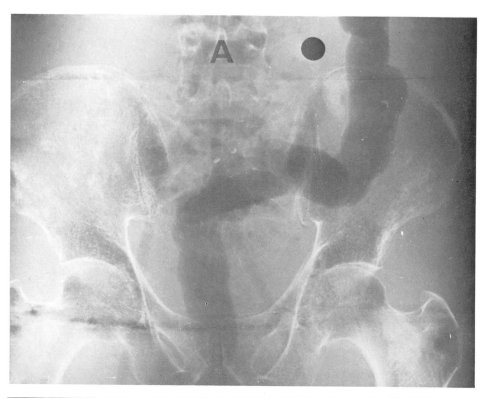

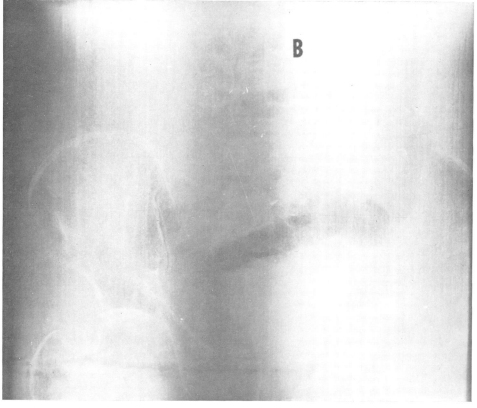

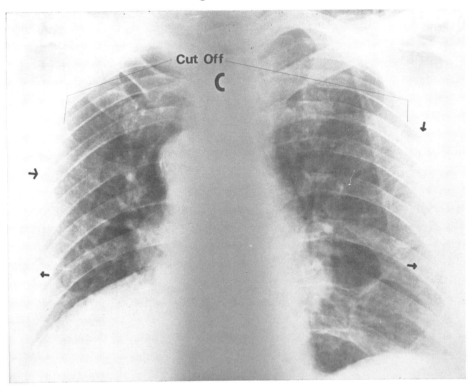

← Figure 149. *A* was exposed using a properly positioned stationary grid. *B* and *C* show the radiographic effect of a poorly positioned grid.

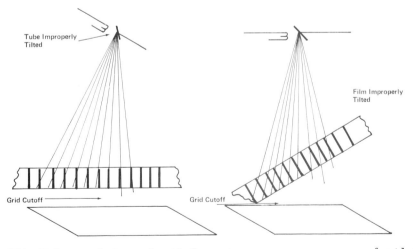

Figure 150. Tube angulation and grid tilt are two very common causes of grid cutoff, especially with portable abdomen or hip examinations. Using the incorrect F.F.D. will also cause serious grid line problems.

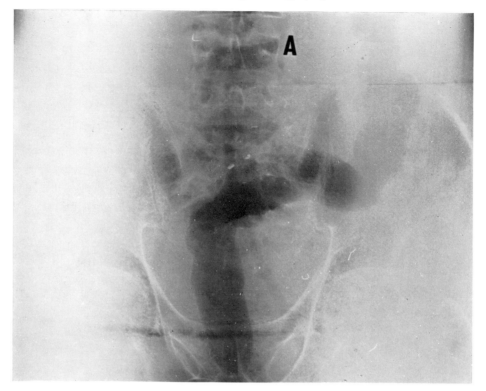

Figure 151. The effect of a focused grid placed upside down.

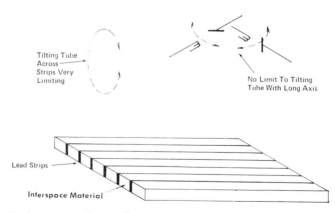

Figure 152. Grid stereo radiography poses no problem as long as the tube is **angled** *with* the lead strips. If the tube is to be shifted across the lead strips, cutoff is likely to occur. Stereo work should not be attempted when cross grids are used.

"core" of the beam. Figure 155 illustrates this situation, the result of which is a marked decrease in radiographic density at the outer edges of the film due to grid cutoff. It must be noted that this condition worsens with (1) shorter focal film distances, (2) large sized films, and (3) with higher grid ratios.

After some reflection on the problem, researchers realized that if the lead strips could be individually angled to varying degrees according to the projected divergence of the primary photons as they pass through the borders of the grid, a more even density distribution would be gained. Figure 156A shows this concept in diagrammatic form. This type of grid became known as a focused grid because its lead strips are "focused" (angled) to accommodate the natural divergence of the primary beam striking the film at the outer edges.

This arrangement was excellent. However, it is necessary to have focused grids made for ranges of focal film distance as noted in Figure 156B. Thus, a focused grid has the important advantage of permitting a uniform passage of primary photons through its lead strips, across the entire surface of the grid. In addition, the focused grid lead strips are indeed "focused" to accommodate the specific divergence which changes according to the focal film distance used. This requires that focused grids be manufactured and used for specific and compatible divergences relative to the focal film distance.

Cross Hatch Grids

With the concept of linear grids (focused and nonfocused) understood, we may now consider the cross grid family which is shown in Figure 154. By definition, the cross hatch grid is basically two linear grids affixed to each other so that the lead strips are at opposing angles. The most common cross grid has the lead strips opposing each other at 90 degrees, however, a rhombic grid has its lead strips opposing each other at 45 degrees.

Like the linear grid, cross grids can be ordered in focused and unfocused types and with a variety of ratios. The major advantage in cross grids is their relatively high cleanup per ratio as compared to linear grids. The major disadvantage regarding cross grids is that they usually cannot be used in a bucky (except for the rhombic design) and that the amount of intentional tube angulation that can be successfully used is very limited. A 5:1 cross grid will have similar positioning latitude to a 5:1 linear grid; but the CR can be angled with the lead strips using a linear grid as much as is desired without risking cutoff while a cross grid does not permit such deliberate angles in any direction. The student should be careful to keep in mind that, whatever design or construction of grid he

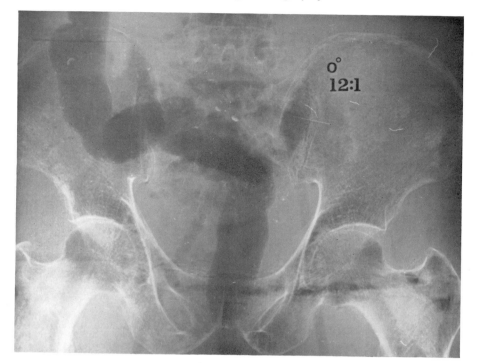

Figure 153A.

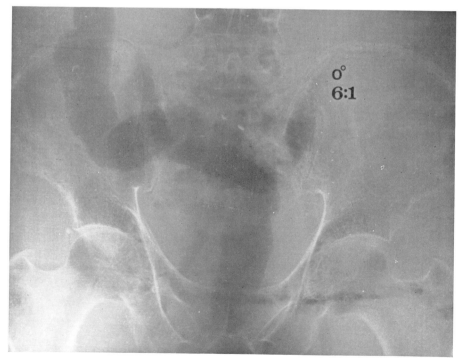

Figure 153B.

Figure 153. As grid ratio increases, the probability of grid cutoff increases.

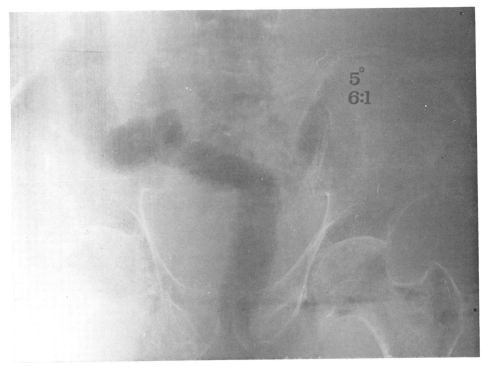

Figure 153C.

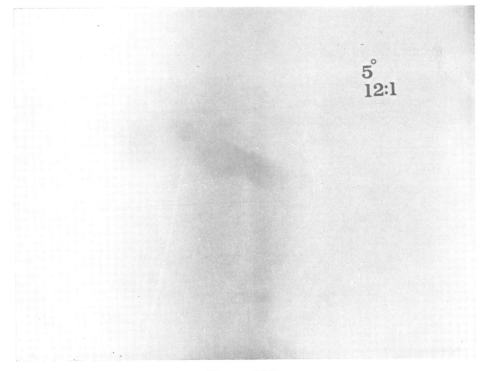

Figure 153D.

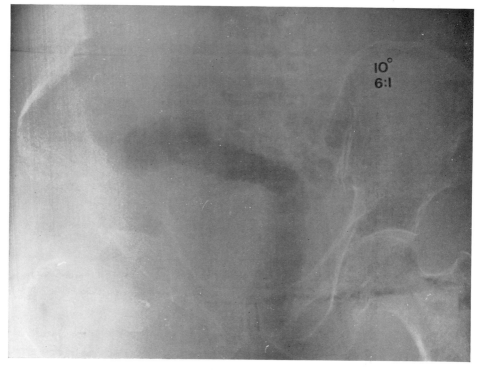

Figure 153E.

chooses, the basic function and rules are always the same regarding positioning latitude, ratio, cleanup, and purpose.

Linear Grid versus Cross Grid

The major advantage of cross grids over linear grids is that cross grids offer better cleanup and equal positioning latitude per ratio than can be expected from the linear grids—providing that deliberate center ray (CR) angulation is not necessary. It is generally accepted that an 8:1 cross grid will approximately equal the efficiency (absorption of S/S) of a 16:1 linear grid, yet the 8:1 cross grid will have wider positioning latitude be-

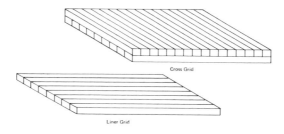

Figure 154.

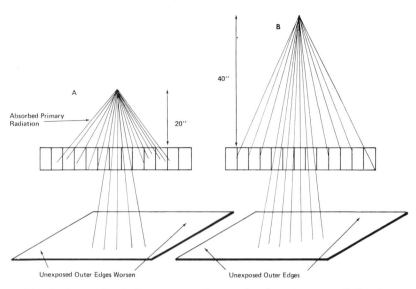

Figure 155. Grid cutoff of this type is the result of an incompatibility between the arrangement of the lead strips and the projected pattern of the photons.

cause of its lower ratio. Figure 157 shows the difference in exposure rates between cross and linear grids and the effect these two grids have on contrast.

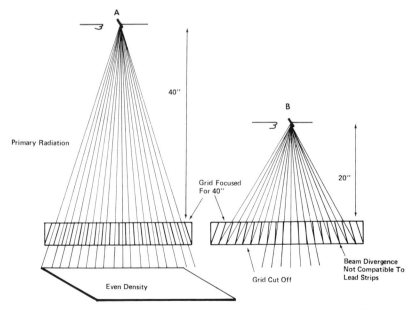

Figure 156. Focused grids are designed in such a way that their lead strips stop the scatter photons but allow the peripheral x-rays to pass through easily. If a focused grid is used when the tube is out of range, grid cutoff will result.

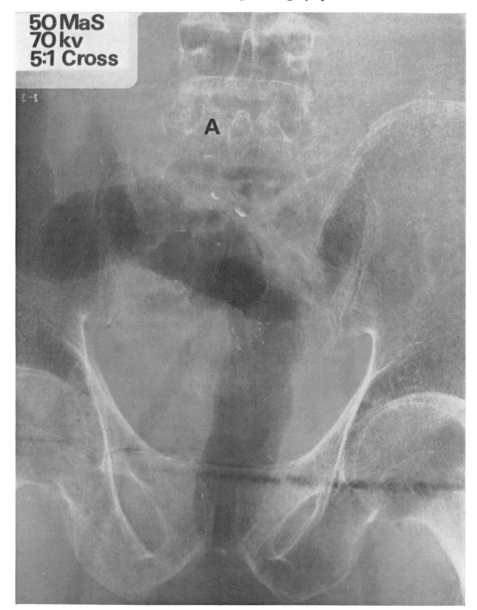

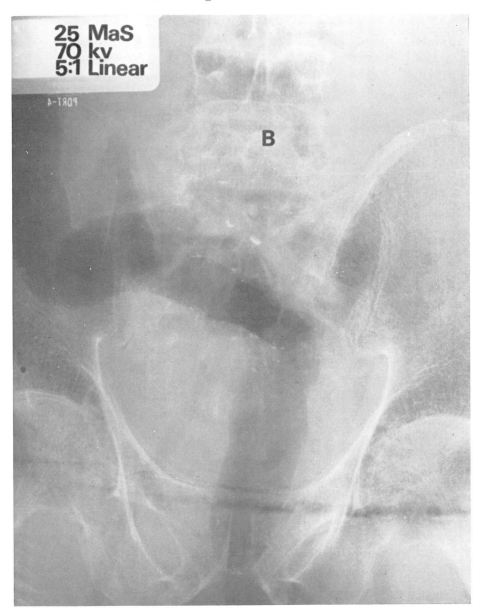

← Figure 157. These radiographs show the effect a cross grid has on the radiographic image compared to a linear grid of equal ratio. Note the exposure values.

Quantity of Lead in the Grid

Earlier in the chapter it was noted that grid ratio is a very important factor in efficiency and cleanup. Yet there is another important contributor to cleanup the technologist should be aware of. It is the factor commonly known as grid weight. It may seem to be a very basic point, but the quality of lead used (purity) to make the lead strips plays an important role in a grid's absorbing ability. The main idea to keep in mind here is that as long as lead content (type of grid weight) is consistent, the ratio will determine cleanup, and as a rule, grid manufacturers maintain a relatively uniform lead composition.

Moving Grids

A bucky assembly is a device that was invented in 1920 to put the conventional linear grid in motion during the x-ray exposure. The advantage in moving a grid during the exposure is that the grid lines can be totally blurred and made invisible in the radiographic image.

A similar effect will result if a photograph is taken of a fast moving object. The faster and smaller the object is, the closer it comes to becoming invisible on the photograph as a result of the blurring effect. With grids, the lead strips are very thin; when put in motion by the bucky, they become invisible in the radiographic image.

In general, bucky assemblies are used with focused grids. With the use of modern high capacity generators, technologists now have the option of using very short exposures, which is almost always an important advantage. However, very short exposures can complicate the use of a bucky assembly somewhat. The problem with fast exposures and bucky assemblies is that occasionally the exposure is so fast that it actually radiographs the grid in motion resulting in grid lines on the film. To correct this situation another breed of grid was designed and has become known as a fine line or micro line grid, which will be discussed later in the chapter. A fine line grid does not have to move during the exposure which permits the technologist to use the shortest exposure he wishes without fear of obvious grid lines.

Types of Bucky Assemblies

There are two major types of bucky assemblies: reciprocating and reciprocatic. The difference is that the reciprocating bucky moves at two different speeds during the exposure and that the reciprocatic bucky moves at one speed. The motion of the reciprocatic bucky is somewhat faster and thus allows slightly shorter exposures to be used without producing grid lines. These bucky assemblies push and pull a grid sideways under the table during the exposure. The total distance moved to either side of center while the grid is in motion is approximately one inch. The excursion of the grid from side to side causes a slight degree of decentering (see Fig. 158) and, as might be expected from such a situation, additional absorption of the primary beam occurs. For this reason, bucky exposures taken with equal grid ratios and lead content will produce slightly less radiographic density than if the same grid was exposed as a stationary grid.

In summary, one can say that grids can be used in one of two ways, stationary and moving. Stationary grids are used mainly for portable and operating room radiography, and occasionally in the radiographic room for various cross table studies. The bucky is used in all table x-ray work except for body parts that measure more than 10 cm and for a small number of special techniques, such as radiography of the paranasal sinuses. The use of a bucky assembly partially determines the standard focal film distance for table work. Because a certain amount of space is necessary for the bucky to function in its place between the film and the under side of the table, object film distance is inevitable. As discussed in Chapter Six, object film distance causes magnification and increases penumbra, but increasing the focal film distance can reduce this problem somewhat. The focal film distance has thus been set at 40 inches to minimize the approximate 2.5 inches of object film distance needed for the bucky. Previous to this, focal film distances were commonly 36 inches.

GRIDS AND THE GENERAL RADIOGRAPHIC EFFECT

Radiographic Density

The radiographic effect can be seen in two forms, density and contrast. Density is affected greatly because so much scatter radiation which had previously reached and exposed the film is now removed from the remnant beam. In addition, small amounts of primary radiation are absorbed by the grid. Contrast is affected because fog caused by S/S radiation is greatly eliminated. In general, 55 percent of the density of a chest exposure and approximately 90 percent of the radiographic density of an abdomen film is caused by S/S radiation. This means that for a chest examination only 45 percent of the density is caused by the primary beam and for an abdomen examination only 10 percent of the radiographic density is caused by the primary beam. If we consider the fact that a 12:1 ratio grid will eliminate up to 95 percent of the S/S radiation from the beam, we begin to appreciate how important the use of grids really is. Figure 159 shows two radiographs exposed with different grid ratios.

There is some question among technologists as to whether it is proper

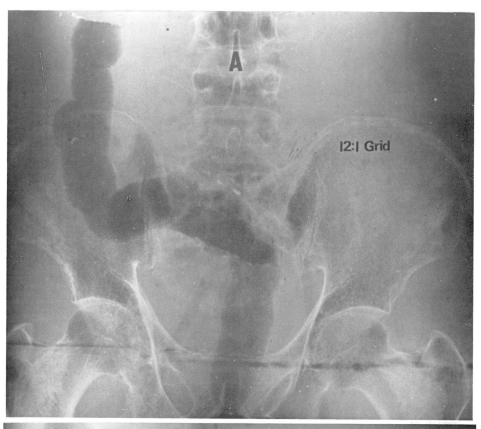

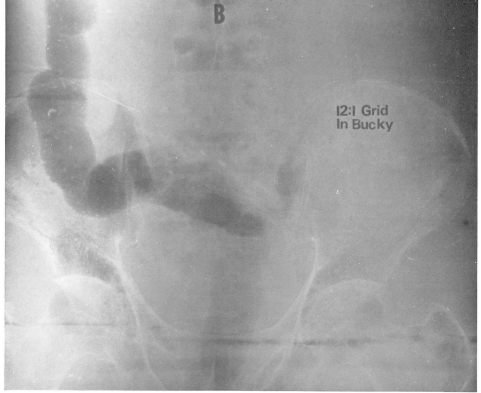

to increase KvP to compensate for the decreased density caused by grids. Although both methods are used, the author prefers increases in MaS. The reason for this is that grids are, after all, used primarily to improve radiographic contrast while absorbing excessive amounts of S/S radiation from the remnant beam. The adjustments needed for additional exposure is only to gain back the lost density, and increasing MaS performs this function very well. On the other hand, if Kv is used to regain the lost radiographic density when a 12:1 grid is used, an increase of approximately 20 to 26 KvP is required. When radiographing a moderate to thick body part, an increase of this magnitude would cause a substantial amount of additional scatter. Thus, if the 95 percent of S/S is removed by the 12:1 grid but the technologist compensates for the loss in radiographic density by increasing 20 to 26 KvP, the net effect of the grid in increasing contrast has been seriously compromised. Another argument against using KvP to compensate for different grid ratios is that the contrast media used for various examinations has an optimum KvP absorption rate. For example, the contrast manufactured by a major pharmaceutical company for doing IVP examinations often recommends a KvP that is not higher than 75. It might be helpful to recap the important results that use of grids have on the radiographic image. Please note the statement below:

> As grid ratio increases absorption of excessive S/S photons increases causing less S/S to expose the film. This, in turn, provides for a higher radiographic contrast.

Selecting the Proper Grid

The fact that there is a wide selection implies no one grid will perform optimally under all conditions. The determining factors for the purpose of grid selection are derived by considering two major conditions. The first is how much S/S will be produced with a given kilovoltage and body part, and the second is the environment in which the grid ratio will be used. It is advisable to study these major factors separately.

Scatter Radiation

The amount of S/S radiation produced as a result of interactions between the primary photons and the body tissues has already been adequately discussed: The primary influencing factor in S/S production is KvP and the makeup of the body part being radiographed.

KILOVOLTAGE: Kilovoltage has an important effect on the ratio of S/S

← Figure 158. Note the density differences in *A* and *B*. Published figures from Liebel Flarsheim Corporation indicate a 20 percent density variation between stationary grid and bucky work. *Courtesy of* Liebel Flarsheim Corporation, Cincinnati, Ohio.

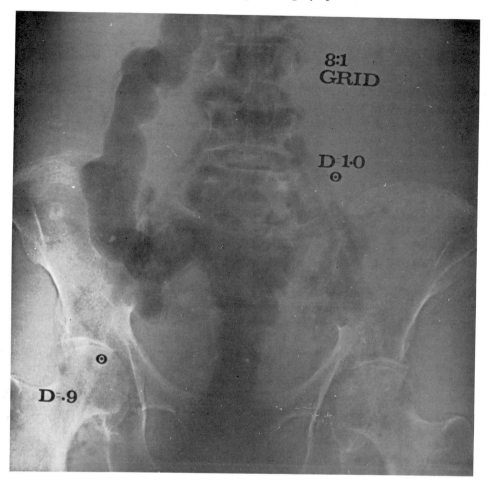

to primary radiation that is produced during an exposure. (It is the ratio of primary and scatter rays in the remnant beam that determines whether fog will result on the radiographic image.) The proper kilovoltage level is determined principally by atomic number, thickness, and body part density. An average lateral lumbar spine exposure requires the use of higher kilovoltages than an average shoulder or knee examination. Not all radiologists and technologists agree on the precise KvP level. However, there is usually general agreement that as the body part's atomic number and density increases, the KvP also should be increased.

One major decision that the technologist must make is to choose the proper grid ratio. The technologist should also keep in mind the major function of a grid is to clean up excessive scatter radiation from the remnant beam and, of course, the primary determinant in cleanup is grid ratio. Since scatter-producing interactions increase rapidly with increases in KvP above 70, more "grid" is necessary to absorb the additional scatter photons. Figure 160 gives a suggested KvP and grid ratio combination. In short, KvP has a strong bearing on the ratio that is to be used be-

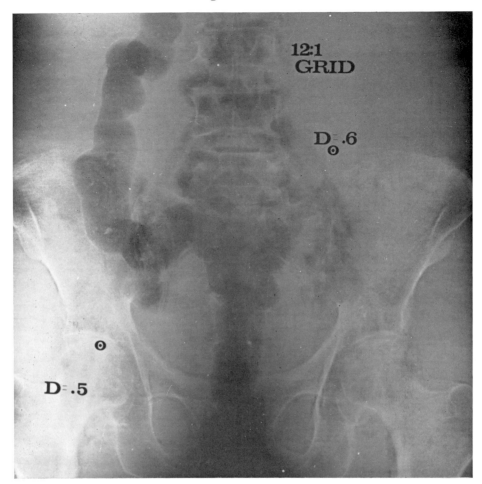

← Figure 159. As grid ratio increases radiographic density decreases, primarily because the scattered photons that would otherwise reach the film and add density are ab-sorbed by the lead strips. Some additional primary photons are also absorbed.

cause Kv strongly influences the type of interactions that will be pro-duced by the body. As KvP increases over 70, disproportionately greater amounts of Compton interactions result as compared to photoelectric interactions; it is the Compton interactions that account for the amount of S/S production that causes fog.

BODY PART: Patient thickness and density also strongly influence the overall amount of S/S radiation the body will ultimately emit. As dis-cussed earlier, patient thickness and tissue density vary with body habitus, body part, and pathology. However, to reiterate, scatter radiation in-creases with body thickness and tissue density. As one can imagine, it would be extremely impractical for the technologist to change the grid ratio for every patient with different body thicknesses and density, so prac-tically all x-ray tables have one type of grid installed. The most common type of grid in modern installations is a 12:1 ratio focused at 40 inches.

Cleanup	Type	Positioning Latitude	Recommended Up To	Remarks
GRID SELECTION ON BASIS OF CLEANUP REQUIREMENTS				
SUPERLATIVE	6:1 Criss-cross	Good	110 KVP	Tube tilt limited to five degrees.
EXCELLENT	12:1 Linear	Very slight	110 KVP (Suitable for higher kilo-voltages)	Extra care required for proper alignment; usually used in fixed mount.
	5:1 Criss-cross	Extreme	100 KVP	Tube tilt limited to five degrees.
VERY GOOD	10:1 Linear	Slight	100 KVP	Reasonable care required for proper alignment.
GOOD	8:1 Linear	Fair	100 KVP	For general stationary grid use.
MODERATE	6:1 Linear 5:1 Linear	Extreme	80 KVP	Very easy to use.

Figure 160. *Courtesy of* Liebel Flarsheim Corporation.

The Environment

The technologist must be equally careful when choosing a particular grid depending on the environment in which the examination is to be accomplished. The term "environment" is used here to describe the conditions under which the exposure is to be made, for example, whether the exposure will be in the patient's room using portable equipment, in the operating room, or in the radiographic room for cross-table exposures.

Without question, the key idea for the technologist to keep in mind during such potentially troublesome situations is the amount of positioning latitude the grid he will choose has to offer. As was pointed out earlier, positioning latitude is dependent largely on grid ratio. The perfect grid under any circumstance would be one that has very wide positioning latitude and very good cleanup. Unfortunately, these are mutually exclusive characteristics because high grid ratios are used for superior cleanup and efficiency and wide positioning latitude can only be gained with low ratio grid. From the radiographs shown earlier, it was noted how much of the image can be lost because of improper positioning of the grid. It is here that the greatest compromise in grid selection must be made. Because of the high probability of grid cutoff occurring when high grid ratios are used for the less than ideal exposure environment, cleanup is usually sacrificed for practicality. With this in mind, a 5:1 or 6:1 ratio is usually the most popular choice.

Even a "routine" portable hip examination can turn out to be a nightmare of unnecessary repeats if the grid used offers too little positioning latitude. Through experience, it will become painfully clear that if a light radiograph is obtained and the cause of the light film is improperly diagnosed as underexposed, additional trouble is close at hand. For example, the author has seen student technologists too quickly guess that their portable abdomen film was too light because they did not use enough exposure. For the repeat film they doubled the exposure, but as luck would have it, the grid was positioned properly for the second exposure and, to their amazement, the repeat film was now too dark. This meant that they had to take a third film. To make matters worse, it is standard procedure at this point to claim the portable unit is not working properly, leaving the whole situation in that much more disarray (see Fig. 161). To reduce the possibilities of this occurring, it is the author's opinion that technologists should use a grid not greater than 6:1 for portable or operating room work. (A 5:1 grid absorbs approximately 80% of the scatter and a 6:1 absorbs approximately 85% of scatter [see Fig. 162].)

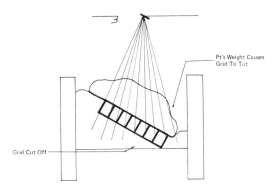

Figure 161. It is difficult for the technologist to see how the grid is aligned with the CR under these conditions, but an intuition can be developed with experience.

If circumstances make maximum cleanup esssential, a low ratio (5:1) cross grid is suggested. Because of their construction, cross grids produce unusually high cleanup per grid ratio. To some extent cross grids do offer the best of two worlds. A 5:1 cross grid will provide the cleanup equal to a 10:1 linear grid, yet because of its low ratio it has wide positioning latitude. The drawback to using cross grids, as was mentioned earlier, is that deliberate angles of center ray cannot be accomplished successfully.

FOCAL FILM DISTANCE: Focal distance should also be taken into account when choosing grids for a particular purpose. During earlier discussions on focus grids, it was emphasized that although grids are not focused for a *specific* focal distance, like 37 or 42 inches, they are limited to *ranges of* focal distances. For example, a grid rated for use at 40 inches can also successfully be used at 35 to 50 inches. When a focused grid is used at a wrong focal film distance, cutoff occurs and is most likely to show itself at the edges of the film (see Fig. 149C). As higher grid ratios

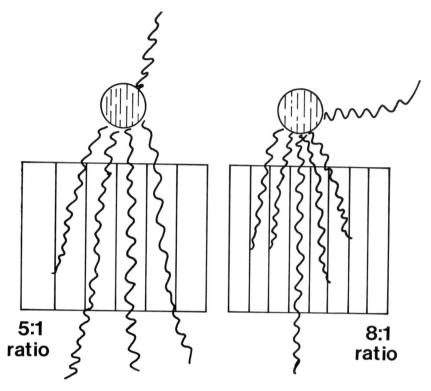

5:1
ratio

8:1
ratio

Figure 162. Compare the placement of the lead strips between the 5:1 and the 8:1 ratio grids. Because of this important factor, the 8:1 grid will allow fewer S/S rays to reach the film and contrast increase as density will decrease. Positioning latitude decreases with the 8:1 ratio grid in this illustration.

are used under these improper conditions, cutoff will worsen. Larger film sizes will also cause this problem to worsen. Parallel grids should be used at the longest focal film distance possible to avoid this peripheral grid cutoff condition. The reason for this is that the lead strips are perfectly vertical, and using increased focal film distances tends to produce a beam whose photons are approaching the film at a more vertical plane.

Chest radiography requires a grid with slightly different construction characteristics than those noted above. A very important factor in good chest radiography is to use the shortest exposure time possible. Because of this, a bucky assembly would be of no real value, since the very rapid exposures (one-sixtieth of a second and less) would radiograph the grid in motion resulting in visible grid lines on the film. To avoid this, bucky assemblies were abandoned altogether and a special grid with very finely cut lead strips (usually 100 to 110 lines per inch) was substituted. These fine line grids have the same basic design as the conventional grids previously discussed, except the lead strips and interspace materials are cut very thin. For example, a standard grid has between 60 and 80 lead strips (lines) per inch so that they can be relatively thick, but these produce a rather obvious distraction to the radiologist when viewing a number of films per day because the lead strips are so visible. With regard to fine line grids, to fit 100 to 110 lines per inch the lead strip has to be extremely thin—when radiographed the lead lines are virtually invisible at normal viewing distances even though this grid is not moving during the exposure. The obvious choice for a grid in chest radiography would be of a high ratio because most often high KvPs (over 100) are used. A desirable installation for chest radiography would be a 12:1 ratio grid focused at 72 inches with 110 lines per inch.

As important as it is for the technologist to use higher grid ratios with high KvP values in thick, dense body parts, it is important to remember to reduce ratio when lower KvPs are used. If this general rule is not followed, the lower penetrating beam will not have sufficient energy to pass through the high ratio grid and unusually high radiographic contrast will result because of underpenetration. To prevent this from happening, students should keep in mind the x-ray beam must not only penetrate the patient adequately but also penetrate the grid as well.

GRID CASSETTES: Grid cassettes are used as a matter of convenience. In a busy department, it is a nuisance to tape and untape stationary grids to cassettes. However, grid cassettes are very expensive, and because a large number of grid cassettes would be needed for a department, there is justifiable reluctance to purchase any more than necessary. In construction, a grid cassette is a standard cassette of any size with a grid built into the bakelite. Any kind of grid design can be purchased as a grid cassette.

SUMMARY

We often become frustrated when there are so few standards and constants for choosing proper exposure conditions including grids. Categorical answers come hard in the medical field because the end result is so dependent upon the patient who can, and usually does, provide an extremely wide range of variables with which we are expected to work. For this reason, it is often best for the student to understand well the various *concepts* surrounding a particular problem before any attempt is made toward a solution.

There is no one grid type, Kv level, body habitus, type of film, or processing condition that will work equally well for all conditions, and it is up to the technologist to make the final decision correctly by evaluating the particular exposure problem at hand.

CHAPTER TEN

TOMOGRAPHY

THROUGHOUT THE AUTHOR'S own experience, tomography has proved to be both an interesting challenge and a very beneficial diagnostic tool. In recent years much more sophisticated equipment has been made available with startlingly improved radiographic results. These more complicated units as well as the more simply designed tomographic units will be discussed along with their relative values later in the chapter. We should first look closely at some of the basic concepts involved in tomography. The terms "tomography," "planography," and "body section" are all synonyms and are thus often used interchangeably.

WHY DO WE USE TOMOGRAPHY?

The purpose of tomography is to make visible those structures that are not clearly seen radiographically because of their location in the body. Very often structures of interest are located beneath very dense body parts and can therefore not be seen by conventional radiography. Figure 163 compares a few tomograms made at crucial levels compared with routine radiographs of the same area. There is no specific organ or body part that may not be tomographed. Virtually any area of the body can be successfully "cut," as long as it has some reasonable thickness and the specific body part under examination is somewhat poorly visualized by surrounding structures. The cervical spine is an excellent example of a thin structure that is very successfully tomographed. The inner ear is perhaps one of the most difficult structures to demonstrate by conventional radiography, but can be nicely demonstrated with the use of tomography. It is easy to understand the value of tomography after reviewing these radiographs.

The Basic Concept

The importance of immobilizing the patient to prevent blurring is obvious. When the patient is allowed to move a body part under examination during the exposure a marked decrease in sharpness results, but motion can be occasionally used to the technologist's advantage. Tomog-

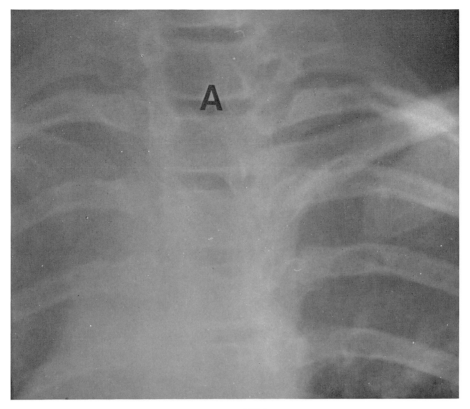

Figure 163A.

Figure 163. These radiographs show the value of tomography over conventional radiography when superimposed body structures are to be demonstrated.

raphy puts to use the concept that motion makes body structures invisible by blurring. During a tomographic study, an attempt is made to selectively blur unwanted structures while maintaining sharpness of the specific body part under examination. With tomography, the motion is not generated by the patient but rather by the x-ray equipment. In Figure 164, a metal rod has been drawn and accompanied by arrows showing its tilting motion. The arrows indicate motion along its entire length except at the pivot point. This point will be referred to henceforth as the fulcrum. If an x-ray tube is attached to one end of the rod and a film to the other end, we can begin to see something that resembles a functional tomographic mechanism (see Fig. 165). If a patient is placed as illustrated, one can begin to see that motion will occur at all levels of the body except at the fulcrum. If the technologist is using equipment that provides a method by which the fulcrum can be raised or lowered at will, virtually any level of the body can be made visible while all other structures lying

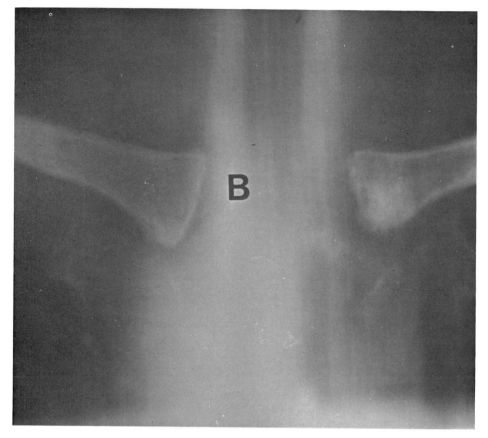

Figure 163B.

outside *that plane* will be practically invisible. Thus, it is the fulcrum that establishes the objective plane so that a specific level of the body can be tomographed. It should be pointed out that the area of clarity is not limited to the fulcrum point alone, but to the entire horizontal plane extending from each side of the fulcrum. Hence, all structures lying at the level of the fulcrum will be of equal sharpness. In short, the purpose of tomography is to improve visibility of specific body parts of interest that are surrounded by other body parts. The increased visibility is accomplished by deliberate blurring of all structures lying above and below the objective plane.

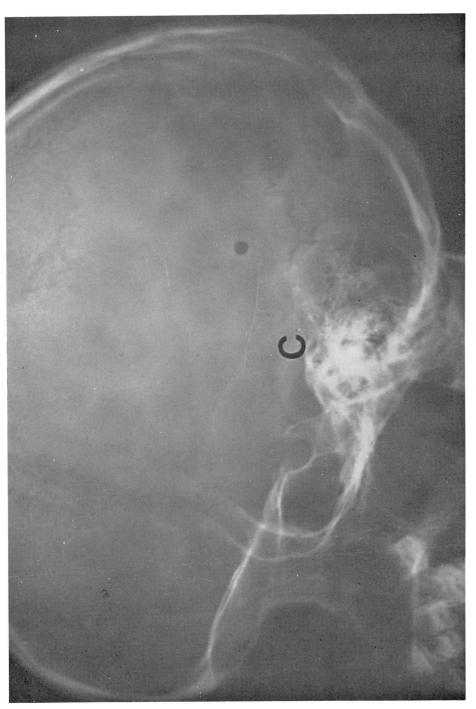

Figure 163C.

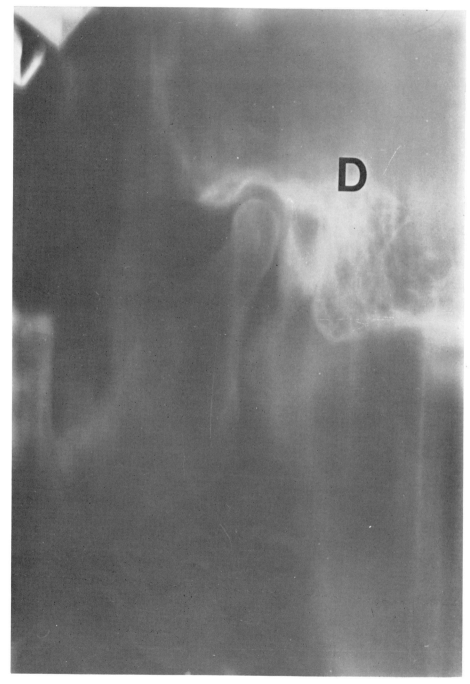

Figure 163D.

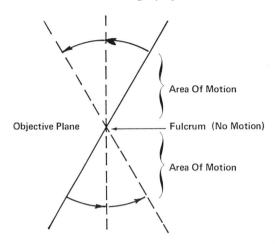

Figure 164. The basic concept of the fulcrum and resulting objective plane.

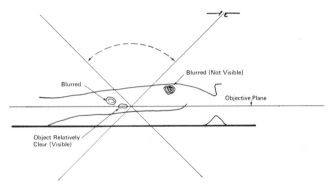

Figure 165. The basic components of a tomographic unit: a pivot point (fulcrum), a connecting shaft holding the tube and film in place, the x-ray tube, and bucky.

MAKING ADJUSTMENTS FOR CUT THICKNESS

There is another dimension to the objective plane—thickness. It is necessary that thickness of the objective plane be changed from time to time and this is handled quite easily by controlling the distance the tube travels *during the exposure*. In reviewing Figure 166, one can see that the area of sharpness (the objective plane) becomes thicker as the excursion of the tube decreases. The distance the tube travels *during the exposure* is known as arc. A very short arc, such as 10°, would produce a thick objective plane (cut) known as a zonogram, whereas a 50° arc would produce a very thin cut. Figure 167 shows a zonogram of a kidney compared with a cut made using a 50° arc. The value of zonography is that, if done properly, an entire organ or at least a large part of a structure can be seen as a unit on the film, as opposed to seeing the structure in

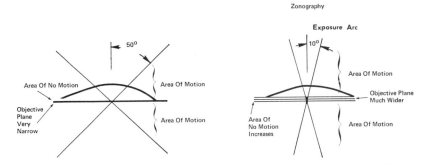

Figure 166. A 10 degree arc is commonly used for doing zonography. It is the exposure arc (the arc of the tube *during* the exposure), not the total distance the tube travels, that determines thickness of the objective plane.

many separate radiographic slices. As a result, the composition of the zonogram structure can be seen with more continuity. Often, zonograms of the kidney are obtained during a drip infusion study. On the other hand, it is common to use very thin cuts for x-ray examinations of small intricate bony structures as well as for bone examinations, and tomograms of the inner and middle ear region are often accomplished with a 40 or 50 degree arc.

In summary, not only the level of an intended cut can be adjusted by raising or lowering the height of the fulcrum but the thickness of the cut can be adjusted as well by changing the FFD providing this adjustment in FFD will cause the exposure arc to change also. The greater the arc the thinner the objective plane will become, and vice versa. Figure 168 shows that changing the FFD will not necessarily change the exposure arc depending on the unit being used. Because the general rule of beam geometry holds regarding FFD, short focal film distances for tomography are seldom, if ever, used. In fact, many feel a 48 inch FFD is optimal for obtaining image sharpness.

EXCURSION SPEED

Excursion speed may be described as the quickness with which the tube and film move during the exposure. There is no relationship between excursion speed and the level or thickness of the objective plane. There is, as you might expect, a direct relationship between excursion speed and exposure time. With the more simplified tomographic equipment, there is a series of switches located under the table and arranged in such a way to both initiate and terminate an exposure (shown in Fig. 169). Clearly, if the excursion speed is set by the technologist to be fast, the exposure will be short because the x-ray tube will pass the on and off switches more quickly.

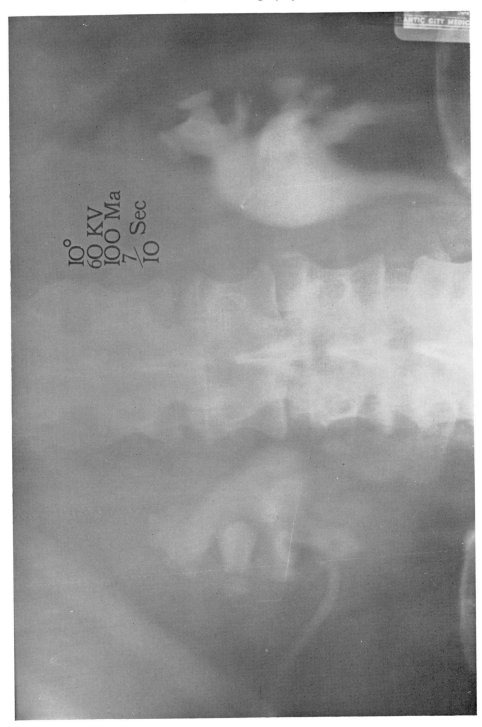

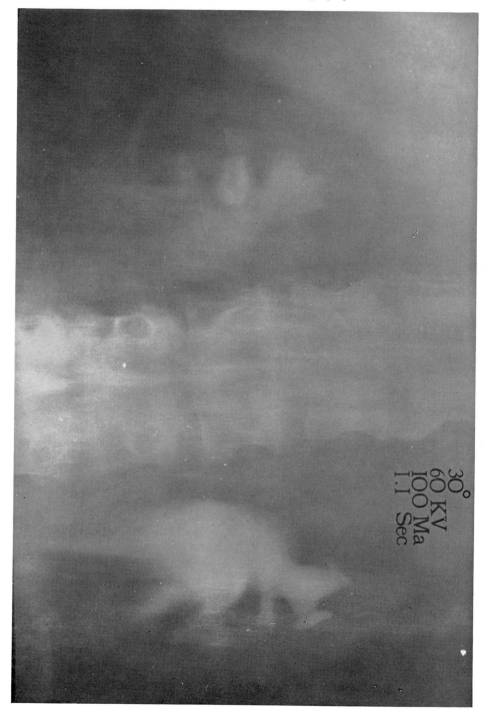

50°
60 KV
100 Ma
2.2 Sec

Figure 167C.

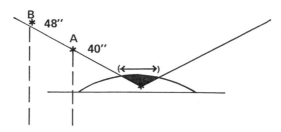

Figure 168.

Although intentional motion is vital to tomography, only certain movements are permitted. The term "unstable linkage" is used here to describe any situation that would allow unwanted motion of the tube or film. Poor (unstable) linkage most often occurs with lightweight equipment common to dual purpose units. Such units if properly calibrated, however, can produce good results, but if they are used improperly, important stress points in the tube and bucky drive mechanism soon begin to give way and loosen. If this process continues, sharpness of the objective plane will decrease resulting in a slow erosion of tomographic quality.

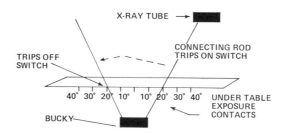

Figure 169. Although this design is certainly not a standard among manufacturers, it illustrates that the technologist does not initiate the *actual* exposure and that the x-ray tube usually travels some distance before the exposure begins.

Poor linkage can be seen radiographically in the form of excessive smearing and unsharpness of the objective plane, but because there is always some blurring present on the tomographic image it is difficult or impossible to tell how much is from "natural" causes and how much is the result of loose mechanical connections. Figure 170 shows a typical dual-purpose, lightweight tomographic unit compared to one of more substantial and sophisticated design.

Figure 171 shows schematically the type of excess motion that commonly results from unstable linkage. Three points might be valuable to consider in attempting to reduce unnecessary tube motion. (1) Use a slow speed for the majority of radiographic work and especially for cutting

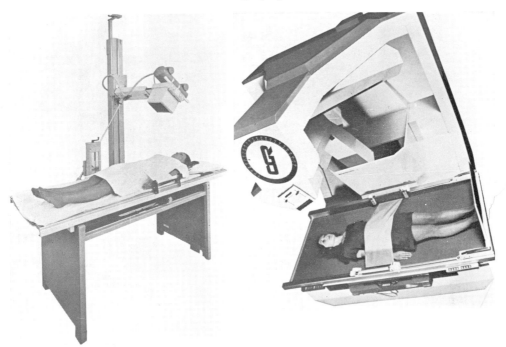

Figure 170. The variation in tomographic equipment design. *Courtesy of* C.G.R. Corporation.

small, difficult to visualize body parts. If the tube moves too quickly through its excursion there is what appears to be a jerking motion initially from which the tube has difficulty recovering as it moves over the patient. Often the tube will still be wobbling slightly after it has made contact with the first turn-on switch. (2) Be sure all knobs that are to be loosened are properly loosened so that once in motion, the tube and film can continue without unnecessary resistance. Any friction or resistance between moving parts can give rise to an unsmooth tube film excursion resulting in decreased image sharpness. (3) Place the body part under examination as close to the film as possible. There is some feeling, incorrectly held by technologists, that the rule of using the shortest objective film distance possible does not apply in tomography, however, this is not the case.

TYPES OF EXCURSION PATTERNS

As long as the prerequisites for a smooth tube film excursion are met (movement without wobbling, jerking, etc.) during the exposure, a more complex excursion pattern will yield less smear and streaking of the tomographic image. With this in mind, manufacturers have designed tomographic equipment that will move in a rather complicated pattern. Figure

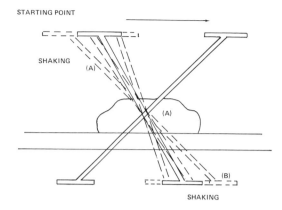

Figure 171. This type of unstable linkage can cause image blurring of the objective plane. When this cannot be corrected mechanically, longer exposure times are advised to reduce tube speed.

172 shows tracings of motions that have become commonplace in modern tomography.

The simplest motion is known as "linear." Although it is still one of the most often used, it does not produce the quality of structure visibility as does the more complex patterns. The problem with linear tomography is that it produces an excess of streaks and smearing that are longitudinal to the direction of the tube motion (see Fig. 173). It can be seen that such blurring certainly detracts from both the overall appearance of the image and more importantly its diagnostic value. This streaking occurs at the points immediately above and below the objective plane as noted in Figure 174. It might be of interest to note that with thin to very thin cuts, these striations will usually appear more pronounced when compared to

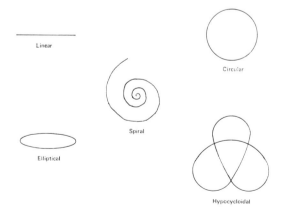

Figure 172. Tracings of commonly used tomographic excursion patterns.

tomographs made with thicker cuts. One way to reduce this type of excess blurring is to arrange the body part so that the long axis of the structure under examination is diagonal and preferably perpendicular to the direction of tube motion (Fig. 175). Figure 176 compares images produced with linear and elliptical patterns. A notable decrease in streaking can be seen as the tube pattern becomes more complex. These more complex excursion patterns are very valuable, especially for the smaller complex structures such as the inner ear.

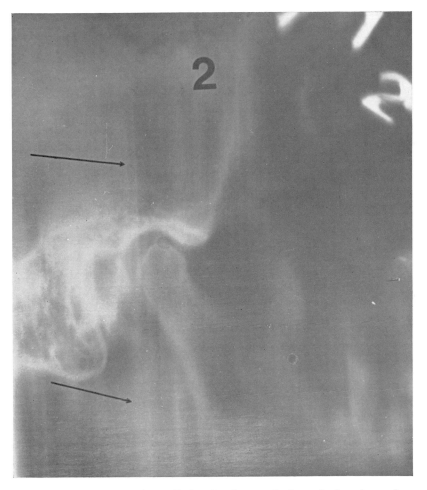

Figure 173. This streaking (also called redundant and ghost shadows) can become so noticeable and obvious it obscures the structures lying in the objective plane.

THE BOOK CASSETTE AND MULTIPLANOGRAPHY

All tomographic equipment has adjustable fulcrum heights as noted earlier. However, a method is available that produces more than one

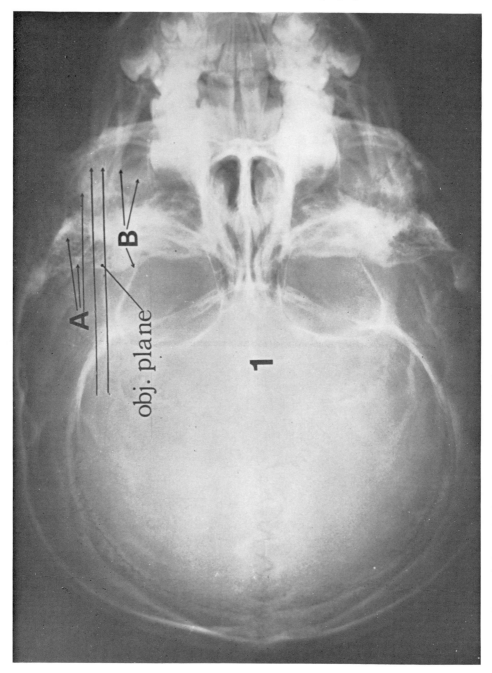

Figure 174. A skull in the lateral position. As the beam is passing through this body part (t.m. joint), other *prominent* body structures are projected to the film as well. These prominent structures are not totally blurred because they are so close to the border of the objective plane.

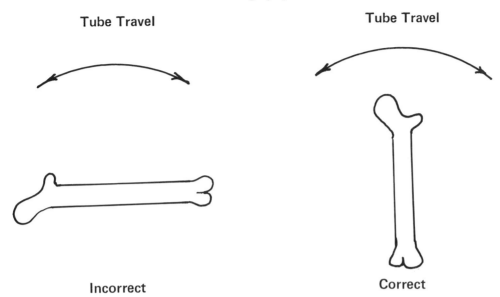

Tube Travel   **Tube Travel**

Incorrect   Correct

Figure 175. For linear tomography, the body part should be positioned with its long axis in such a way that the tube *will not* move parallel to it.

objective plane with a single exposure. This is known as multiplanography and has questionable value for two important reasons. (1) A device called a book cassette is used (as noted in Fig. 177) which is very thick and causes a great increase in object film distance. It is composed of as many as seven individual sets of intensifying screens that are arranged to be 0.5 to 1 cm apart. The object film distance caused by the thickness of this cassette ranges anywhere from 2.5 inches for the first film to 5 inches for the film that is at the bottom of the cassette. (2) Because the beam loses a moderate amount of intensity as it reaches the film stacked at the bottom of the cassette, it is necessary that the bottom few screens have fast speed screens, and this, of course, decreases sharpness of detail even further.

The combination of these two factors has resulted in decreased interest of multiplanography because of the lack of sharpness. The advantage of multiplanography is that it is a very quick way to produce many different levels of the body part with one exposure. Thus, it can be used effectively as a scanning tool by which approximate levels or structures can be located. Once the structure is located, conventional tomography could be used for the final diagnostic cuts. Thus, the trial and error procedure for finding the correct level of a body part would be greatly reduced and there would be a significant decrease in patient dose as well. Multiplanography is not a popular technique today.

Plesiotomography utilizes the same book cassette concept but it is

Understanding Radiography

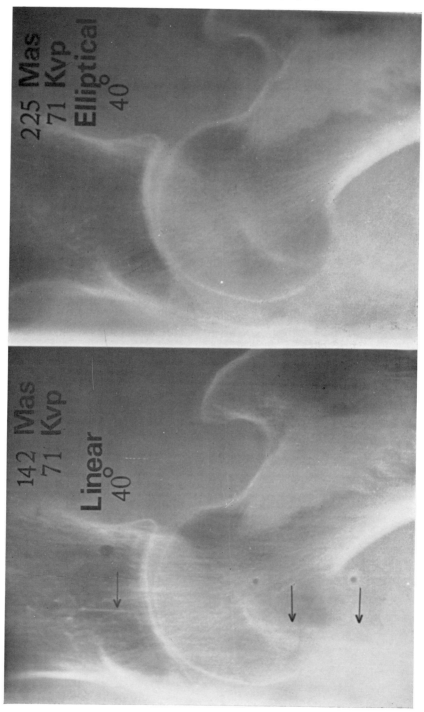

Figure 176. As the pattern of tube motion becomes more complex, ghost or redundant shadows are reduced or eliminated.

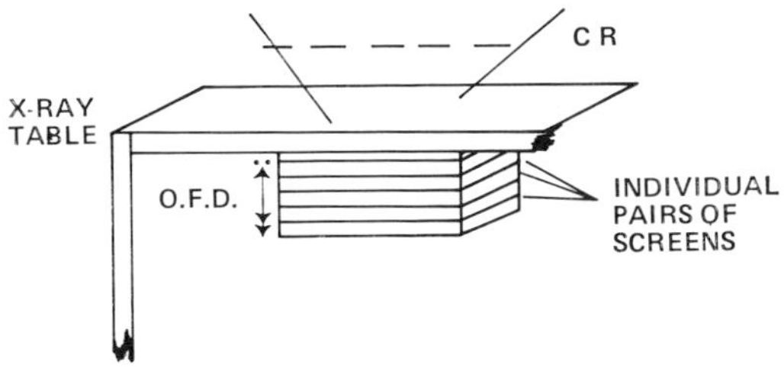

Figure 177. A book cassette as designed and used, revealing one of its most important disadvantages—increased O.F.D.

much thinner and mainly used for radiographing smaller parts such as the middle ear. The standard book cassette noted above commonly produces cuts 1 or 0.5 cm apart, whereas the plesiotomography cassette produces cuts approximately 1 mm apart. Also, there are usually only three or four films used in the plesiotomography cassette as compared to seven with the standard book cassette. More recently, tomographic equipment has been used in conjunction with special procedures, but the expense of this equipment restricts its practical use in many departments.

THICK VERSUS THIN CUTS

The technologist and radiologist have an important option when doing tomography work that can mean the difference between a study of optimal quality or something considerably less. It involves the choice of whether thick or thin cuts are to be obtained. It has been the author's experience that diffuse, loosely constructed structures are often better visualized when moderate to thick cuts are used. However, better results can be noted if dense and unevenly constructed body parts are cut thinly. If a dense but thin body part, such as bone, is to be examined, a thick or thin cut may be used to produce excellent visibility. When equipment that is not of the sophisticated multipattern design is used for too thin a cut, the results are often unsuccessful because the zone of semisharpness which lies immediately above and below the objective plane dominates the visibility of the structures of the objective plane. The result of this is an image with excessive amount of streaking and little structural detail. With multidirectional equipment, however, thin cuts can be made with much less streaking, thereby allowing the objective plane to dominate the image instead of vice versa.

ZONOGRAPHY

Zonography is the technique using tomography equipment that produces a very thick cut on the order of 6.5 cm, as compared to 1.19 mm which is commonly used in routine work. Zonograms are important be-

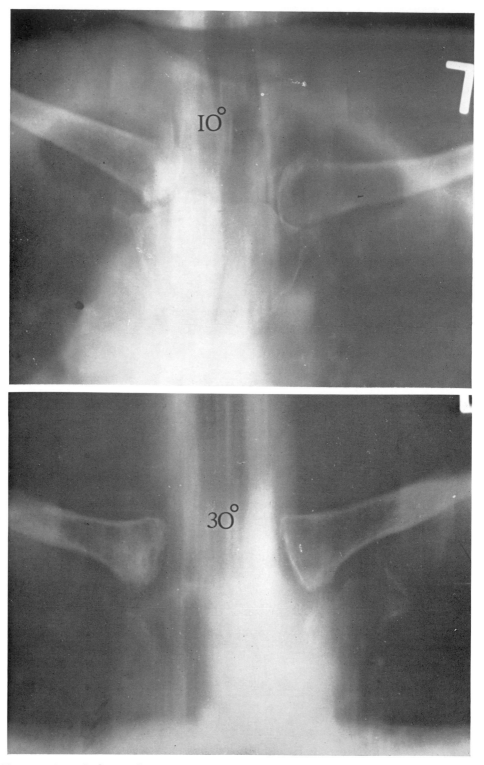

Figure 178. Thick cuts have important advantages over thin cuts when the structure under examination is to be viewed as a consolidated unit.

cause they can produce such a thick cut that often the entire body part under examination would be included in the objective plane. Zonograms are accomplished, as noted earlier, by using a short exposure arc or amplitude. A popular arc used for this function is 10°. Figure 178 shows the effect that zonography has on visualizing a body part compared to other exposure arcs.

CHOOSING A STARTING POINT

Nothing can be more frustrating to the technologist if he cannot locate the body part tomographically and the book cassette is not available. To begin with, no attempt at tomography should be made until an adequate set of routine films are obtained. To find a structure in a patient's chest, for example, the first thing to do is to place the patient properly on the table keeping in mind that the body part to be planographed should be as close to the table as possible. The reason for this, of course, is to maximize sharpness of detail by reducing the object film distance. Oblique positions can also be helpful in this regard. A pair of calipers can be used to localize the level of the structure to be tomographed (see Fig. 179). If properly marked and if the tomographic unit is cutting at the level indicated on the scale, such a device can be used quickly and easily to avoid wasted time in taking many unnecessary scout films. This type of localizer is used for mainly palpable structures, however. A good knowledge of topographic anatomy aids greatly in estimating the position of internal body structures.

For planograming of structures set deep within the chest or abdomen, "eyeballing" the routine film is the only alternative. Figure 180 shows how this can be done effectively: Such measurements are approximate because there is at least a moderate amount of shifting of internal structures between the time the patient is standing for an erect chest, for example, than when he is in a recumbent position on the tomographic table. Also

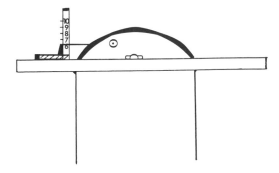

Figure 179.

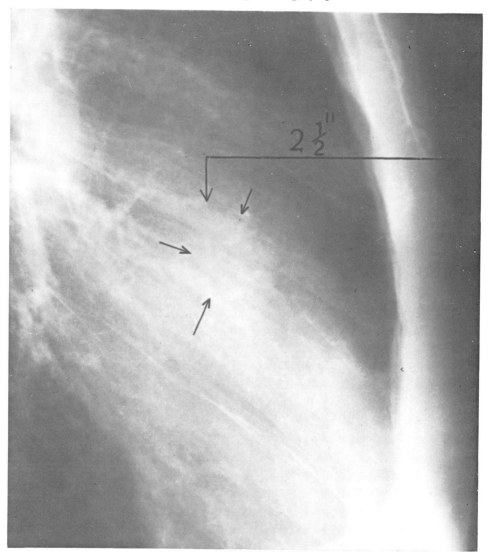

keep in mind that the patient's breathing motion should be stopped for each tomographic exposure made. There is a common practice to simply have the patient stop breathing immediately upon command, but the command is sometimes given at different points of the patient's breathing. The problem with this, of course, is that it takes only little differences in phases of respiration to affect the immediate position of a structure such as a calcified node in the chest or perhaps a duct system of the gall-bladder.

The patient's comfort should be an important consideration from both a professional point of view and a technical one as well. Because a considerable amount of time usually elapses before a tomographic study

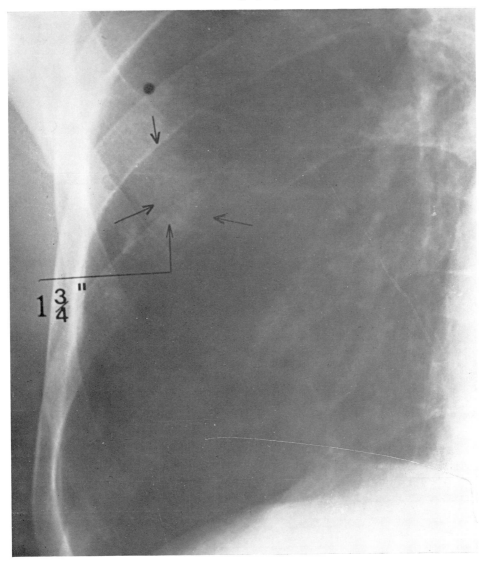

← Figure 180. When measuring the position of the object under examination, the technologist should remember that when the patient lies prone there might be a slight shift in location toward the chest wall. In a lateral chest film, the lesion may be 2.5 inches from the anterior surface; when the patient lies prone, the lesion may be 0.75 to 1 inch from the anterior wall.

is completed, the patient should be kept in as comfortable a position as possible. An appropriate amount of table padding is certainly advisable. If the patient is comfortable he is more likely to cooperate and hold still for longer times. Since the objective plane changes with each adjustment up or down with the fulcrum level, slightly different absorbing rates often occur, producing variations in radiographic density from one film to the

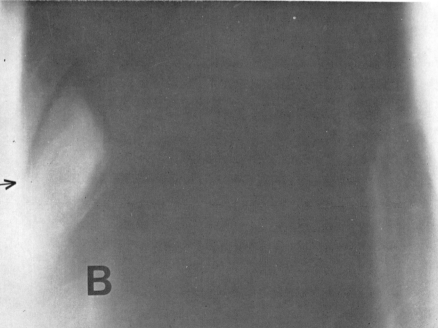

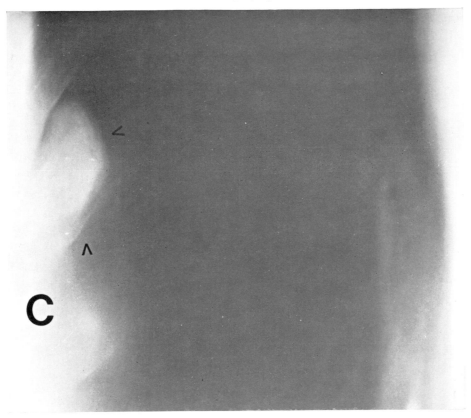

← Figure 181. Variations in density are inevitable when changing fulcrum levels, especially when tomographing areas with high subject contrast.

next. Figure 181 shows this situation. There is nothing that should be done to compensate for this as long as the technologist is reasonably sure that the x-ray equipment is producing the proper amount of exposure each time.

Establishing Exposures for Tomography

Although it is often said that an increase in exposure of 50 percent is required over what was used for a routine film, such adjustments from conventional techniques to tomography actually vary depending on the arc used. A zonogram, for example, will require less exposure when compared to the same body part using a 30 or 40° arc (see Fig. 182).

Types of Linkages

Earlier in the chapter excursion patterns were discussed briefly. It is of some importance that we look into these in more detail now. There are three basic types of linkage systems as noted in Figure 183. (Please

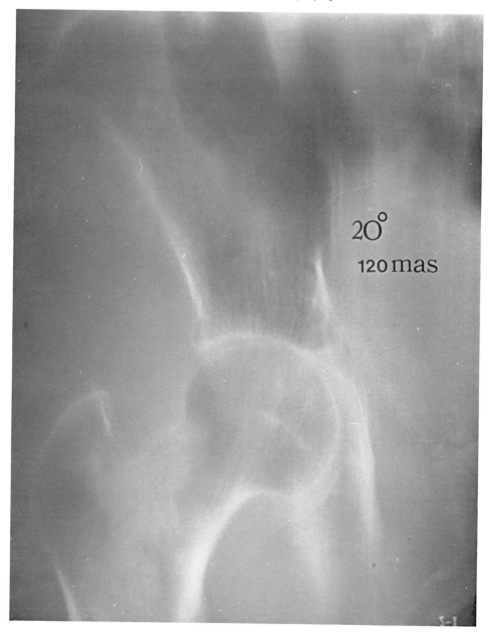

Figure 182A.

Figure 182. Because of the many variables involved, no categorical technique compensations for various arcs can be specified. However, the exposure values used here can be used as a guide.

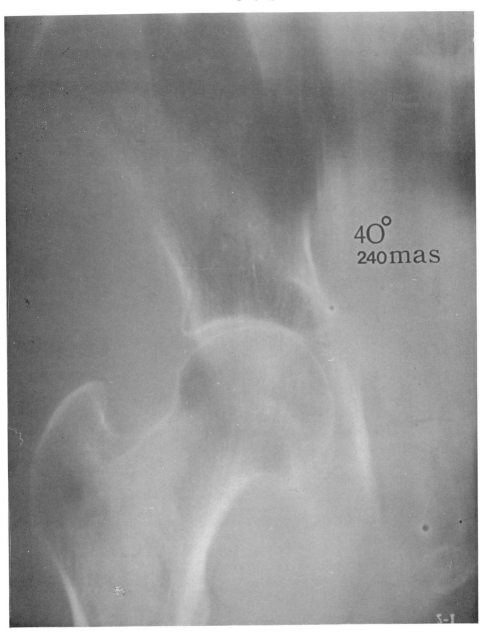

Figure 182B.

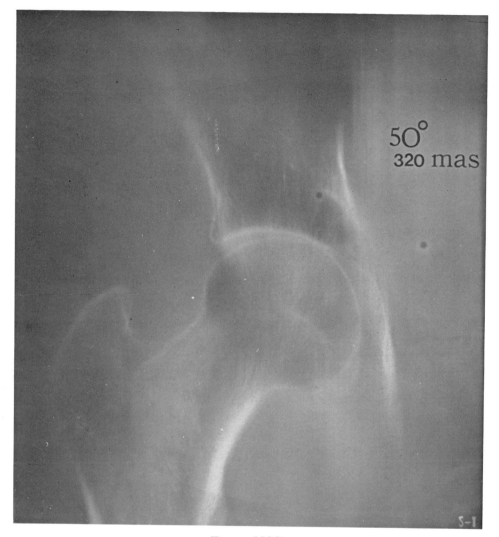

Figure 182C.

note that the ratio between (a), (b), and (c) is important.) Linkage is a term that refers to the method of mechanical connections between the tube and the film. Very often the type of linkage used depends on the manufacturer, and any of these can be used for either linear or pleuri-directional motions. If the tube and film move parallel to each other, we have a line-to-line type of linkage. If the tube moves in an arc but the film moves on a horizontal plane, we have an arc-to-line linkage, and if both the tube and film arc, an arc-to-arc linkage is being used.

The linkage is important to tomographic quality because it influences the sharpness of the objective plane. This one factor, when compared to

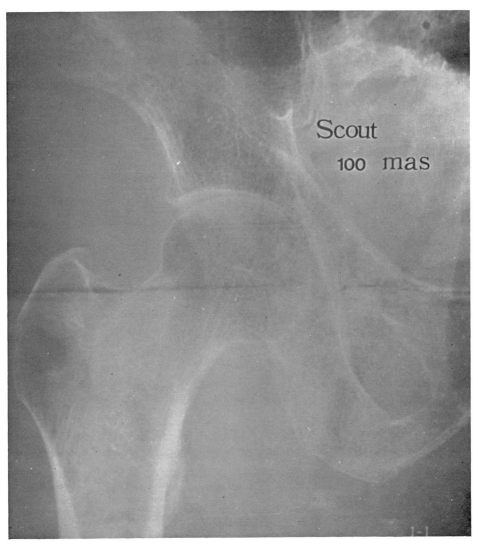

Figure 182D.

the thickness of the cut, the use of book cassettes, and the use of multi-directional patterns, has relatively little effect on the quality of the tomographic image. In general, the arc-to-arc linkage is considered to be the best of the three. The reason for this is that when the tube and film are in motion during the exposure the relationship between the object and the tube and the object and the film with regard to distance is constant. With arc-to-line and the line-to-line types, the object film distance changes somewhat during the exposure and this causes a reduction in the sharpness of the objective plane.

LINKAGE FOR LINEAR TOMOGRAPHY

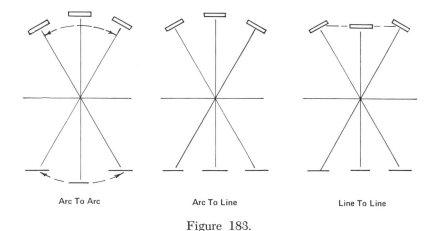

Arc To Arc   Arc To Line   Line To Line

Figure 183.

Pluridirectional Tomography

An important disadvantage regarding (linear) unidirectional tomography has been noted. As tomograms became more popular, there was strong interest in reducing the "smearing" that was common with the simpler linear mechanisms. With linear tomography the smearing (ghost shadows) cannot be avoided. However, with a more complex excursion pattern of the x-ray tube during the exposure these ghost shadows can be virtually eliminated. Figure 176 shows two radiographs that were exposed under the same conditions except one was exposed with a pluridirectional tomographic unit and the other with a linear unit. When you view these radiographs you may come to the conclusion that unless the tube moves in such a way that all structures in the objective plane are transversed ghost shadows will not be eliminated from the tomographic image. As a result, manufacturers began experimenting with these more complex designs and today they are used extensively throughout the country.

As pluridirectional motions were developed, four basic patterns have emerged (Fig. 172 shows these patterns schematically). As one might guess, the most complex of these provides the greatest opportunity for the tube to transverse all structures thus reducing smear and ghost shadows substantially. Smearing, of course, is a kind of half blur and half sharpness that occurs at the transition points which lie between the objective plane and the areas immediately above and below. In other words, the more abrupt a change there is between the sharpness of the objective plane and the total blurring of structures superior and inferior to the plane, the better the objective plane will be. In addition, the need to position the body part under examination in such a way that the tube will move across its long axis is not necessary for pluridirectional tomography. The primary danger with this type of equipment is excess heat at the anode, and very long exposures.

Balancing Exposure Factors for Tomography

A moving tube and film should not preclude the possibility of obtaining good contrast and density. It is the author's opinion that the KvP should not be increased when going from conventional radiography to tomography. There are some who feel that the kilovoltage should be increased; considering the extra thickness of tissue that must be penetrated when the tube is angled for a 40 or 50° arc, there appears at first to be some justification. However, contrast is extremely important when doing tomographic work since thin cuts greatly reduce subject contrast, and so the technique factors should be adjusted so that they produce as much subject contrast as possible. With this in mind, it would be more realistic from a technical point of view to keep the KvP as low as possible and adjust for the density with MaS. This would be especially true with very thin cuts. Figure 184 shows two tomographic images that were made at 10° and 50°, each demonstrating a different radiographic contrast. You will note that the 50° arc cut being thinner has a notable decrease in radiographic contrast.

The Selection of Exposure Angle (Arc)

Figure 185 shows some commonly used arc settings and approximate cut thicknesses they produce, but it should be pointed out that these values may vary slightly between different manufacturers' equipment. There has been no mention, so far, regarding matching the thickness of cut used with the increment of cuts for the tomographic study. Suppose a rather diffuse and moderately large structure was noted in the patient's chest and, as a result, tomograms were ordered. With this situation, it may be important to evaluate the presence of small calcium deposits in the structure as well as the general appearance of the structure itself. If the structure measured 4 cm in size and the technologist made a series of cuts using 1 cm increments, it should also be kept clearly in mind that, if the thickness of the cuts is not matched with the increments from one level to the next, it is possible that some of the structure being examined will not be radiographed at all. Figure 186 shows what can happen if the thickness of the cut obtained is not appropriate or matched with the increments that were used. In short, either the body part will be cut too "fat", or in the case of making thin cuts, with large fulcrum increments, some of the structures will be missed altogether.

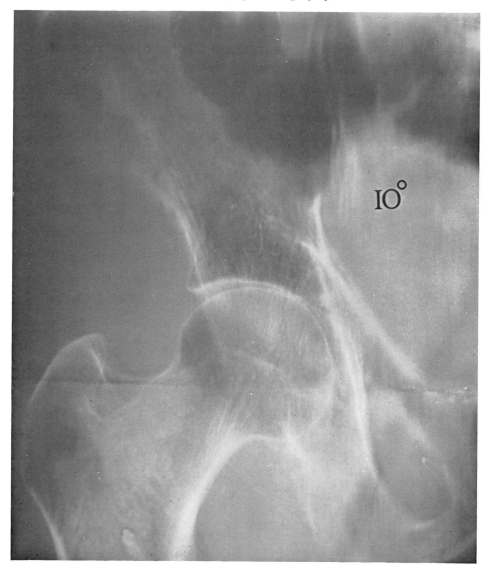

In summary, tomography is a very valuable tool which can be used to improve visualization of structures that are superimposed by other body parts to an extent where they cannot be adequately examined using conventional radiography. The technique utilizes a fulcrum (or pivot point) which provides a level where no tube or film motion is evident. The fulcrum establishes what is known as an objective plane. All structures lying in this plane will be seen in the radiographic image with sufficient sharpness for a radiological diagnosis. The level of the objective plane is adjustable and can be raised or lowered to any point in the body depending on

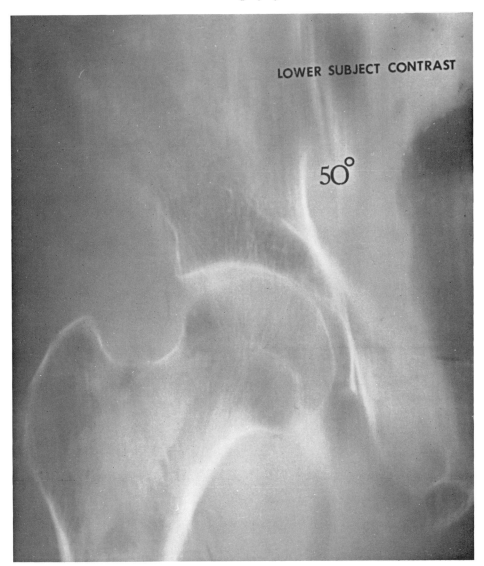

← Figure 184. Radiographic contrast decreases as the thickness of the objective plane decreases, because as thinner "slices" of the body are obtained there are fewer objects present to absorb the x-ray beam.

the wishes of the technologist and radiologist. The thickness of the objective plane is dependent on the arc or amplitude, which in turn determines the distance the tube travels *during the exposure*. The greater the exposure arc or amplitude, the greater the distance travelled during the exposure; a thicker cut will result.

There are many types of excursion patterns available ranging from

EXPOSURE ARC	THICKNESS OF CUT
(Degrees)	(mm)
10	6.5
20	3.2
30	2.12
40	1.52
50	1.19

Figure 185. The actual thickness obtained with a given amplitude may vary slightly between units of different tomographic manufacturers. Thickness of cut is a function of the exposure arc and not the complexity of the tube motion.

the simplest rectilinear to the most complex pluridirectional patterns. The major advantage of the more complex types of tube and film motions is that they can reduce the smearing (also called phantom images) of the tomographic image which, of course, increases diagnostic value substantially. An additional increase in exposure is required for tomography over conventional films, but generally these changes should be made by adjusting the MaS and keeping the kilovoltage as low as possible to improve or at least maintain subject contrast. Almost always the exposure must be increased when going from conventional radiography to tomography, but this should be compensated for by increases in MaS. The degree of exposure adjustment necessary for tomography depends on the arc that is to be used. As the exposure arc increases, more tissue is radiated and more exposure will be necessary to produce satisfactory density.

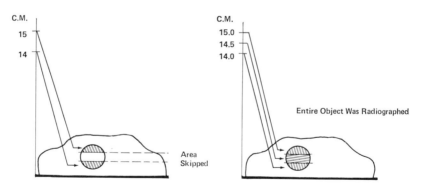

Figure 186. If this concept is not taken into consideration, an important area of an object could be missed unknowingly.

CHAPTER ELEVEN

CONVERSION FACTORS IN RADIOGRAPHY

IN THE FINAL CHAPTER of this text we will discuss the all-important matter of how to review or critique a film for technical quality. As you will see, there are a number of variations possible that can be made by the technologist to correct a technically inadequate radiograph. However, we must first learn how to make adjustments using the four common prime exposure factors and recognize what practical effect these adjustments have on the final radiographic image.

Suppose, for example, a film you have taken is too light because of inadequate penetration. First, we must know or be able to predict how the image will change if we increase the exposure by 5, 10, or 15 kilovolts. In other words, we must know how to increase the KvP so a third repeat exposure and all of the related work in setting up for it will not be necessary. If the lack of sufficient density in our imaginary radiograph is because of insufficient MaS, we must know how much additional MaS is needed to produce the desired effect. In brief, conversion factors and technical adjustments should be made only after we have made a technical diagnosis of the poorly exposed radiographic image.

The second use for conversion factors comes when the technologist is presented with a situation that will not allow the use of the technique posted. Under these conditions, it would be necessary to convert or change the posted exposure factors that are ordinarily used to something that is more compatible to the immediate situation. Experience plays an important role here and, as you watch other technologists improvise and convert their exposure factors for various technical situations, it might be well to ask for a brief explanation at an opportune moment so you can benefit from their experience and expertise.

Any number of such situations may arise, but we will begin with some of the more basic. The first problem might well involve a very sick patient who is not coherent and thus not able to follow breathing instructions adequately. The posted technique for the particular exposure to be made is 200 Ma, 2.5 sec, and 72 KvP. The patient is moving about on the table and so it is clear the exposure time will have to be reduced to

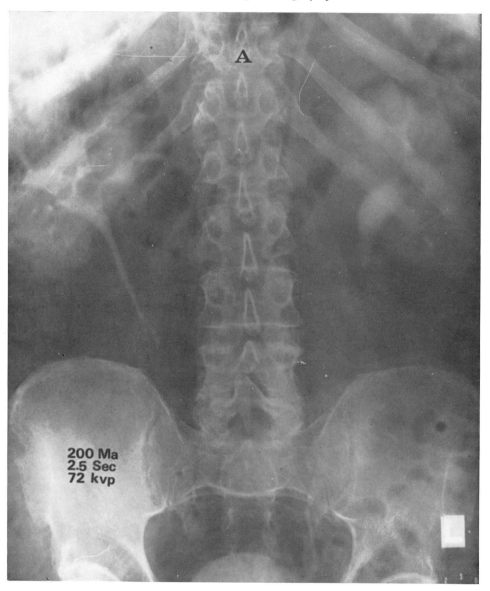

200 Ma
2.5 Sec
72 kvp

minimize patient motion. Through experience, the technologist who will be responsible for the case knows that any exposure much more than one-half of a second would probably cause too much blurring from patient motion. A situation now exists where the posted technique must be converted to one that will be compatible and will better accommodate the situation at hand. In this example, we will assume the x-ray control panel cannot be used beyond 400 Ma. In Chapter Five we discussed the concept of Ma and time reciprocity, and with this in mind we can proceed with some possible solutions for the problem.

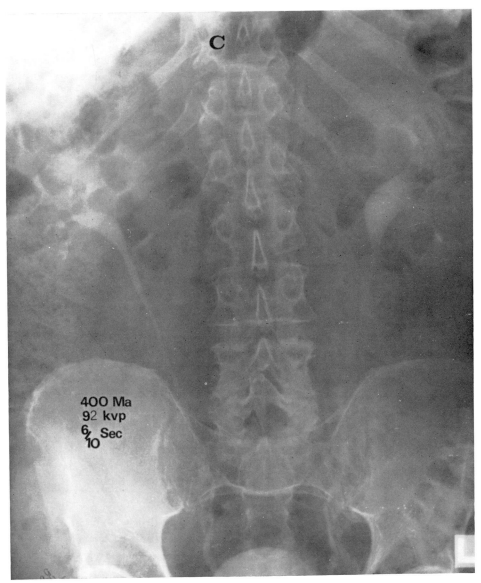

Figure 187.

You will recall the technique chart was for 200 Ma, however, that will have to be changed to 400 Ma. This allows us to reduce the exposure time from 2.5 sec to 1.25 sec, however, it is feared that this will still be too long an exposure time. Decreasing the focal film distance will certainly increase radiographic density; however, it will also increase magnification and the penumbra (which would not be to our advantage). Kilovoltage can increase radiographic density without producing any negative effect on sharpness or magnification. With this in mind, we will increase the kilovoltage by 12 and this will permit us to lower the exposure time to 0.6 sec. Thus, our new technique to accommodate this situation is 400 Ma, 0.6 sec, at 92 KvP. Figure 187 shows two radiographs, one exposed using the posted technique showing excessive blurring, and the other exposed with the converted technique. There might be other possibilities for finding an acceptable conversion for this situation, but this seems to be the most appropriate.

Another problem might be to try to increase the visibility of detail by producing more contrast in the image. The study is an IVC examination and we want to make the bile ducts as visible as possible radiographically. An experienced technologist would be able to mentally review a list of the items that would potentially improve radiographic contrast. The posted exposure, in this case, is 200 Ma, 0.5 sec, 90 KvP, and the patient's abdomen measures 25 cm A.P. The concept of producing good radiographic contrast was discussed thoroughly. The technical constraints are that the control panel does not go beyond 400 Ma and it is felt that the patient's breathing would become a problem if the exposure time is increased to over one second. In this situation, the reciprocity law by itself would not be of any benefit. The basic problem here is that radiographic contrast is poor. First, coning will be checked and made as tight as possible to reduce the number of excess scatter interactions. Since KvP is so important in determining subject contrast, it is felt that it should be lowered to 70 KvP to produce more differential absorption and less Compton interactions which produce large quantities of scatter. If the Kv is lowered to 70, the resulting radiograph would be too light so we must do something to build up the density lost by reducing the kilovoltage. The first step will be to increase the Ma to 400 (times two), and the Kv will be reduced to 80. If we increase the exposure time to 1 sec (times two) the Kv can be lowered to 70. The converted technique then is 400 Ma, 1 sec, at 70 KvP. It should be mentioned, also, that compression can be used in such a situation, if possible medically, to help increase radiographic density. In fact, depending on whether the patient is really flabby, sufficient compression might allow a further reduction of KvP to approximately 66 without danger of insufficient radiographic density. One could also use a very fast pair of screens (rare earth) if available which would allow a further reduction of exposure time to 0.5 sec (beware of quantum mottle).

Portable and operating room work often requires a good deal of technical imagination and ingenuity. Orthopedic work can be especially trying and portable grid work can be very demanding. Let us discuss a circumstance where a cross bed lateral hip must be performed to check for postoperative alignment of the broken fragment. The patient, however, was in bed in a way that caused increased object film distance, resulting in a

$\rightarrow$

Figure 188. A series of radiographs exposed with longer times. It would be very helpful for the technologist, when calculating exposures for repeat films, to note the increments of times and the respective density values obtained.

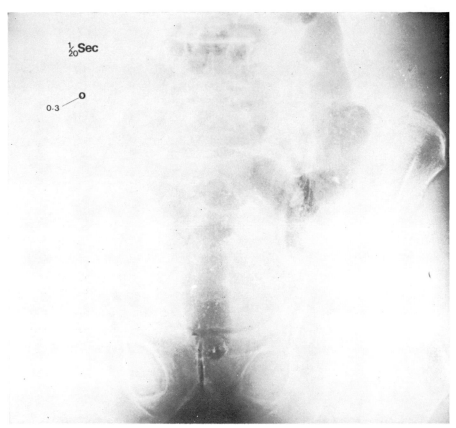

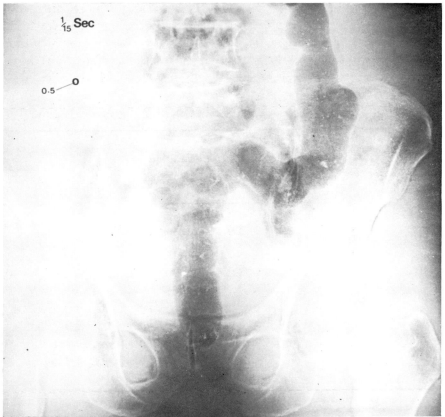

Figures 188A and B.

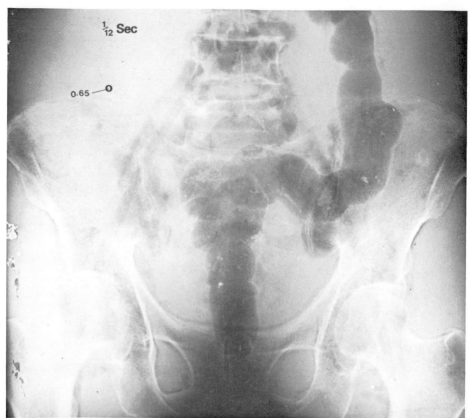

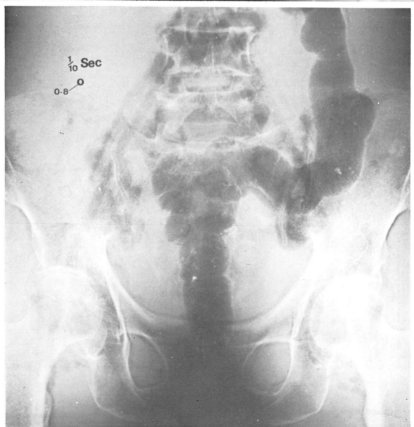

Figures 188C and D.

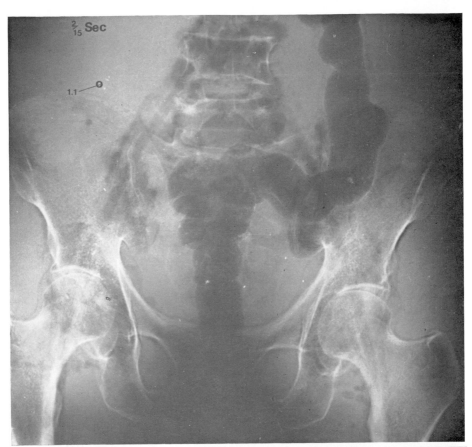

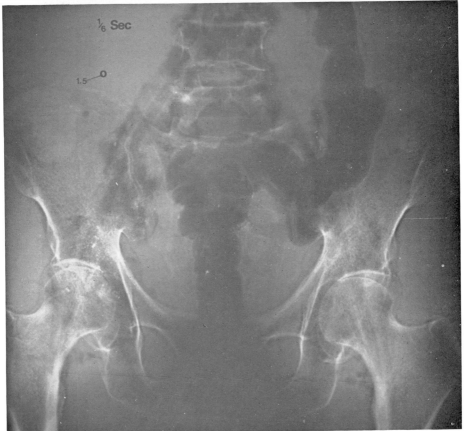

Figures 188E and F.

considerable amount of penumbra and magnification. You will recall that magnification greatly increases the effect of penumbra by the divergent rays emitted from the target. Since, in this case, object film distance cannot be reduced for the exposure, its effect on the radiographic image can be reduced substantially by increasing the F.F.D. The object film distance was, in fact, 6 inches and the portable unit was originally set for 200 MaS at 80 KvP at 40 inches focal film distance. The 6-inch object film distance could be effectively compromised by using 60-inch focal film distance, but this would cause, of course, a reduction in radiographic density and the technologist must convert his original exposure factors accordingly. In this situation, the portable unit has no Ma station above 200. Recalling the inverse square law, if the focal film distance is increased by 50 percent, the radiographic density will be reduced to one half of what it was originally. Since the Ma and time factors must be held constant in this particular situation, it is necessary to increase the kilovoltage to 94, which would make our converted technique 200 MaS, 60 inches F.F.D., at 92 KvP.

Another requisition may come along involving the use of a grid where

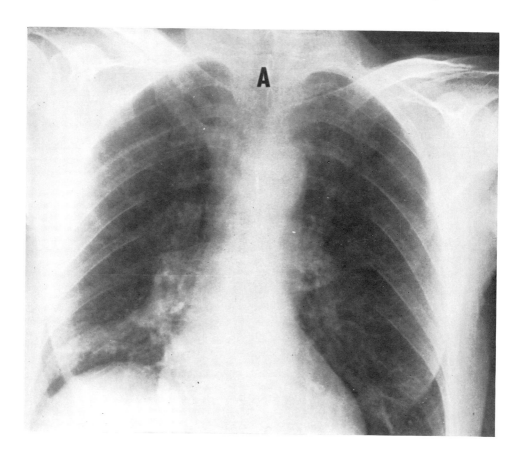

it was not used previously. A chest film, for example, was made on a very heavy patient nongrid and had too much fog caused by excessive secondary scatter radiation. The technologist chose a 5:1 grid for the second exposure to reduce the number of scatter rays reaching the film. The original exposure was 5 MaS at 80 KvP and since the grid will absorb much scatter, the radiographic density will be greatly reduced. The new technique using a 5:1 grid was increased subsequently to 16 MaS at 80 KvP. Figure 189 shows these two radiographs for comparison.

Another situation may arise when doing extremities using direct exposure technique where the patient cannot hold still for the long exposures needed for nonscreen radiography. The technologist chooses to use a cassette with par speed screens, so the original exposure must be reduced drastically to compensate for the intensification factor of the screens. The original technique (Fig. 190B) was 50 KvP, 100 Ma, at 3 sec and was converted to 100 Ma, $\frac{1}{15}$ sec, at 50 KvP (Fig. 190A).

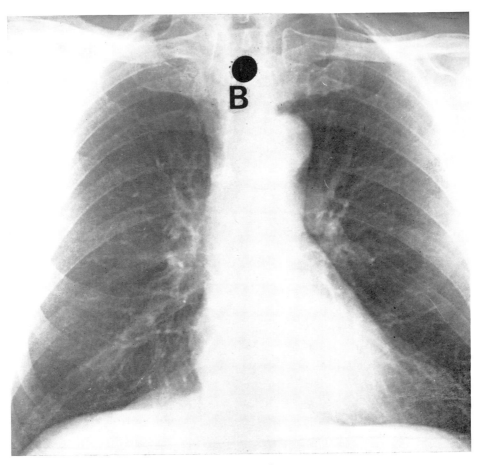

← Figure 189.

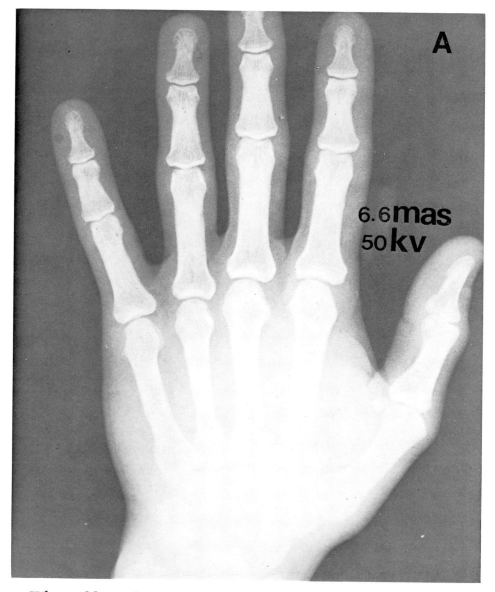

6.6mas
50kv

A

When additional contrast is required as a result of too much scatter fog reaching the film, one may always use a grid cassette or a stationary grid taped to a cassette, which is then placed in the bucky tray. If a second grid is used in this way to supplement the one already used in the bucky tray, additional absorption of the remnant radiation is obvious. To keep from overloading the tube one may often successfully use a very high speed, rare earth screen. It is worth noting here that when two grids are used in this way one can easily afford to increase KvP substantially without fear of scatter fog detracting from radiographic quality.

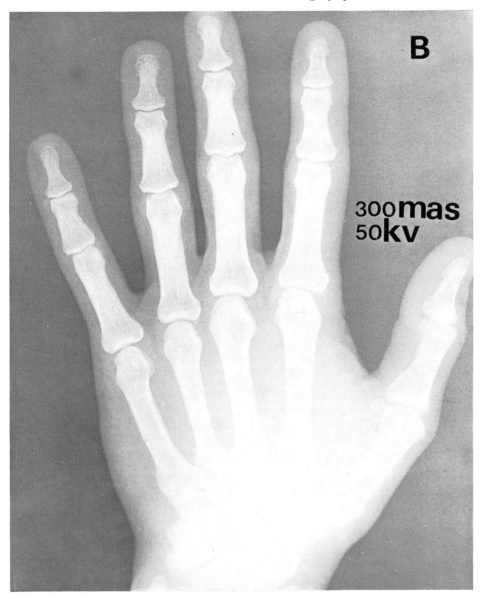

Figure 190. The nonscreen (*B*, exposed in cardboard) exposure required is approximately forty times more than the screen (*A*, par speed screen) exposure.

As you will recall, high KvP at low MaS exposures is also considerably less damaging to the tube than vice versa.

CONVERSION FACTORS

There is no question that the converted techniques noted above could have been arrived at through lucky guesses; however, a more reliable method of making these technical changes would result from using or at least becoming familiar with various conversion factors that will be presented in this chapter. It is appropriate, at this point, to remind the student that a technologist is often measured by the consistency with which he produces good radiographic examinations. To produce an occasional "pretty" or exceptional film is certainly to one's credit, but it is also simply a stroke of fortunate luck and does not necessarily indicate a talented technologist.

Very often technologists have a tendency to emphasize a few of these rare "Roentgen Rembrandts" and forget the majority of the marginal cases they have produced throughout a given week or month. The author has worked with many talented technologists who have developed their abilities wisely, and also those who have learned to become good technologists through practice and effort. In any case, the most valued technologist is the one who can produce good radiographs consistently. Fortunately, there are reliable methods that can be used to calculate new conversion factors mentally during the examination if necessary. Although there are many such conversion formulas to know depending on the situation, only a few are used with any degree of regularity. If one becomes familiar with these few it will prove very beneficial to both the technologist and the patient. To make the problem of conversions more manageable the author will introduce the concept of the conversion factor or "C" factor. This can be used effectively for practically any radiographic situation similar to those noted above. The reader will take note that KvP conversions (see Table II) are not written merely in increments of 10. Because of the effect exposure latitude has with decreasing and increasing changes in KvP, it is misleading and incorrect for the technologist to do so because it often results in an incorrect choice of exposure factors.

Increasing the KvP, for example, will give much different results depending on the KvP level of the original film. An alternate method is to think in terms of increasing or decreasing the exposure by increments of 15 or 7 percent (see Fig. 191). In Table III, you will note formulas that can be used to find the new time or Ma when the other has been changed. Other conversion tables and formulas have been included to aid in finding the correct conversion factors.

Two final points must be made before concluding this chapter. First the contrast of the film has a very important role in calculating conversion factors. You will find that high contrast film will yield more dramatic radiographic changes in the image when the same conversion factor is used. For example, if we used a high contrast film and a low contrast film and had an original technique of 15 MaS at 70 Kv, then increased the

TABLE II
KvP — DENSITY CALCULATIONS

New KvP→	Density +50% Column A	Density +100% Column B	Density −25% Column C	Density −50% Column D
Initial KvP	(Plus 7%)	(Plus 15%)	(Minus 7%)	(Minus 15%)
50 :	53	57	46	42
52 :	55	59	48	44
54 :	57	62	50	45
56 :	59	64	52	47
58 :	62	66	53	49
60 :	64	69	55	51
62 :	66	71	57	52
64 :	68	73	59	54
66 :	70	75	61	56
68 :	72	78	63	57
70 :	74	80	65	59
72 :	77	82	66	61
74 :	79	85	68	62
76 :	81	87	70	64
78 :	83	89	72	66
80 :	85	92	74	68
82 :	87	94	76	69
84 :	89	96	78	71
86 :	92	98	79	73
88 :	94	101	81	74
90 :	96	103	83	76
92 :	98	105	85	78
94 :	100	108	87	79
96 :	102	110	89	81
98 :	104	112	91	83
100 :	107	115	93	85
102 :	109	117	94	86
104 :	111	119	96	88
106 :	113	121	98	90
108 :	115	124	100	91
110 :	117	126	102	93
112 :	119	128	104	95
114 :	121	131	106	96
116 :	124	133	107	98
118 :	126	135	109	100
120 :	128	138	111	102
122 :	130	140	113	103
124 :	132	142	115	105
126 :	134	144	117	107
128 :	136	147	119	108
130 :	139	149	120	110

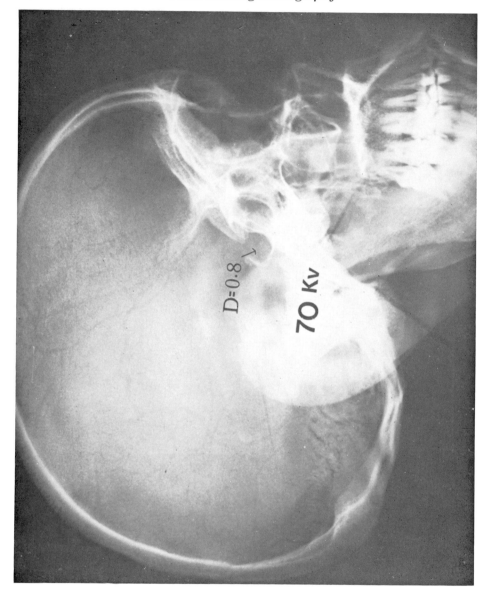

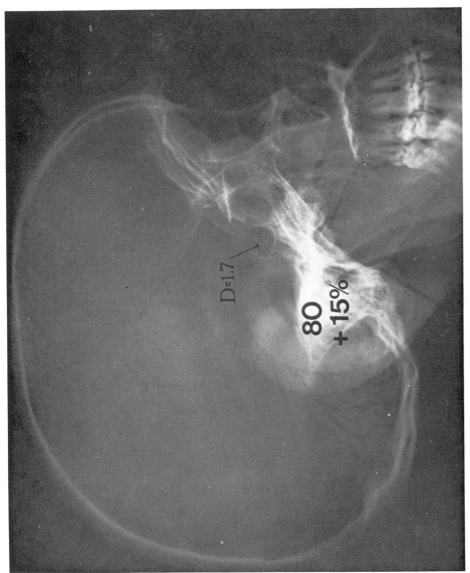

Figure 191. These radiographs show the effect in density with a 15 percent increase in KvP. If more subtle changes in density are required, the author's experience is that an 8 percent increase in KvP will prove adequate.

TABLE III

CONVERSION TABLES

It must be kept in mind that the conversion factors below were established under controlled conditions. Because of the tremendous number of variables involving film type, processing conditions, type of screens, various KvP levels, filtrations, generator calibrations, and the variety of factors affecting body habitus, it is simply impossible to have categorical conversion factors that will work equally well under all situations. However, because medical x-ray film has wide latitude characteristics, the conversion factors noted below will provide diagnostic radiographic images.

I.   How to find the exposure time and Ma from a given MaS value.

A. $\dfrac{\text{MaS}}{\text{The Ma Desired}}$ = Exposure Time

PROBLEM: If MaS is given as 71, what would the time be if 200 Ma is desired?

$$\frac{71 \text{ MaS}}{200 \text{ Ma}} = \frac{35}{100} = \frac{7}{20} \text{ sec}$$

Please note that if 7/20 is not on the timer selector, the available time can be estimated by using the following logic: Since 10/20 equals 1/2 and 5/20 equals 1/4, 7/20 would be close to the midpoint of 1/4 and 1/2 so the nearest time to this point can be used.

B. $\dfrac{\text{MaS}}{\text{The Time Desired}}$ = Milliamperage

PROBLEM: If 71 MaS is given, what would the Ma be at a desired time of 1/2 second?

$$\frac{71 \text{ MaS}}{1/2 \text{ sec}} = 140 \text{ Ma}$$

Because of the many time stations available, it is usually better to choose the desired Ma and calculate for the exposure time.

II.   How to find the new Ma when the exposure time has been changed.

$$\frac{\text{Original Ma} \times \text{Original Time}}{\text{New Time}} = \text{New Ma}$$

PROBLEM: Originally, the exposure was made at 100 Ma at 1/4 sec. What would the new Ma be if the time is changed to 1/20 sec?

$$\frac{100 \times 1/4}{1/20} \times \frac{25}{.05} \times 500 \text{ Ma}$$

III.   How to find the new exposure time when the Ma has been changed.

$$\frac{\text{Original Ma} \times \text{Original Time}}{\text{New Ma}} = \text{New Time}$$

PROBLEM: If the initial exposure is made at 500 Ma at 1/2 sec, what would the time be at an exposure of 200 Ma?

$$\frac{500 \times 1/2}{200} = 1.25 \text{ sec}$$

IV.   How to find the new Ma or MaS or time when the distance is changed.

$$\frac{\text{Original MaS} \times (\text{New F.F.D.})^2}{(\text{Original F.F.D.})^2} = \text{New MaS}$$

TABLE III—*Continued*

CONVERSION TABLES

Note that time, Ma, or MaS values can be interchanged whenever these calculations are made.

PROBLEM: If the exposure is 5 MaS at a distance of thirty inches, what would the exposure be at a distance of forty inches?

$$\frac{5 \times 40^2}{30^2} = \frac{8000}{900} = 9 \text{ MaS}$$

V. How to find the new exposure when only a slight change in density is needed.

To increase: Original MaS × 1.25 = New MaS
Original KvP × 1.08 = New KvP

To decrease: Original MaS × 0.75 = New MaS
Original KvP × 0.93 = New KvP

VI. How to change the radiographic density.

Original MaS × 1.5 = 50% increase
Original MaS × 2.0 = 100% increase
Original MaS × 3.0 = 150% increase
Original MaS × 0.75 = 25% decrease
Original MaS × 0.50 = 50% decrease
Original MaS × 0.25 = 75% decrease

VII. How to change the radiographic density (also see Figure 191 and Table II).

Original KvP × 1.08 = 50% increase
Original KvP × 1.15 = 100% increase
Original KvP × 1.23 = 150% increase
Original KvP × .93 = 25% decrease
Original KvP × .85 = 50% decrease
Original KvP × .78 = 75% decrease

VIII. How to find the new MaS when coning is sharply increased (from open field to a tight spot).

Original MaS × 1.35 = New MaS
Original KvP × 1.08 = New KvP

Note that these conversions will vary depending on the amount of S/S radiation present in the original exposure.

IX. Alternate method for calculating new time, MaS, or Ma when the distance is changed.

New F.F.D.^{2} ÷ Original F.F.D.^{2} × Original MaS = New MaS

X. How to make quick estimates when focal film distance is altered.

25% increase, from 40 to 50 inches: Original MaS × 1.5 = New MaS
Original KvP × 1.08 = New KvP

50% increase, from 40 to 60 inches: Original MaS × 2.25 = New MaS
Original KvP × 1.23 = New KvP

100% increase, from 40 to 80 inches: Original MaS × 4 = New MaS
Original KvP × 1.3 = New KvP

10% decrease, from 40 to 36 inches: Original MaS × 0.8 = New MaS
Original KvP × 0.94 = New KvP

25% decrease, from 40 to 30 inches: Original MaS × 0.55 = New MaS
Original KvP × 0.85 = New KvP

50% decrease, from 40 to 20 inches: Original MaS × 0.25 = New MaS
Original KvP × 0.72 = New KvP

XI. How to find the new MaS, time, or Ma when changing screens (at 65 KvP).

Direct exposure to detail speed: Original MaS × 0.043 = New MaS

Direct exposure to medium speed: Original MaS × 0.022 = New MaS

TABLE III—*Continued*

CONVERSION TABLES

Direct exposure to fast speed:	Original MaS	× 0.011	=	New MaS
Direct exposure to rare earth:	Original MaS	× 0.006	=	New MaS
Detail speed to nonscreen:	Original MaS	× 23	=	New MaS
Medium speed to nonscreen:	Original MaS	× 46	=	New MaS
Fast speed to nonscreen:	Original MaS	× 87	=	New MaS
Rare earth to nonscreen:	Original MaS	× 170	=	New MaS

Please note that most screens are made with speed factors in multiples of 2: A fast speed screen is two times faster than medium, so the exposure must be decreased to one-half of the original; a medium speed screen is two times faster than detail speed, so the exposure must be reduced to one-half of the original value.

XII.   How to change grids (the conversion factors will vary slightly depending on the KvP, part thickness, density, and general body habitus).

At 75 KvP:

From no grid to 5:1 grid:	Original MaS	× 1.5	=	New MaS
From no grid to 6:1 grid:	Original MaS	× 2.0	=	New MaS
From no grid to 8:1 grid:	Original MaS	× 3.0	=	New MaS
From no grid to 12:1 grid:	Original MaS	× 3.5	=	New MaS
From 5:1 grid to 6:1 grid:	Original MaS	× 1.5	=	New MaS
	Original KvP	× 1.08	=	New KvP
From 5:1 grid to 8:1 grid:	Original MaS	× 2.0	=	New MaS
	Original KvP	× 1.15	=	New KvP
From 5:1 grid to 12:1 grid:	Original MaS	× 2.33	=	New Mas
	Original KvP	× 1.30	=	New KvP
From 6:1 grid to 8:1 grid:	Original MaS	× 1.5	=	New MaS
	Original KvP	× 1.08	=	New KvP
From 6:1 grid to 12:1 grid:	Original MaS	× 1.75	=	New MaS
	Original KvP	× 1.11	=	New KvP
From 8:1 grid to 12:1 grid:	Original MaS	× 1.5	=	New MaS
	Original KvP	× 1.09	=	New KvP
From 12:1 grid to 8:1 grid:	Original MaS	× 0.75	=	New MaS
	Original KvP	× 0.95	=	New KvP
From 12:1 grid to 6:1 grid:	Original MaS	× 0.55	=	New MaS
	Original KvP	× 0.88	=	New KvP
From 12:1 grid to 5:1 grid:	Original MaS	× 0.43	=	New MaS
	Original KvP	× 0.75	=	New KvP
From 8:1 grid to 6:1 grid:	Original MaS	× 0.66	=	New MaS
	Original KvP	× 0.93	=	New KvP
From 8:1 grid to 5:1 grid:	Original MaS	× 0.5	=	New MaS
	Original KvP	× 0.85	=	New KvP
From 6:1 grid to 5:1 grid:	Original MaS	× 0.75	=	New MaS
	Original KvP	× 0.95	=	New KvP

exposure by 50 percent, the high contrast film would show a greater effect on the image than would the low contrast film, as indicated:

Exposure	Radiographic Density	
	High Contrast Film	Low Contrast Film
15 MaS	0.8	0 8
22.5 MaS	1.4	1.1

The second point that must be made regarding the effect a conversion factor has on the radiographic image is dependent on where the original density of the film *lies* on the H & D curve. If all other factors are constant, it is very likely that the same conversion factor will produce a different effect on the density of the image (see Fig. 192).

When the density of the original film lies close to the toe or shoulder of the H & D curve, the technologist should be mindful that a slightly greater adjustment in technique for the repeat is usually necessary.

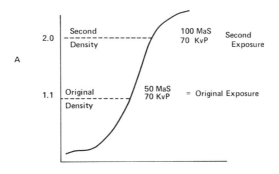

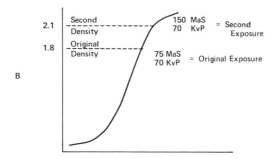

Figure 192. X-ray film is not able to record densities increasing beyond 1.5 in proportion to increasing exposure values. In *A* and *B* the MaS was doubled from the original density. However, *A* does not produce the same effect in density as *B*. This is one important reason why it is difficult and misleading to make categorical statements about the effect any one specific conversion factor will have on the radiographic image.

Experience has taught the author that more often than not technologists are too conservative when estimating how light or dark their original radiograph appears and consequently make too small an adjustment in exposure factors. It is very important for students to learn to make accurate estimates in determining how light or dark the original film is from what is considered to be optimal. Only then can he hope to make consistently accurate technical adjustments for the repeat films. It is also strongly recommended that he learn to express these differences in density in terms of percentages, such as this film is 50 percent darker than it should be or this film is 50 percent lighter than optimal for this exam. Once this information is determined, the exposure may be increased or decreased by 50 percent according to the tables. With this system, the technologist has better control over his finished work, and radiography becomes something more than a mere guessing game. It might be helpful to have a small library of films exposed deliberately with different factors. A densitometer can then be used to get a reading that represents their overall densities. During film critique classes, these films may be presented to train and test abilities in correctly estimating how light or dark the images are from what is thought to be optimal.

CHAPTER TWELVE

FILM CRITIQUE

V ERY SHORTLY after Roentgen's discovery of x-radiation, many physicians began experimenting with this amazing phenomenon for possible applications in the medical field. By the early 1900s the usefulness of x-radiation in helping determine fractured bones had been documented, but as time went on physicians, in cooperation with physicists, began to realize that this new medium could be used for purposes other than the mere visualization of bony fragments and dislocations.

By 1910, the use of x-rays was almost commonplace, and experimentation with other medical applications had, by that time, built to almost a fever pitch. More and more medical facilities had acquired x-ray equipment for the first time, and those facilities that were fortunate enough had additional equipment installed for newly developed uses and research. It was indeed a rare moment in medical history. As the use of x-rays continued to grow, it became clear that physicians who had a special interest in this aspect of medical care had little time available for the more traditional methods of practicing medicine and before too long the A.M.A. recognized the need and the practicality for a new medical specialty known as Roentgenology. The roentgenologist, as he was known in those days, performed all the required procedures necessary to produce a roentgenographic image until the demands in interpreting the plates caused him to seek people to help with the more functionary duties of making the exposure and processing the plates. From these humble beginnings the x-ray technologist has evolved into today's highly trained, sophisticated technologist who has continued to grow with the changing demands of the medical profession. Not only has the diagnostic application of x-ray technology continued to grow but many subspecialties have evolved as well; this has, in turn, increased the demands of their technical knowledge even further.

In the realm of modern radiographic technology, the technologist is playing a more dominant role than he had in the past. Today many radiology residents receive abbreviated theory and little or no practical training regarding the writing of technique charts and the actual application of such accessory equipment as intensifying screens, grids, and processing film.

More and more, the responsibility for the truly technical aspects of radiology are being delegated to technologists. Most radiological technologists welcome this responsibility because it offers more opportunity to grow as professionals and to develop their departments to produce a higher quality radiographic study and interpretation by the radiologist.

Today the radiologist and the technologist work as a team to achieve the end result of producing a high quality radiographic diagnosis for the patients referred to them by private physicians. With this responsibility, cooperation, and recognition of the technologist as a thoroughly trained radiation worker, it is important for us to function in such a way as to not only maintain this growth but further develop this trust by conscientiously learning the basic concepts of how and, equally important, why certain radiographic problems manifest themselves and how to make the appropriate corrections when necessary. This is, above all else, the fundamental purpose and responsibility of today's radiologic technologist. With this idea firmly fixed in our minds, we will begin a review of the various characteristics we should want to produce in the radiographic image.

The immediate task at hand will be to review some of the important terms that will directly pertain to the matter in this chapter. The first of these is density, a general term used to describe the blackness of a radiograph or a specific area. Density is measured with a device known as a densitometer, which measures the tones of radiographic density in quantitative terms such as 1.5^D. The diagnostic tone range for most studies lies between $.25^D$ and 2.0^D.

Radiographic contrast is the difference between two or more tones of density in a radiographic image. The optimal amount of contrast and density desired in an image is purely subjective and cannot be given a specific value. Contrast *scale* refers to the tonal differences of a series of densities ranging from the darkest to the lightest tone. You will recall that if there is an abrupt difference in the tones (the tonal range from black to white happens very quickly) you are looking at a short (high) scale contrast. If, however, the differences between the various neighboring tones is subtle and one observes total range of these tones from black to white very gradually, you are viewing a long (low) contrast scale. It should be made very clear that for routine studies neither extreme is optimal, but rather something in between is desired. Figure 193 shows samples of these variations in contrast scale.

Definition or sharpness is another concept that must be understood. Blurred structures, of course, cause a loss in diagnostic information and

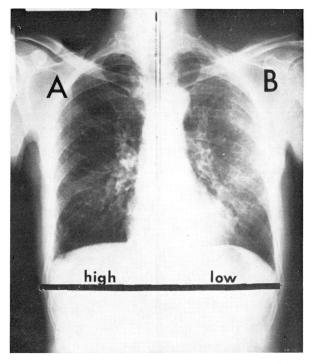

Figure 193. Two films with different contrast characteristics (*A* is a high contrast film, *B* is a low contrast film) were cut to fit into the cassette. One exposure was made, and both halves were processed at the same time to eliminate any variables.

should be avoided. It is not by coincidence that contrast, density, and sharpness were reviewed first. A radiograph, to have optimal detail, must have sufficient contrast to see or differentiate the various structures of the body, it must have enough density or overall accumulation of black metallic silver to present the various body structures, and finally it must possess definition so the structures being viewed are sharply defined or delineated in the image. If the various structures did not have well defined borders, each would "melt" (blur) together making their evaluation or diagnosis impossible. Thus, it is these three characteristics that are considered prerequisites for good radiographic quality.

The terms detail and image quality should be reviewed as well. The remnant x-ray beam is a media that transfers an impression (via variations in intensity) onto the film's emulsion. If all possible conditions are properly set, they will be recorded by the film, and processing will make visible the desired body structures that are to be examined. If, however, something is amiss with either contrast, density, or sharpness, the required reproduction of the body part will not be seen adequately. Detail is the ease with which the viewer can see small structures in the image, and quality is a more general term used to describe the overall diagnostic value of the radiographic image.

VISIBILITY AND DEFINITION OF DETAIL

The radiographic image can be analyzed from two distinct points of view. These two areas are visibility of detail and definition of detail. In order to critique (analyze) a film effectively one must *first* be able to tell the difference between visibility problems and definition problems. It was pointed out earlier in the text that visibility of detail refers to the ability to see the structures under examination. Figure 194 shows two radiographs that have very little visibility of detail. One image possesses very poor contrast and the other image is too dark to make visible the structures or detail that has been impressed into the film's emulsion by the projected image. Visibility of detail then is how well one can see or visualize body parts that *are present* in the emulsion.

The Effect of Contrast on Visibility of Detail

With the understanding that visibility of detail means the ease with which detail can be "seen" in the image, a brief review of all the factors that affect visibility is in order. With this as a starting point, contrast should be considered among the most important. In Figure 193, there is no question as to which radiograph has greater visibility of detail and diagnostic value.

Many factors affect radiographic contrast, and it would be misleading to list them in order of importance because each of the items shown in Figure 195 can cause equally urgent technical problems. It would be helpful, however, to list them in the order of probability of occurrence so when one begins to critique a radiograph with poor contrast an appropriate starting point can be more easily found. Throughout the text, much time was given in describing various causes and remedies of poor radiographic contrast. As a supplement to these discussions it is advisable that Figure 195 be reviewed.

Trying to find the cause of poor contrast can be an extremely difficult problem to isolate because contrast has so many contributing factors. On the other hand, if a radiograph is too dense, usually the technologist need only know how much of a reduction of exposure (MaS) is needed, or, if overpenetration is suspected, the KvP should be reduced to achieve the required results.

The technologist must first realize that contrast is indeed poor, then a technical diagnosis must be made that will help him find the cause of the poor contrast. To do this properly, a basic knowledge of x-ray principles

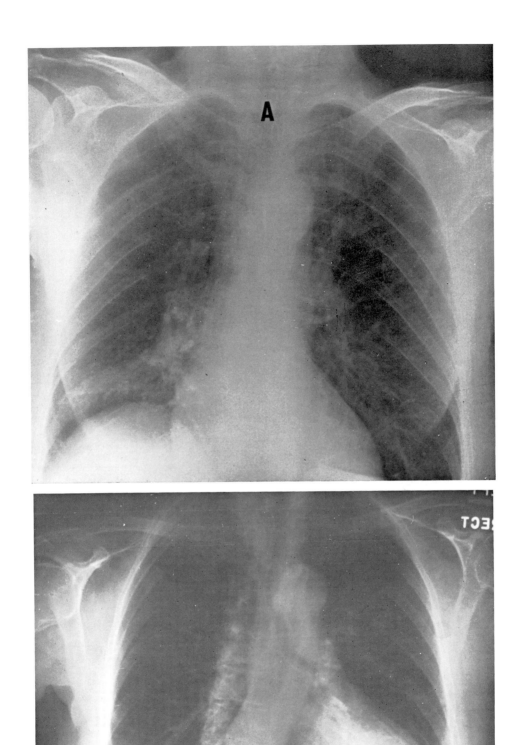

Figure 194. The total diagnostic value of the films is diminished as a result of poor visibility of detail.

MAJOR FACTORS THAT AFFECT RADIOGRAPHIC CONTRAST

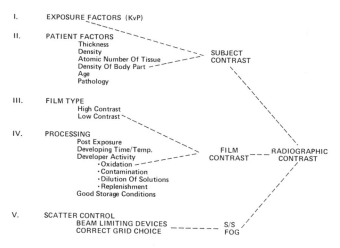

Figure 195.

must be well in hand and a few additional years of experience and practice is certainly helpful. The technologist must train himself to become a good technical diagnostician and to do this effectively. Later in this chapter, radiographs taken during a routine working schedule will be critiqued and compared with the corrected exposure settings. An important point to remember when making corrections for poor radiographic contrast is that frequently there is more than one factor contributing to the undesirable image so that it is necessary to mentally review more than one or two items.

In the case of density, the technical diagnosis and solution to the problem is usually more straightforward than with contrast. It would not be good practice to increase the F.F.D., for example, because of the focused grid and the limitation with the height of the ceiling. Also, changes in focal film distance would affect the sharpness of the radiographic image as well as density. Similarly it would not be appropriate to reduce KvP because of the possible effect on radiographic contrast as well as the possibility of insufficient penetration. More realistically, corrections for density are made by changing the time or the Ma (MaS). In the section covering conversions we saw the effect MaS changes had on density.

There are adjustments that could be made with processing, film, or screens; however, these items should be held as a constant and not used as additional factors to correct exposure problems. Changing developer activity through replenishment or by temperature could lead to more complicated problems. For the purpose of correcting poor contrast or density, one should be keenly aware of the fact that the vast majority of poor quality radiographs are the result of improper exposure factors and not processing, grids, film, or screens.

The following summary of probable causes for poor radiographic quality due to visibility of detail may be helpful.

Artifacts	Dirty screens
	Static
	Fingerprints on the film
	Scratches from the processor
Body habitus	Elderly patient—absorption is poor
	—poor subject contrast
	Pathology
	Obese patient—poor subject contrast
Poor contrast	Excessive S/S reaching the film
	KvP set too high
	Excessive base fog
Density	Incorrect exposure factors
	Chemical fog
	Using wrong film/screen combination
	Improperly positioned grid—light film
Penetration (KvP)	Body too thick or dense for KvP—poor penetration
Processing	Chemical fog
	Under replenishment
	Over replenishment
	Contaminated solutions
	High developing temperature

Sharpness of Detail

Throughout the discussion of visibility of detail it was presumed the body structures have sharply defined borders or edges and these must be seen as such in the radiographic image, but because of contrast or density problems these structures are not seen easily. This concept is illustrated in Figure 196, a photograph of dashes sharply impressed in the page, yet the image of a circle cannot be seen as such in B because the overall density and contrasts do not allow easy visualization of the dashes. Also, we may use the example of trying to visualize a ship sailing in the midst of a heavy fog. The structure of the ship with its sharply defined lines are certainly present yet because of the overall fog its structures are not seen. The problem is not because of poor definition. In radiography the same type of condition exists when a radiographic image is too dense or has too much fog to see the various structures that have been placed in the image by the projected beam.

In the case of sharpness or definition of detail we are concerned with actual delineation and sharpness of the borders of each body structure. Figure 197 shows examples of good sharpness and poor sharpness. This

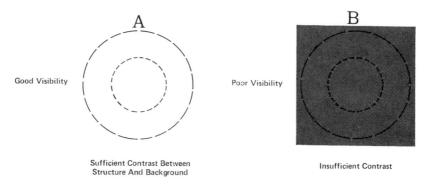

Figure 196. The geometric aspects of the image are optimal. However, they are not visible in film because the density and contrast portion of the technical image is poorly balanced.

factor (sharpness of detail) is controlled by those factors that affect or influence the projected (or geometric) beam, which include intensifying screens. It is the projection (geometry of the photons in the beam) that determine sharpness of detail. The actual detail or information is carried from the patient by the remnant beam and is implanted into the film's emulsion. At that moment the geometric structure (borders) of the image is present, and it is the responsibility of contrast and density to make the various structures optimally visible. In Figure 198 you will be able to see this as you look at the simulated x-ray photons pass through a wire mesh and carry the impression of the structure to the film. This information containing the structural and border lines of the body part under examination is known as the geometric or projected image and is totally responsible for definition of detail.

Thus there are two separate components to every x-ray beam, those related to visibility of detail and those related to definition of detail. When critiquing a radiograph one must first be able to differentiate between the two. Further, one must keep in mind that visibility problems are directly related to radiographic contrast and density and problems related to definition of detail are related to radiographic blurring or distortion.

AN APPROACH TO FILM CRITIQUE

If one would watch a radiologist as he interprets a film and reaches a diagnosis, one would begin to realize that he uses a somewhat fixed regimen. While viewing a PA chest examination, for example, he may first look at the bones of the right side then the left and go on in each progressive step until he has seen the entire film.

$\rightarrow$

Figure 197. *A* reveals acceptable sharpness of the structures in the body. In *B* the borders of the body parts are poorly defined. This unsharpness at first may seem minimal, but much important information could be lost as a result.

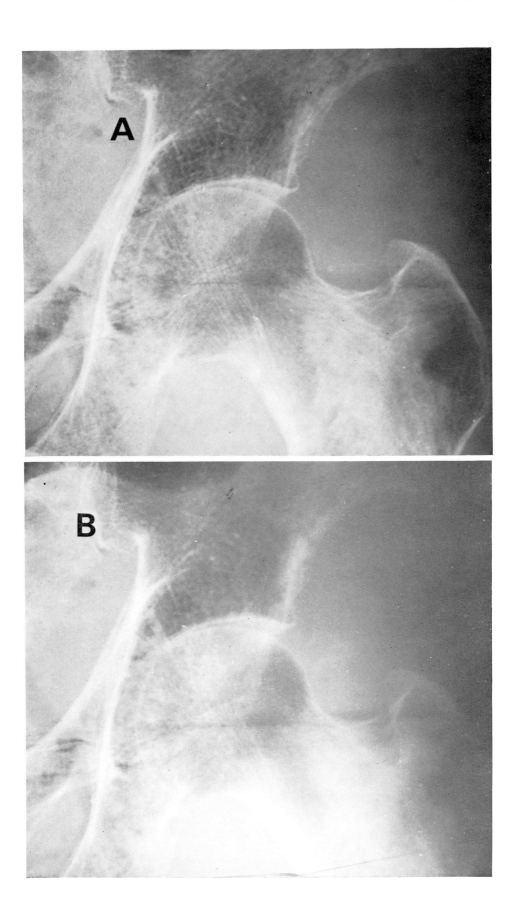

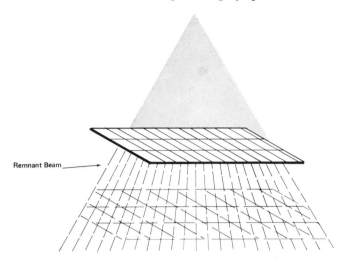

Figure 198. Additional evidence that there are indeed two separate components to the x-ray beam: (1) the projected or geometric image as seen here and (2) the variation in beam intensity and quantity which can be called the technical beam.

In practice, the radiologic technologist should develop a similar regimen with film critique that can be followed easily and quickly but at the same time provide enough information so that important observations are not overlooked. The remaining part of this chapter will deal with the problem of establishing a fixed regimen for critiquing a film and to present some actual examples of radiographs with technical problems to show how effective this method can be to help identify a technical problem and make a decision as to the correct technical adjustment.

There are three basic steps involved in this approach to film critique. The first step is the determination as to whether the radiographic problem is that of definition of detail or visibility of detail, the second is to determine the cause of the problem; the third is to choose the appropriate corrective action.

How to Identify a Definition Problem

Before going any further, it would be valuable to gain an understanding as to how these definition problems may present themselves in the radiographic image. Definition problems will usually manifest as blurring, although shape and size distortion can also be frequently seen. Magnification most often occurs when doing portables, operating room procedures, or off-table grid work, and often cannot be corrected entirely because of the nature of the situation. Sometimes blurring can be so uniform that it is difficult to notice because of the lack of any neighboring sharp lines with which comparisons can be made, so the technologist must be careful not to overlook this subtle but important problem. Figure 199A shows the problem of such a blurring situation. Most often it appears in chest films because of the patient's involuntary motion. In fact, the most frequent cause of blurring is patient motion, and the second is poor screen contact. Figure 199B is the same patient using a quicker exposure and observing the patient with more care while the exposure is being made. The difference by comparison is obvious yet if the blurred film were viewed alone and one was not trained to look for this problem it might very well not be picked up.

Causes for Blurring

The focal spot size influences sharpness or blurring of the radiographic image. The focal spot or target is, of course, where the beam originates. Many photons coming from different areas of the focal spot pass through the same body part but in slightly different directions, causing a kind of criss-crossing action. Only a little imagination is needed to understand how this would affect an examination of a complex structure such as the lung fields of the chest. Thus, one should use the smallest focal spot possible. When object film distance is increased, enlargement (magnification) occurs of all body parts including blurring itself. Figure 200 shows one film taken at a normal object film distance. Also note that the object film distance of the immediate structure under examination increases with increasing angles of the x-ray tube. In summary, problems with definition of detail are caused by how the beam is projected through the patient and eventually to the film. Screens complicate the projection problem even further under normal conditions by emitting divergent light patterns from each crystal, and if film screen contact is poor, the projection of the screen's light rays makes blurring even more obvious. Please note the outline below regarding *probable* causes for definition of detail problems.

Definition of Detail (Projection)

Problems	*Causes*
Blurring	Patient motion
	Screens
	Large focal spot
	O.F.D.
	Acute C.R. angle
Size Distortion	Long O.F.D.
	Short F.F.D.
Shape Distortion	Angled C.R.
	Angled film

Causes of Problems in Visibility of Detail

It is entirely possible that a radiograph with poor density or contrast characteristics could result from one of a number of items, including outdated film, processing problems, defective grids, screens, etc. To further

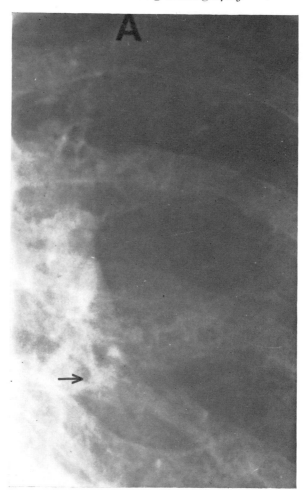

illustrate the complexity of the situation, it is important for the technologist to be able to organize these items mentally while critiquing a film for technical quality so the possibilities of making the correct adjustment will be significantly increased. For example, in the case of a radiograph that looks too light one should be able to determine whether the problem is because of insufficient quantity of radiation or because of insufficient penetration. Each of these problems requires different corrective action. The light film could also be caused by inactive developer solution, perhaps through poor replenishment rates or low temperature. The poor density might be the result of an improperly used grid resulting in grid cutoff. On the other hand, the light film might have been caused by pathology in the patient. An equal number of possibilities would be present if we were looking at a radiograph that was too dark.

When diagnosing poor contrast, a similar number of possibilities loom.

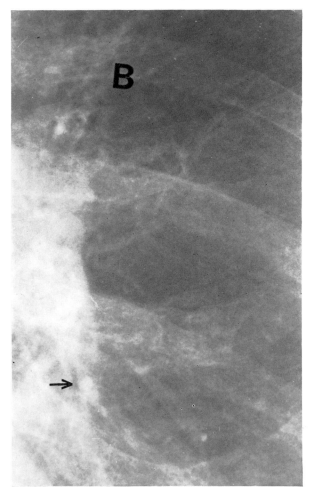

← Figure 199. Although there may appear to be only subtle differences in blurring, small fracture lines, small vessels in an arteriogram study, or small bone trabeculi would be almost impossible to see.

A flat gray-looking film could be caused by too high a Kv that produced homogenic penetration of the body part and increased the number of scatter photons. It is a possibility that the developer is too inactive (weak) to reduce the exposed silver bromide crystal properly. The grid ratio might be too low allowing excess numbers of scatter to reach the film. The developer solution might also have been too warm causing indiscriminate reduction of the unexposed crystals producing unwanted densities (chemical fog). The film might be outdated or stored for a time in too warm an area. The patient may simply have very little subject contrast and predictably a low radiographic contrast would result. As

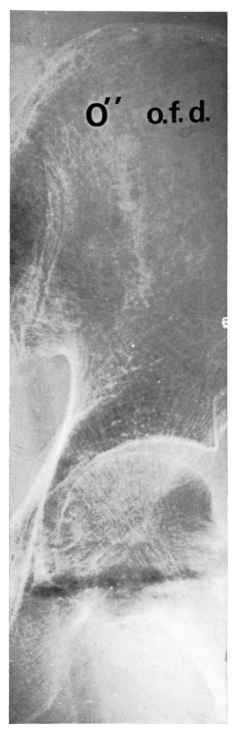
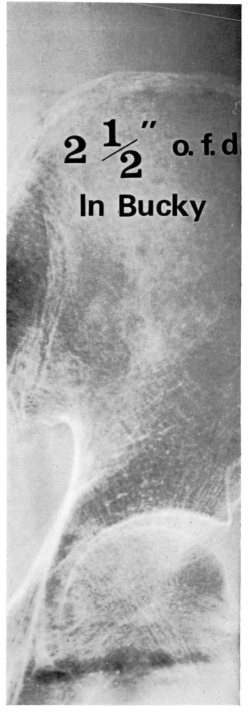

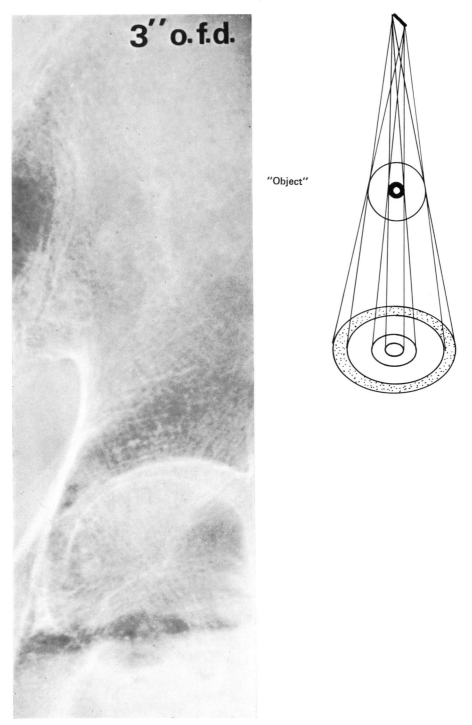

"Object"

← Figure 200. As object film distance increases, there is a notable decrease in sharpness of the borders of body parts to the point where they cannot be distinguished from the surrounding area.

these items come to mind, the technologist must be successful at diagnosing the problem so that the proper adjustment can be made for the repeat film. In short, the film must be analyzed technically by establishing an organized logical outline which progressively reduces the possibilities of causes and eventually will help us arrive at which factor actually caused the poor image.

The Correct Procedure

The three steps used in film critique are the following (see also Fig. 201): (1) Identify the problem; (2) Determine the cause of the problem; (3) Choose the appropriate corrective action. The first step is to determine whether or not the problem is that of definition or that of visibility of detail.

AN EFFECTIVE APPROACH TO FILM CRITIQUE

STEP ONE   Determine whether you are observing a visibility problem or a projection problem

STEP TWO   Review the status of the image regarding the eight items below:

Projection Problems
1. Blurring
2. Distortion of
   —size
   —shape

Visibility Problems
3. Artifacts
4. Body habitus
5. Contrast
6. Density
7. Penetration
8. Processing

STEP THREE   (A) Determine the correction
(See Table IV)

(B) Choose the new exposure factors
(See Chapter 11)

Figure 201. The three primary steps for film critique.

The second step in film critique is to mentally review the items that are likely to cause problems with either definition or visibility of detail. For example, if the problem is a definition problem we might want to look at exposure time because of patient blurring or perhaps we would examine a cassette for poor screen contact. The eight categories in Figure 201 can be memorized quite easily if arranged in alphabetical order. After reviewing this list mentally, and keeping in mind the information discussed in previous chapters, you should be able to tell whether you are dealing with a visibility problem of a definition problem, and determine the root cause of the problem.

We must then go to step 3 to determine which course of action is most likely to correct the problem, including the choice of new exposure factors.

With the information previously given, it is possible to determine which adjustments should be made to correct the appearance of the radiographic image, i.e. whether the KvP should be adjusted, whether to use a higher or lower grid ratio, or whether the temperature of the developing solution is incorrect. Once the decision has been made, the actual exposure factors that are to be used can be obtained with the information given in Chapter Eleven. Thus, the final step to be taken in our established film critique procedure is to actually decide on which of the items in step 2 need to be corrected. This can be done successfully only if you know how far the original radiograph is from what is desired. For example, assuming the radiograph is too light and a simple adjustment in MaS is needed, before one can choose the new exposure factors he must know how light the image really is. He must be able to say to himself the radiograph is one-half the density it should be or perhaps it is three-quarters the density it should be.

If the film is one-half the desired density the exposure factors must be increased by two times or doubled, but if the image is approximately 75 percent of the density which is desired, the exposure must be increased by 25 percent. Another situation may be that the image is too low in contrast, which of course could be the result of a number of problems. It might be because the exposure produced too many scatter photons causing radiographic fog. It might also be a problem of poor subject contrast as a result of an old patient with very poor absorption tissue characteristics. In this case we will assume that the KvP is to be reduced to help increase subject contrast. The next step is to choose the correct KvP that will provide the desired effect. It has been the author's experience that a reduction in KvP to increase subject contrast requires an adjustment of at least 15 to 20 KvP. This, of course, would reduce radiographic density and so the MaS must be increased accordingly. However, before such an adjustment is made, once again the technologist must ask himself specifically how much of a kilovoltage decrease is necessary to cause the desired effect. A fairly accurate estimate must be made as to how this change will affect density; by reviewing Chapter Eleven, the correct adjustment can be made with relative ease.

We will now critique two examinations using our three-step film critique procedure. Refer to Figure 202: Here we have a technical prob-

TABLE IV

POSSIBLE REMEDIES FOR POOR RADIOGRAPHIC QUALITY

General Radiographic Problem	Specific Radiographic Problem	Possible Solutions
Artifacts	foreign body in screens	These generally show as white marks on the film. Cassette screens should be replaced if the surface is pitted.
	static	Try to increase humidity. Use a fine mist to spray walls of darkroom. Use antistatic solution on screens during the static season (winter).
	fingerprints	Try to reduce moisture on hands. Occasional washing or cleansing with alcohol soaked cotton balls will help. Avoid contact with film except on edges.
	scratches	These can result from dirty rollers and/or misadjusted guide shoes; occasionally, too much pressure when feeding film will cause tiny plus density scratches. Usually, processor scratches will be in the direction of film travel.
Blurring	patient motion	Use compression, restraints, shorter exposures. Reduce to at least 1/10 or 1/15 of a second for adults, 1/120 of a second for children. Use super-high speed screens. Reduce F.F.D. only if absolutely necessary. Use lower ratio grid.
	poor film contact	Do a screen test. New felt straps or hinges might be needed. Replace cassette if frame is warped.
	too much O.F.D.	Increase F.F.D. if possible.
Poor Contrast	excessive S/S radiation	There are many ways to reduce S/S radiation reaching the film. All, however, become more or less desirable depending on the particular situation. One must not lower the KvP at the risk of inadequate penetration. The formula for finding optimal KvP can be used as a guide.
		Coning is very important as shown in Chapter Nine. With flabby patients, compression can be very helpful. Of course, increasing grid ratio will absorb extra S/S. One should keep in mind that flabby, fat patients often do not require nearly the MaS or KvP that is indicated on the chart through mere cm measurement. This type of patient requires a reduction in KvP to yield positive results. Muscular people, of course, often require more KvP than indicated by measurement. Even with muscular people, the base KvP formula will usually be sufficient.
	chemical fog	Chemical fog is the result of overactive developer solution. There are three ways to correct the situation. If processor is overreplenishing, reset the switches. Often the incoming water temperature will rise and cause the developing tank to overheat through conduction; reduce water temperature to approximately 60 degrees for quick cooling and readjust to 85 for routine operation. Also check

TABLE IV (*Continued*)

POSSIBLE REMEDIES FOR POOR RADIOGRAPHIC QUALITY

General Radiographic Problem	Specific Radiographic Problem	Possible Solutions
		for exhausted solution or improperly mixed developer solution. Inactive developer will produce poor contrast. Basically the same factors that affect developer activity must be reviewed: developer replenisher, temperature, and one additional factor, oxidation. If too low, increase replenisher rate to at least 100 cc of fixer per 14 $\times$ 17 and 55 cc developer per 14 $\times$ 17. In low volume areas these rates should be as much as 50 percent higher. Consult the chapter on automatic processing. The developer temperature should be increased to 92–95°. If the developer is oxidized, new solution must be mixed.
	body habitus	Subject contrast relates directly to radiographic contrast. If the patient is elderly and dehydrated with poor muscle tone, reduce the KvP by *at least* 10 to 15 and adjust MaS for density. Certain pathologies require KvP adjustments. These are so varied that space does not allow a detailed description. One should know which diseases cause tissue to break down, requiring lower KvP. For muscular patients, you are forced to use at least base KvP. Very little can be done in these cases except to use a heavier grid and careful coning.
	film fog	Occasionally, film will be stored incorrectly, resulting in high levels of base fog and mottle. One must check the film's base fog if poor storage conditions are a possibility.
Density	exposure factors	Through experience one must develop an intuition about the exposure factor used for the body type and related situation (grid/screens/film etc.). Over penetration might be part of the problem. Excessive developer temperature, over replenishment, or simply too much MaS could be the problem. The wrong film or screen may have been used. As we will see later, the general appearance of the film as well as background information must be used to review all possible causes. Most often, it is a matter of over penetration and/or too much MaS.
Distortion	angulation of body or CR increased O.F.D.	This probably occurs most when doing portable work. One must be sure to position the body part correctly with the appropriate angle of CR. When the object film distance is increased, the F.F.D. should be increased proportionately. Maintain an acceptable ratio of O.F.D. to F.F.D. whenever possible, i.e. 2.0 inch O.F.D. to 40 inch F.F.D. or 3 inch O.F.D. to 60 inch F.F.D.
Patient	elderly patients	Often, body habitus has a very strong influence on the appearance of the radiographic image. Elderly patients have a very low hydration level which decreases subject contrast

TABLE IV (*Continued*)

POSSIBLE REMEDIES FOR POOR RADIOGRAPHIC QUALITY

General Radiographic Problem	Specific Radiographic Problem	Possible Solutions
		greatly. Use a beam that will produce more photoelectric interactions, which would improve or amplify the existing poor subject contrast (decrease KvP 15 to 20 KvP).
	demineralized bone	Occasionally, the bones of elderly people become demineralized and thus offer little stopping power to the beam. The bones on radiographs look almost as dark as the surrounding soft tissue, which produces even less subject contrast (lower KvP by 15 to 20 KvP).
	muscular patients	Another problem relative to body habitus is when the patient has very firmly developed muscle tissue. The water content in this case is high which makes penetration difficult and, in addition, produces an excess of S/S radiation. Both situations combined cause a light and very low contrast radiographic image. The solution is difficult, because the increase in KvP needed to penetrate the highly absorbent muscle tissue helps produce an increased scatter ratio. A very careful balance of KvP and MaS is needed. Extra grid ratio is helpful. Sometimes using a grid cassette in a bucky will help in extreme cases. The use of rare earth screens sometimes can compromise for a decrease in KvP. Proper coning is important as well.
	obese patients	Patients with a high fat content often produce a very flat, gray looking film. One should keep in mind that fatty tissue is easily penetrated and, although the physical size of the patient is great, one can often reduce the KvP by ten and not loose too much density. This would also reduce the ratio of S/S production producing a higher contrast image. Finally, compression can be used with some degree of success. Very often, radiographs of fat patients are simply over exposed and a reduction of technique in general, preferably in KvP, will produce a better image. Coning is important also.
Penetration	too thick or dense body tissue (too much absorption)	It is very difficult for untrained technologists to determine whether under or over penetration is the problem. Too little penetration produces an image that has poor contrast with very light density. The bony parts when viewed are almost undistinguishable, as they have very similar density to the surrounding soft tissue. An increase of at least 10 to 15 KvP is usually necessary. Increasing by 20 KvP is sometimes indicated in extreme cases. Coning is important to help reduce S/S as much as possible.
	too little absorption	The solution and resulting image is more successful than with the above condition. Here one may reduce the KvP by 15 and in extreme cases by 20. One must be careful, of course, not to use a KvP that will produce inadequate penetration.

TABLE IV (*Continued*)

POSSIBLE REMEDIES FOR POOR RADIOGRAPHIC QUALITY

General Radiographic Problem	Specific Radiographic Problem	Possible Solutions
Processing	chemical fog (overactive developer)	Chemical fog occurs when the reducing agents begin to work on the unexposed silver bromide crystals. This produces an over-all density in areas of the film that are supposed to be white. Check for high developer temperature. The easiest way to identify chemical fog is to look at the area of the film containing the patient's name and x-ray number. If it appears gray, then it is probably caused by overactive or contaminated developer. Also, the replenishing rates should be checked for possible replenishment.
	underactive developer	This situation produces an image with very little contrast and low density. The film usually has a "washed out" look. Too low a developer activity is usually the cause. The solution is simple once the cause is determined. The developer temperature for automatic, 90 sec processing should be about 95°. Replenishment should be about 55 cc per 14 × 17 film (developer), and if the solution is exhausted and oxidized it must be replaced with fresh solution. Check for diluted developer.

lem with poor contrast and increased density. The patient was to have an IVP examination, and the scout film is very dark and extremely low in contrast. The patient measured 26 cm and was quite flabby. As you can see from looking at the scout film (A), a major adjustment in technique is necessary. The lack of contrast makes it impossible to see any detail whatever.

Our first step in film critique is to determine whether we are dealing with a definition of detail problem or a visibility of detail problem. The structures appear to be sharp; upon close inspection of the radiograph, the bone markings of the pelvis appear to be well defined, and since there is no evidence of distortion we can direct our attention to the factors involving visibility of detail. We can now review the six items under visibility of detail which would help us isolate the problem still further. Artifacts can be eliminated immediately. Body habitus is certainly a problem because of the patient's size and fat content. Contrast is certainly too low, and there is too much density on the film. At first the film appears to be overpenetrated, but the KvP used was 90 and this would not cause overpenetration of a part of a patient of this size. Processing does not appear to be the problem because the patient's identification name plate has good image characteristics. (Fig. 201 shows a checklist that can be used in the classroom.) Also, the other films processed at the same time looked normal. Thus, with step 2, we have eliminated more possibilities and have come

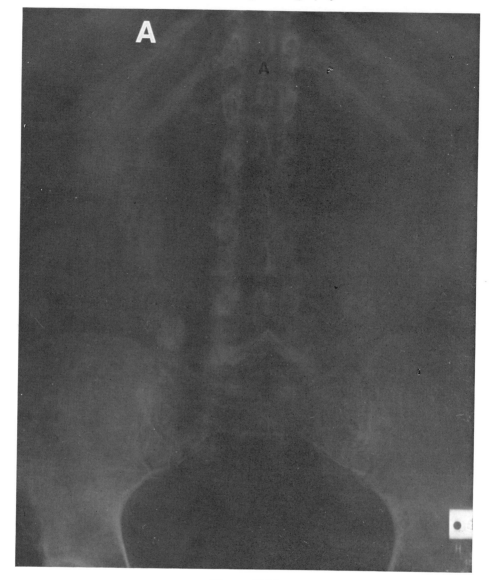

Figure 202A.

Figure 202. By using an established, logical, simple approach, film critique can be accomplished easily. Proper use of the various conversion factors will help make step 3 easier.

closer to the root of the problem. It appears in summary that the patient is the basic problem because of his body habitus, but because of the factors used, contrast and density were unacceptable. Although this may seem obvious, the established procedure or regimen helped us to review

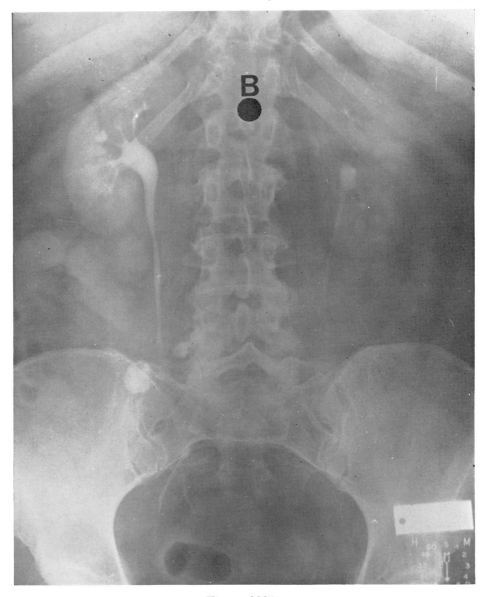

Figure 202B.

and eliminate many other problems very quickly that *could have* contributed to this situation.

The third step in our program requires us to review all of the items in step 3 that involve body habitus. We may also wish to review the sections on radiographic contrast and density for additional information that may help us decide on the most appropriate corrective action. The original technique was 400 Ma, one second at 90 KvP. We estimate through

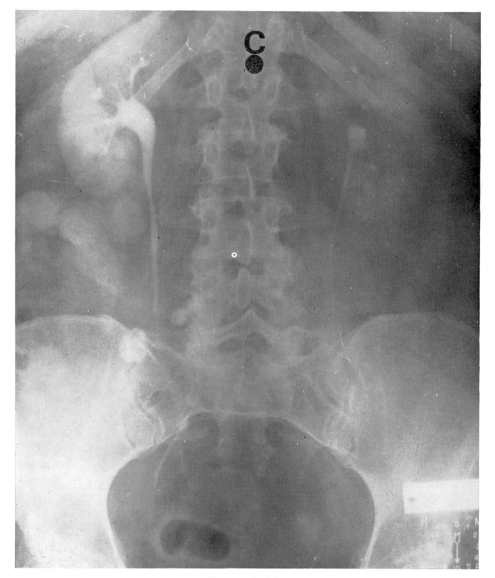

Figure 202C.

experience that the density is approximately four times darker than it should be and reduce the exposure to one quarter and obtain film B. We note, however, that the contrast is still low because of poor subject contrast and probably excess scatter as well. We know subject contrast can be improved by decreasing KvP, also less scatter will be produced. We know further that a reduction in KvP of 15 percent will reduce radiographic density by one-half. With this in mind, the third film was exposed using 76 KvP at one-half second and the results are noted in film C. Thus,

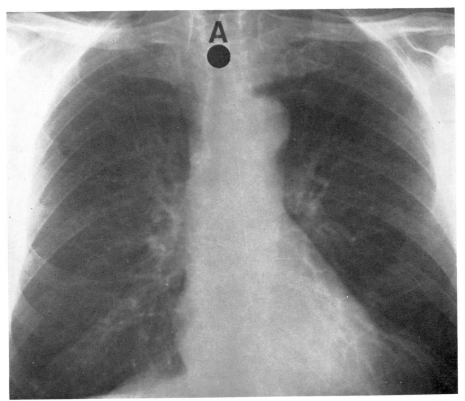

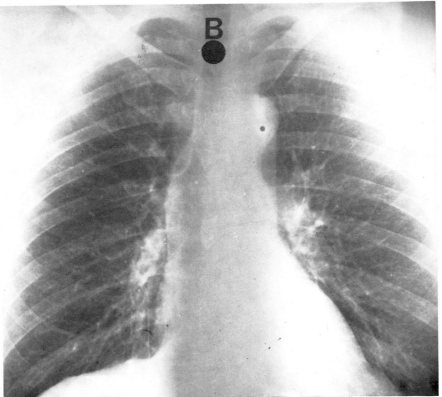

Figure 203.

by using the three step system we have effectively, without skipping any important factors, reviewed many possibilities and systematically came to the conclusion that subject contrast and overexposure were the primary problems. By using the information presented in Chapter Eleven, we were able to make the necessary conversions and affect the optimal image for that particular patient's examination.

A chest examination is shown in Figure 203. By inspection of film A we note that there is a great deal of density. There is no evidence of distortion or blurring so we can once again direct our attention away from those factors that affect definition of detail and distortion. As we go down the list of six categories listed in step 2, all can be eliminated quickly except for excessive density and possibly overpenetration. At first one might suspect overpenetration (too much Kv for a given body part) but it is very common to use 110 KvP for grid chest studies. This leaves us with a simple case of overexposure as a result of too much MaS. We must now look again at film A and determine how much darker it is over what we consider optimal and in this case we decide the image is approximately two times darker than it should be. The exposure time was thus reduced by one-half and the results of the second exposure can be noted in film B. The initial exposure factors were 300 Ma, 110 KvP at $\frac{1}{15}$ second. The final exposure used was 300 Ma, 110 KvP at $\frac{1}{30}$th of a second.

CHAPTER THIRTEEN

RADIATION PROTECTION

BACKGROUND RADIATION

B ACKGROUND RADIATION constitutes any ionizing radiation coming from the sun, from within the earth, and from our own bodies. Background radiation is broken into two basic categories: external radiation and internal terrestrial radiation.

External sources are cosmic rays which come to the earth's atmosphere in two forms: galactic and solar. Galactic move through outer space with unknown origin. They strike the atmosphere and produce scattered particles, where they expose the general population. Solar cosmic radiation is emitted directly from the sun during high energy explosions on its surface. Cosmic radiation can carry energies up to 2.5 billion electron volts. These particles then move toward the earth with a myriad of energy levels and at an infinite number of projected angles. Cosmic radiation dose to the gonads averages 26 to 28 millirems per year measured at 5 cm in a tissue equivalent in the U.S.A., with Denver at about 50 mrem and Leadville, Colorado (10,000 ft. altitude) at about 125 mrem per year (measured in 1979).

Internal or terrestrial background radiation is emitted from within the earth's crust and from various radioactive elements that find their way into building materials (the cement, concrete blocks, and bricks). Other forms of internal radiation coming from the earth's crust are uranium, platonium, actium, thorium, but the most prominent are carbon 14 and potassium 40. It is probably no surprise that the amount of background radiation present varies in different parts of the world. There are notable differences, for example, within the U.S.A. alone, and there are striking variations around the entire earth. The intensity of scattered cosmic rays also varies with altitude. For example there is about an 100 percent increase in cosmic radiation present at an altitude of 1,000 feet from that at sea level. The position of an area of the earth relative to the sun also has an important effect on the intensity of the scattered cosmic radiation. There is a 10 to 20 percent increase in cosmic radiation for an area at the equator compared to an area located at approximately 50 degrees latitude. The average amount of cosmic radiation to the earth is approximately 44 milliroentgens per year,[1] and with the protection provided by buildings, gonadal dose is about 28 mrem yearly.

---

[1]N.C.R.P. Report #45

A striking example of internal background radiation variances can be seen in areas such as Brazil and Kerala, India; individuals in these areas are exposed to between ten to twenty times the average dose of internal radiation compared to other parts of the world.

An individual living in the United States receives on the average 100 millirems per year from both internal and external sources. We are all aware that radiation has some effect on the number of genetic mutations of off-spring generations, and there are estimates that from 4 to 10 percent of the genetic mutations[2] in this country are the result of background radiation. However, many of these mutations develop in the womb and cause secondary physiological problems during gestation which we suspect result in either miscarriage or fetal death and, therefore, never become full-term deliveries.

In addition to the 70 to 100 millirems per year, many individuals are involved in receiving additional doses of radiation from medical and dental sources. Artificial radiation includes those forms of ionizing radiation produced by man such as medical, dental, and nuclear power plants. The average individual receives from 70 to 100 millirems per year from medical and dental x-rays,[3] which accounts for by far the greatest amount of exposure to individuals in the U.S.A. today. The average amount of radiation received by an individual based on general population figures for both dental/medical and background radiation can be seen in Figure 204. To put this figure in perspective, an average PA and lateral chest examination using 120 KvP with rare earth

Approximate doses to background radiation	
TYPE	DOSE PER YEAR
Cosmic radiation	26 - 28 m rem
Internal/terrestrial	60 m rem
Body radiation	25 m rem
(Some resulting from food eaten)	
Approximate maximum total background	111 m rem per year
Background	111 m rem per year
Medical dental (artificial)	70 to 100 m rem per year
Average Population	
Approximate average total maximum radiation to average population	180 to 211 m rem per year

Figure 204. A summary of average approximate doses from background radiation. These vary greatly depending on actual location and altitude.

imaging system generates a skin dose of about 50 millirems of radiation to the patient. There are very interesting and informative facts available on the subject of background radiation obtainable from the federal government in a handbook identified as "Background Radiation" NCRP report #45, available from the superintendent of documents, Washington, D.C. 20002.

---

[2]Sheele, Wakely: *Elements of Radiation Protection;* Charles C Thomas, Publisher.
[3]Graham, Thomas: *Physics for Radiological Technologists;* W. B. Saunders.

MAXIMUM PERMISSIBLE DOSES AND THEIR CALCULATIONS

Maximum permissible dose is the maximum amount of absorbed radiation that can be delivered to an individual as a whole body dose or as a dose to a specific organ and still be safe. The term "safe" in this context means that there is no evidence that individuals receiving the maximum dose mentioned will suffer harmful immediate or long-term effects to the body as a whole or to any individual structure or organ of the body. One should keep in mind that whenever an individual is exposed to ionizing radiation there is always some degree of damage that could theoretically occur. Analogy might be made here to an individual smoking a cigarette only once a month. There is no evidence that with this frequency of smoking physical damage could result; however, with increasing frequency of smoking, the probability steadily escalates by virtue of its accumulative effect. Unfortunately, there is no established threshold for either cigarette smoking or radiation under which damage will not occur or over which damage will surely result. One of the primary issues is that radiation damage tends to accumulate with each exposure, and that safe limits have been arrived upon through statistical analyses based on scientific and medical data over generations.

Maximum permissible dose (MPD) refers to the maximum amount of radiation that has statistically been shown not to contribute to destructive changes to the body. The maximum permissible dose for radiation workers can be seen in Figure 205. MPD standards are divided for two segments of the population: radiation workers and the general public. Radiation workers constitute any group of people who work in a radiation controlled area. A radiation controlled area is any area or section inside or outside of a building that is under the direct supervision and survey responsibility of a radiation health physicist. The MPD for the general population (nonradiation worker) is 1/10 of the MPD that has been established for the radiation worker shown in Figure 205.

MAXIMUM PERMISSABLE DOSES

	Maximum 3 month dose (13 weeks)	Maximum yearly dose
Whole body, gonads, blood forming organs, lenses of eyes°	3 rem	5.0 rem
Skin of whole body	10 rem	30.0 rem
Hands, forearms, head, neck, feet, ankles	25 rem	75.0 rem
General public		0.5 rem
		(over total gestation period)
Pregnant women		0.5 rem

°Because of the very small dosages involved, it is not practical to calculate MPD on a weekly bases, assuming the guidelines in the thirteen week period are observed.

Figure 205. Maximum permissible dose as prescribed by the National Committee on Radiation Protection (NRCP).

As you can see from reviewing Figure 205, a maximum permissible dose varies with particular body part. It differentiates the amount of maximum permissible exposure to the body as a whole compared to exposure to certain individual parts of the body. The maximum permissible dose to the whole body is the same as certain specific structures that have high x-ray sensitivity, such as blood-forming organs and lenses of the eye. However, since radiation has an accumulative effect over time, a method had to be derived to determine the maximum amount of radiation an individual could safely receive during the course of his life in addition to what he could safely receive during any one year of his life. The simple formula below can be used to calculate what is considered a safe life-time dose.

$$\text{Life-time MPD} = 5 \ (\text{AGE OF INDIVIDUAL} - 18)$$

The above MPD formula includes any form of background or artificial radiation.

It is very important to remember that the effect radiation has on the body is dependent on how a given dose had been fractioned out to the body over a period of time. For example, the maximum permissible whole body dose for a year is 5 rems. However, there is a considerably greater biological effect to the radiation if 5 rem is received all with one exposure rather than spread out over a period of time. With the concept of dose fractionization in mind, the MPD figures shown in Figure 205 require some qualification. Even the small dose of 5 rems per year is undesirable if received at one time, so the federal government has assigned MPD exposures for three month periods of time which are shown in addition to the yearly MPDs in Figure 205.

Basic Units of Radiation Measure

The basic unit of measure of radiation is the roentgen, defined as a quantity of radiation that produces 2.58^{-4} coulombs per kg of air. The amount of radiation that will produce such an effect is arbitrarily named roentgen in recognition of its discoverer, Wilhelm Konrad Roentgen. Quantities of radiation that do not generate the effect equal to one roentgen or rad (R) are expressed as milliroentgens (a milliroentgen is equal to 1000th of 1 roentgen); for example, 1/2 of a rad would be expressed as 500 millirads. It is important to keep in mind that the measure of roentgen or milliroentgen is a statement of quantity and not quality of radiation. With this in mind, the same dose of 5 roentgens could have very different biological effects depending on the quality (energy level) of the radiation used.

It should be pointed out that there are differences in the result when measuring quantities of radiation in air compared to the same dose absorbed in the body. Although 1.0 roentgens measured in air is actually equal to an absorbed dose of 0.87 rads, the formula used to determine or convert roentgens from air to absorbed dose (rad) is not significant here since, in general, we consider 1 roentgen to be equal to 1 rad, with a rad being the unit of measure of absorbed dose.

Roentgens to Rads to Rems

At this point we have discussed the roentgen and the rad as standard units of radiation, and that one roentgen is considered to be equal to one rad. However, the MPD is based on rems, so we will now discuss briefly how these three terms tie together. We have already learned that the units of measure so far discussed relate only to quantity of radiation. In order to determine the actual biological effect of a given dose of radiation we must take into account the energy level of the radiation as well as the quantity. To accomplish this we refer to a term called the quality factor. When a quality factor of radiation is figured to a given quantity of radiation we can arrive at its biological effect. The various energy levels or quality factors for different types of radiation have been subdivided into categories shown in Figure 206. With this we can define a rem as being equal to a rad times the quality factor.

Type of radiation	Quality factor
X-rays, gamma rays, electrons, beta rays	1.0
Neutrons and photons up to 10 Mev	5.0
Neutrons and photons	10.0
Naturally occurring alpha particles	10.0
Heavy recoil nuclei	20.0

Figure 206.   Listing of radiation quality factors.

With the table shown in Figure 206 we can determine that a dose of x-radiation equal to 0.5 roentgens will have an absorbed dose of 0.5 rads. The biological effect of this amount of x-radiation is determined by multiplying 0.5 (rad) times 1.0 (radiation quality factor), which yields 0.5 rems.

A dose of alpha radiation of 0.5 roentgens is considered to be 0.5 rads. The biological effect of this particular dose of radiation can be determined by multiplying 0.5 rads times 10.0 (radiation quality factor), which will produce 5.0 rems. Once the various doses under evaluation are converted to rems, they can be gauged against the standard MPD for whole body or whatever part of the body is of interest at the moment.

If an individual were exposed to various energy levels of radiation during a single dose, the individual dose received for each type of radiation must be calculated separately with its respective quality factor and the product of each separate dose and related quality factor added together to obtain a total radiation dose expressed in rems, as shown in Figure 207.

The MPD for children is the same for the public, 1/10 of radiation workers.

When a combination of dose energies are received:
(150 mr x-rays) $150 \times 1 = 150$ mrem
(10 mr alpha particles) $10 \times 10 = 100$ mrem
                        150 mrem
                    + 100 mrem
                    _____

Total =   250 mrem

Figure 207.   Calculating doses for exposures with multiple types of radiation.

RADIATION AND ITS BIOLOGICAL EFFECT

With the understanding that any amount of ionizing radiation is potentially damaging to tissue, the following section identifies a broad spectrum of doses and related biological effects. The net biological effect resulting from radiation exposure is primarily determined by the following points:

1. the quality of radiation used
2. the amount of dose absorbed
3. the overall period of time or fractionization through which the dose was distributed
4. which tissues or organs were involved
5. over what area of the body the dose was given, i.e. whole body or specific area

The Reaction of Tissue to Radiation

The reaction of tissue to radiation can be divided into three basic categories: local, general, and genetic. An example of local reaction is erythema dose. Erythema is a reddening of skin over the radiated field. Typically, erythema is temporary and within a few days or weeks the skin heals and returns to its original tone. Some examples of general reaction to radiation are nausea, vomiting, diarrhea, and even leukemia. Large amounts of radiation required to produce these symptoms could even lead to death. Genetic effects are considerably more difficult to evaluate since they often do not appear for two or three generations after the original dose has been received, and also because many of these pregnancies are terminated before full-term, making it extremely difficult to evaluate the direct effect of the radiation dose; more specific information will be discussed later in the chapter.

As noted previously, genetic damage can lie latent for two or three generations and then appear in the form of a malformed offspring with a range of severity from mild and barely noticeable to very obvious and requiring institutional care. The one responsibility technologists should always keep in mind is their own potential role in keeping the patient's dose during childbearing years to the gonadal area as low as absolutely possible, and in doing so protect the patient from producing this potential genetic effect. This can easily be accomplished by providing patients with proper shielding, especially those patients in childbearing years.

Since a child has partially developed reproductive organs, these patients in particular are vulnerable to ionizing radiation. Children are, in general, more sensitive to ionizing radiation than adults. Perhaps the familiar term of "child-bearing" years is short sighted and instead should be expanded to include newborns through age 50.

When ionized by radiation, the chromosomes can be damaged. In short they become ionized and in doing so they take on different characteristics; combined with other ionized chromosomes they form chromosomes of very different characteristics. Sometimes these damaged chromosomes can combine and become camouflaged, and coexist with healthy chromosomes maintaining a latent or dormant level which can reappear and play a primary role in developing mutations of later generations.

Basic Biological Considerations

Although many variations are apparent with respect to biological effects toward radiation, there are some general points that remain constant throughout. (1) Generally, as a quality of radiation increases, the more potentially damaging is the effect to human tissue. (2) The amount of total accumulated dose is directly proportional to the potential tissue damage. (3) Generally, the biological effect diminishes if the total dose to be delivered is fractioned over a number of smaller doses rather than the entire dose delivered at one time.* (4) There is a workable trade-off between the total amount of dose relative to its biological effect depending on the degree of fractionization, which is to say that one can produce less damage with a larger dose given over a long period of time than a smaller dose given all at one exposure. (5) The type of organs and tissues that are radiated has a great deal to do with the net effect of the radiation (to verify this one need only review the NPD charts for skin, distal extremities, and blood-forming organs). (6) The size of the area radiated weighs heavily in the general biological effect of the tissue; as the radiated area increases in size the biological effect increases proportionately.

Sub-, Mid-, and Supralethal Doses

Massive doses of radiation can be categorized into three subgroups known as sublethal dose, which refers to the quantity of radiation to the whole body that produces general effects and symptoms that do not directly lead to death. Midlethal dose range refers to a dose that produces a 50 percent chance of recovery, and a supralethal dose is a whole body dose that results in death in sixty days.

---

*Fractional doses cause less tissue damage because less dose is used per exposure, and because the exposed tissue had an opportunity to rebuild during the entire treatment plan.

General Symptoms of Supralethal Doses

(From 500 rem and above)

1. Nausea
2. Vomiting
3. Prostration within one to two hours
4. Rising fever
5. No appetite
6. Onset of diarrhea
7. Dehydration
8. Rapid drop in white blood count
9. Death within about two weeks after exposure

General Symptoms of Midlethal Doses

(From 300 rem to 500 rem)

1. Biological effects are as noted above except with somewhat less severity
2. Death occurs to approximately 50 percent of the population radiated with such a dose within a period of time extending from one to three months

General Symptoms of Sublethal Doses

(From 100 rem to 300 rem)

1. Mild nausea
2. Vomiting
3. Malaise for about 12 to 24 hours
4. Loss of hair after about 3 to 5 weeks from the exposure (in many cases this is transient and new hair will regrow)
5. Cataracts develop typically in 2 to 6 years if the eyes were involved
6. Noticeable drop in blood count, which usually recovers (there are often some permanent although manageable biological effects that occur with sublethal doses such as anemia; with all of those exposed to sub- and midlethal doses there is at least a nine time increase in incidences of leukemia)

The above symptoms are known as acute radiation syndrome and become apparent when a whole body dose of 25 rads or more is received at a single time.

Long-Term or Chronic Doses

Another form of radiation to the body is known as chronic dose. This usually involves a much lower dose received either continuously or at frequent intervals over a long period of time. Continuous doses produce such effects as cracking and drying of the skin and in more advanced cases ulceration; the result in instances of cancer and leukemia are high. An example of continuous radiation exposure would be recorded increases in cancer among workers in uranium mines.

As was noted before, there is no absolutely defined amount of radiation that automatically causes a certain biological effect; however, there appear to be general thresholds levels of radiation as indicated below that usually account for somewhat predictable effects.

Reaction	*Dose*
Cataracts	200 rads
Depressed sperm count	100 rads
Reddening of skin (erythema)	200 rads
Temporary sterility	200 rads (varies with age and sex)
Epilation (loss of hair)	500 temporary but becomes permanent as dose increases
Decreased number of lymphocytes & leukocytes	25 rads or more (whole body dose)

General Reaction of The Body To Radiation

The sequence of biologic effects develops through four typical phases, although the severity of the dose makes these three stages more apparent. They become less apparent and practically undetectable as the dose decreases to the MPD levels.

1. The initial effect, when transient clinical signs are often noted.
2. Latent period (where symptoms are not seen after radiation was received)
3. Manifest effects (when symptoms first appear; with moderately low doses the symptoms often last from a few days to two months)
4. Recovery/tolerable damage/death (this is the period of time extending after the manifest period is apparent where the individual totally recovers from the symptoms, dies, or maintains some permanent but manageable damage such as anemia)

RADIATION MONITORING DEVICES

There are a number of different radiation devices that can be used for measuring exposure rate. Although these devices are very accurate in detecting exposure rate, they are generally not useful for determining the wavelength or energy of the radiation being surveyed. Although a few of the devices we will be discussing shortly are capable of identifying the presence of radiation at both high and low levels, we will see that each device has its own characteristic that makes it more accurate for certain specified circumstances.

Gas-Filled Devices

The first group of instruments we will discuss are generally known as gas-filled detectors. Two very common such devices are called the G.M. meter (Geiger-Meuller) and the "cutie pie." Both utilize a gas-filled chamber (air), but operate in slightly different style.

The "cutie pie" type has two separate metal plates in its air chamber, with one having a positive and the other a negative charge. These plates are electrically connected to a small circuit and energized by a battery. When the air chamber portion of the device is exposed to ionizing radiation, the air molecules inside the chamber begin to ionize into separate positive and negative ions. These ions separate and travel toward the oppositely charged attracting plates. These free ions produce an electrical current in the device which is measured by an ampere meter connected in series to the electrical circuit. If the air chamber is placed in a field of relatively low activity, only a few air molecules will be ionized, and in turn, a relatively low voltage will register on the meter. Conversely, if the device is placed in a field with high activity, a proportionally high number of ions will be produced and the ampere meter will respond according to the increased current in the circuit of the detector. The amount of exposure is directly related to the amount of ions that will form. The scales on these devices have been calibrated to read in either roentgens per minute or per hour or in milliroentgens per minute or per hour depending on the dose value. Fields of low activity will result in the reading of milliroentgen per hour, and fields of high activity will read on a scale as mr/min or r/min, as shown in Figure 208.

The Geiger-Meuller (G.M.) meter works on a principle similar to the cutie pie. It has its own gas-filled chamber, and it has a positive and negative pole connected to an internal circuit powered by a large cell battery. However, the voltage that the G.M. meter produces is considerably higher than the cutie pie. The high voltage in the G.M. meter is an important characteristic because it causes a kind of amplification of the ions that are released by the radiation exposure. In fact, when the chamber is radiated the voltage in the circuit is sufficiently high so that the free ions are driven at a rate of speed and collide with other air molecules that were unaffected by the exposure to radiation. The accelerated ions collide with neighboring air molecules that were unaffected by the exposure and produce an increased number of free additional ions, called the "avalanche" effect that dramatically increases the amount of current in the circuit of the instrument. With the increased current the ampere meter, of course, reads a much higher value than would have occurred if only the irradiated ions were involved.

These characteristics of high voltage and the avalanche effect makes the G.M. meter able to detect very small amounts of radiation as compared to the cutie pie. Because of its low voltage design, the cutie pie would not have been able to produce sufficient ions to register on the amp meter in the circuit. The G.M. meter is considerably more accurate than the cutie pie when reading low dose rates per hour. On the other hand, the cutie pie is more reliable when measuring areas of high radiation intensity, because its low operating voltage characteristics do not produce the avalanche effect that would likely produce hyperstimulated and inaccurate reading by the high voltage of the Geiger-Meuller meter.

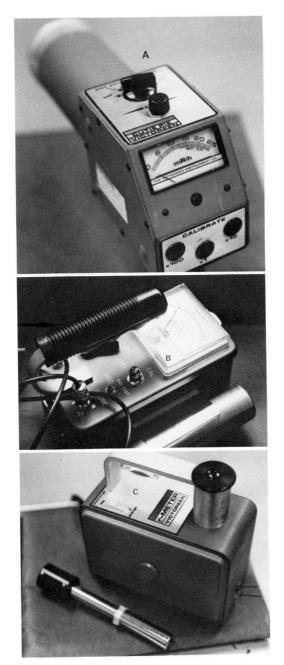

Figure 208A. Cutie pie;

Figure 208B. G.M.; and

Figure 208C. condenser R meter. The condenser R meter is commonly used for measuring *total* exposure values of x-ray and therapy equipment, as opposed to taking radiation activity readings as a general survey function for nuclear medicine type isotopes, which is the case with cutie pie and G.M. meters. The condenser R meter has different "thimble" chambers so that the total exposure ranging from rays to linear accelerator types of radiation can be measured accurately. The condenser R meter is also called an *integrating device* since it integrates exposure time with the amount of radiation it receives to produce a total exposure dose, not simply an exposure rate.

Scintillation Detectors

The next group of radiation monitoring devices are known as scintillation detectors. These devices have a fluorescent material similar to those used in intensifying screens. When exposed to a field of radiation, the fluorescent material is ionized and in the process raises some of its electrons to a higher level in the valance band. These electrons soon lose their excess energy and drop back into their original orbit and give off a representative amount of energy in the form of light. A photomultiplier tube is located in back of this fluorescent material. A photomultiplier tube is a device that can electronically amplify the amount of visible light. The photomultiplier tube amplifies the intensity of the light from the fluorescent material and is registered on a light meter. The more ionization of the fluorescent crystal that occurs the more light will be emitted to the photomultiplier tube, which will be measured by the light meter. Fortunately the light meter and photomultiplier tube can be adjusted and calibrated to produce readings that are directly proportional to the amount of radiation originally received.

Each individual photon or particle of radiation falling upon the fluorescent material produces an emission of light, which is then amplified or exaggerated in intensity by the photomultiplier tube. Scintillation counters are especially flexible since the type and size of fluorescent material used can be changed to handle various energies and activity levels that are to be measured. For example, a material known as anthracene is used as the fluorescent material for low energy emissions such as beta particles, while a material known as sodium thallium-activated iodine is frequently used to detect higher energy level photons such as gamma rays. The quantity of fluorescent material used relates inversely to its accuracy in measuring levels of activity. As a larger piece of crystal is used, the detector becomes more sensitive to lower activity levels such as in the range of 0.1 and 0.2 mr per hour; for higher dose levels in the range of 100 r per minute, smaller sized crystals are used. The meters discussed above are used only to measure exposure rates and not energy level. Exposure rate is a simple ratio of the number of individual photons or radiation particles that are present over time, and must not be confused with total exposure, which is the exposure rate multiplied by the time of exposure duration. See Figure 210.

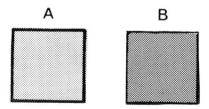

Figure 209. Demonstrates the concept regarding exposure rate and activity concentration. As exposure rate increases for an equal area and time span, the photons or particles would become more concentrated or densely packed, with *A* of lower activity than *B*.

Another, perhaps more practical, method of expressing exposure rate is as the concentration of radioactivity in a given area over a given time. For example, if a given area, such as 5 cm square, contains an intense dose of radiation, the concentration of activity throughout the 5 cm square area will be very high as compared to a situation whereby the same 5 cm square area contains a less concentrated amount of radioactivity. If the amount of area and time is standardized and these ions are counted by a detector either per minute or per hour, the exposure rate would actually be equal to degree of distribution (concentration of the activity in the 5.0 cm area). To quantify exposure rate we have to establish parameters identifying the size of the area sampled and over what period of time the counting took place. See Figure 210.

Calculating *total exposure* from a known exposure rate:

A. A survey meter reading 10 r/hr
   If an individual had been exposed to this rate of activity for 10 minutes, the total exposure received would be 1.66 r.

Exp. Rate	10 r/hr
Time (hrs)	× 0.1666
Total Exp.	1.166 r

B. A survey meter reading 5 r/min
   If an individual had been exposed to this rate of activity for two hours, the total exposure received would be 600 r.

Exp. Rate	5 r/min
Time (min)	× 120
Total Exp.	600 r

Figure 210.   Exposure rates and total exposure.

Personnel Monitoring Devices

Certainly no less important is the process of monitoring the amount of radiation an individual receives. Unfortunately, there are no devices currently available that guarantee that the body has actually received the amount of radiation recorded by the monitoring device. For example, an individual may wear a film badge on the waist band or belt of an uniform. Theoretically, this badge will record the amount of overall radiation given to the whole body, but in fact it is only measuring the amount of radiation received by the badge. This, at first, may seem to be a basic and unimportant simple observation; however, it indicates the importance of how a film badge should be worn, when it should be worn, and on what part of the body it should be worn for various situations.

There are three basic types of personnel monitoring devices commonly used today, and all have acceptable accuracy if used and read properly. In order to be useful there are a few specific characteristics that personnel moni-toring devices must possess to satisfy state and federal regulations as well as providing a very practical function of monitoring radiation to an individual.

1. They must be portable and lightweight.
2. They must be able to measure a wide range of energy levels.
3. They must be able to measure accurately both small and large accumulated doses.
4. The readings must be relatively easy to record.
5. They must be very rugged and practically unbreakable.
6. They must have reproducible reliability.
7. They must be relatively low in cost.

Film Badge

The first and probably most commonly used personnel monitoring device is the film badge. These are usually supplied, maintained, and recorded by commercial companies equipped especially for this purpose.

The film badge is a packet that contains a piece of dental film (different types of film can be used for specific purposes, such as recording various energy levels), a set of copper filters, and a layer of thin lead foil placed behind the film to absorb any backscatter from the body that would distort the reading if it got back to expose the film (Fig. 211).

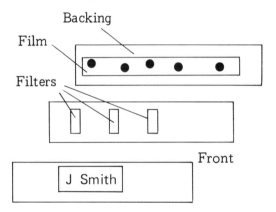

Figure 211.   Schematic of a typical film badge pack.

Readings from this device are accurate and measure the accumulative dose that the badge received. The accuracy range for these badges is from 10 mrems to 500 rems[4] depending on the type of film used. Fortunately, film badges possess all the desirable characteristics mentioned above. They do, however, have some disadvantages as well, which are noted below.

1. They do not respond accurately with equal readings from energy doses from significantly different energy levels. However, for the range of radiation involved in medical diagnostic work, film badges have proven to be very reproducible in their readings and measurements.

---

[4]NCRP Report #48.

2. Because the recording medium is x-ray film, fading (latent image failure) can occur and distort the report coming back to the hospital from the commercial company. To prevent this, film badges should be collected and sent promptly to the commercial companies for processing.
3. The readings produced are somewhat dependent upon the angle the photons strike the film badge packet. For example, if an individual wearing a film badge on the front of the body turns approximately 45 degrees away from the patient, the film badge will accurately record the radiation delivered to the axillary side of the person, and will produce a false low reading.
4. Development techniques are, of course, very critical, and although quality control in the commercial processing companies is extremely high, there is some potential for producing a faulty reading.
5. Immediate readings are not possible since the film badge has to be mailed to the commercial company for processing and evaluation.

A question often arises as to where the film badges should be worn, and the answer lies usually with the preference of the licensed health safety officer. State and federal regulations simply state maximum MPDs for whole body or special sensitive organs and structures and say that the film badge should be worn in a place that shows representative readings. If the film badge is worn at the collar it would receive a similar dose as lenses of the eye and thyroid gland, but the remainder of the body is unaccounted for by badge readings. However, if lead aprons are properly worn and are in good condition, the amount of scatter to the body through the apron would be undetectible.

If there is any doubt as to how much radiation may be penetrating the lead apron, careful reading and calculations can be done by a health physicist with thimble type ion chambers placed behind the lead apron at various positions from a fluoroscope while fluoroscoping a phantom torso at different Kv levels, up to perhaps 115. From these measurements one can be assured that the lead apron is, in fact, a safe barrier from the scattered radiation of the patient. It is certainly important to take these readings while using the maximum KvP that would possibly be requested for a fluoroscopic study as well as for fluoroscopic spot films, so the integrity of the radiation barrier to the person can be accurately evaluated relative to all Kv levels used. Radiation reading from film badges for x-ray personnel who are conducting their activities appropriately in fluoroscopic and routine radiographic situations would normally not exceed about one-tenth of the MPD monthly.

Pocket Dosimeter

The next type of personnel monitoring device that is used frequently is the pocket dosimeter. Pocket dosimeters are very accurate at energy levels from 1 mr to 200 r. There are two types of pocket dosimeters: One model is known as "self reading," meaning that the dose it receives can be read directly through one end of the unit as shown in Figure 212. The second model operates with the same basic principle; however, it must be placed into a small portable

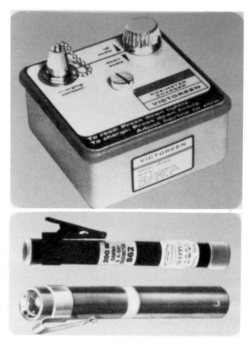

Figure 212. Typical pocket dosimeter along with a charging device. These dosimeters come with varying sensitivities.

box-type device which produces the reading. Pocket dosimeters are approximately the size of a fountain pen and their primary advantage is that they produce accurate readings immediately.

Both of these pocket dosimeters work on the principle of the gold leaf electroscope. With this device an electrical charge is placed on the two leaves inside the dosimeter by a battery or a charge through an A.C. adapter. The radiation causes a certain amount of ionization in the air chamber surrounding the small gold leaf electrodes and this ionization allows some of the pre-charged ions to escape through a small circuit connected in series with a small voltmeter and calibrated in milliroentgens. An additional ability is that they are accurate when measuring very high and intense doses of radiation.

The pocket dosimeter presents some practical problems as shown below.

1. They are fairly susceptible to rough handling and can be easily knocked out of calibration.
2. They must be recalibrated frequently and routinely to maintain satisfactory accuracy.
3. They do not record the presence of low energy beta particles.
4. They are relatively expensive, at approximately $200 each.
5. Because of some of the problems with calibration and their low sensitivity to beta radiation, they should be worn under close supervision of a radiation safety officer, which makes their use not practical for many x-ray working personnel.

6. Records and documentation of readings can often be difficult since their own reading is temporary.

Thermoluminescent Dosimeter (TLD)

Becoming more widely used is the thermoluminescent dosimeter, or TLD. These devices use a thermoluminescent material such as lithium fluoride, which has very similar characteristics to those used in conventional intensifying screens. When this material is bombarded by radiation, it ionizes and throws some of its electrons into a higher energy level in the outer valance band of the atom, where they become trapped. If these fluorescent materials are heated after their exposure to a certain point, the elevated electrons are released from the trapped band, lose their excessive energy, and return to their original orbit and energy level. In the process, their excess energy is converted into visible light. The quantity of light given off by this process can be calibrated on a meter to relate directly to the amount of radiation that was originally used.

The characteristics this particular type of personnel monitoring offers are listed below and are in addition to those already mentioned for the film badge.

1. TLDs, unlike dental film badge packets, are insensitive to humidity.
2. Lithium fluoride is very similar in absorption properties to human tissue and responds linearly with radiation biological effects.
3. The lithium material is reuseable, which reduces long-term cost.
4. Immediate read-outs are possible since the heating and reading devices are portable if put on a small metal mobile cart.
5. TLD readings are said to be approximately 95% accurate from exposure rates of 10 mr to 100,000 mr, as compared to 80 to 85% accuracy with the conventional film badge readings.

STRUCTURAL SHIELDING

All x-ray facilities have to meet rigid specifications, which are established by state and federal regulations. The primary monitor and enforcer of these regulations when dealing with radiation emitted from diagnostic and dental x-ray equipment is usually the state government. For radiation emitted by product materials—such as those with natural sources similar to cobalt, radium, and cesium, and those doses used in nuclear medicine—the facilities are enforced and monitored primarily by the federal government* with, in some cases, state assistance. Radiation produced by x-ray tubes, including linear accelerators, is regulated by the states.

Before installation of a diagnostic or therapeutic facility, the manufacturer of the proposed equipment typically provides an architectural drawing show-

---

*The federal government has contracts with some states requiring those states to provide monitoring and licensing of areas and materials that are usually considered to be federal government domain.

Radiation Detecting Devices

Detector	Types of Radiation Measured	Typical Full Scale Readings	Use	Minimum Energy Measured	Advantages	Possible Disadvantages
Scintillation counter	Beta, x, gamma	0.02 mR/h to 20 mR/h	Survey	20 keV for x rays. Variable for betas	1. High sensitivity 2. Rapid response	1. Fragile 2. Relatively expensive
Geiger-Muller counter	Beta, x, gamma	0.2 to 20 mR/h or 800 to 80,000 counts/min	Survey	20 keV for x rays. 150 keV for betas	Rapid response	1. Strong energy dependence 2. Possible paralysis of response at high count rates or exposure rates 3. Sensitive to microwave fields 4. May be affected by ultra violet light
Ionization chamber	Beta, x, gamma	3 mR/h to 500 R/h	Survey	20 keV for x rays. Variable for betas	Low energy dependence	1. Relatively low sensitivity 2. May be slow to respond
Pocket ionization chamber	X, gamma	200 mR to 200 R	Survey and monitoring	50 keV	1. Relatively inexpensive 2. Gives estimate of integrated dose 3. Small size	1. Subject to accidental discharge
Film	Beta, x, gamma	10 mR and up	Survey and monitoring	20 keV for x rays, 200 keV for betas	1. Inexpensive 2. Gives estimate of integrated dose 3. Provides permanent record	1. False readings produced by heat, certain vapors and pressure 2. Great variations with film type and batch 3. Strong energy dependence for low energy x rays

Figure 213. Summary of various radiation detection devices from NRCP handbook #48.

ing what type of wires are needed for proper operation of the equipment and how these wires must be placed through the walls and floors. At this point, the manufacturer's architectural drawings are then submitted to the radiation health physicist, who draws in his specifications for lead shielding and other related radiation barriers. The architectural plans with radiation safety specifications are submitted to the building contractor, who supplies the owner of the facility with total costs for construction. The construction company must also pay special attention to the various wiring schemes suggested by the x-ray manufacturers, so that they will meet local and state electrical safety regulations.

Fortunately, the regulations for structural shielding are relatively simple and straightforward, but they do vary dramatically between an x-ray facility, a radiation treatment facility, and nuclear medicine labs.

The *controlled area* is a section of the building that is controlled and monitored regularly by the radiation health physicist, as opposed to an *uncontrolled area*, which includes all other sections of the facility such as corridors and public waiting areas that are not involved in the radiation exposure. Two more terms necessary for discussion are primary and secondary barriers. *Primary barriers* for a diagnostic x-ray unit absorb primary x-rays produced by 150 KvP, which is typically accomplished by providing a 1/16 inch thick sheet of lead on all four walls of the exposure room, extending from the floor to a height of 7 feet. A *secondary barrier* absorbs secondary radiation and must be the equivalent of 1/32 inch lead thickness for a typical diagnostic room. The secondary barrier extends from a 1/2 inch overlap of the primary barrier up to the ceiling of the room. Floors in the exposure rooms require no specific shielding whether they are on the ground level without a basement beneath or whether they are on an upper floor. This is because the cement needed to support the weight of an x-ray machine is a minimum of 6 inches thick and this, of course, provides more than adequate absorption for the primary and secondary photons from a diagnostic x-ray tube operating up to 150 KvP.

There are some alternatives to this basic shielding plan that are frequently used; however, the greatest majority of installations are protected as discussed above. One common exception is that if an x-ray room is on an outside construction wall, there is more than adequate absorption provided by the materials used for the construction of an outside wall—such as concrete blocks in addition to whatever material is used for the outside surface such as bricks.

The control booth must also be protected with primary shielding and ideally arranged in such a way that any scatter radiation from the patient would have to bounce twice before entering the control area. In addition, all x-ray generators are now manufactured with the exposure switch located in a position far enough from the entry of the control booth so that a technologist cannot make an exposure by stretching his arm while standing outside the control booth entrance. The glass observation port built in control booths must be a primary barrier. As a general rule of thumb, the lead glass used as a primary barrier is about four times the thickness of lead to be equally effective against radiation.

It might be of some interest to know that there are various grades of concrete used for primary and secondary barriers for therapy treatment rooms. Federal regulations require that the stated thicknesses of the concrete are valid only when the concrete has a consistency to produce 140 pounds per cubic foot.

For all installations of high energy therapy equipment, regulations require that a closed circuit T.V. monitor be used in place of a glass window since the necessary one foot thick glass window provides very little practical visibility of the patient. In addition, there must be provisions and policies in the therapy department for having a spare monitor on hand or assurances to the regulatory agencies that no treatments are performed unless the T.V. system is in operation. There must also be adequate sound contact between the patient in the treatment room and the technologist at the control station.[5]

Converting A Nonradiographic Room

Many different methods are acceptable for providing primary shielding. To prepare an existing room that was not originally designed for x-ray purposes and was not shielded previously, there are a variety of decorative panelling materials or vinyl covered dry wall sections with 1/16 inch lead sheets bonded behind the decorative surface. For new construction sites, many contracting companies use cement blocks with bonded sheets of lead 1/16 inch thick. When the cement blocks are laid properly, the small individual sheets of lead on each block overlap slightly to provide complete protection.

All doors located along primary barriers must contain a minimum of 1/16 inch of lead. Any holes punched or drilled through the protective lead sheeting of a primary barrier must be plugged with special small lead plugs that are available.

Nuclear Medicine Department

Nuclear medicine laboratories do not require structural shielding because of the low energy levels involved, which includes the hot lab itself. However, the nuclear pharmaceuticals that are stored short-term in the hot lab must be placed behind special barriers, which usually take the form of lead bricks stacked on the work countertop. Regulations also require a venting system be installed in the hot lab so that any airborne radioactive particles can be drawn out of the room quickly, carried through a duct system safely, and vented out

---

[5]For cobalt installations 1/4 inch thick lead is acceptable shielding for areas that receive radiation that has been bounced twice from the primary source.

through the roof of the building, where they mix with the air, and the activity level becomes very diffuse. This circumstance would only occur during a radioactive spill. Sensitive monitoring devices such as the G.M. meter should be placed around the hot lab at critical points to continuously monitor the amount of radiation that is present in the room. In addition to this, radiation monitoring devices should be mounted close to each exit of the nuclear medicine area, which would signal any increased accumulation of radioactive material that might have spilled or otherwise accidentally be carried out by the technologist. The nuclear medicine department should also be vented to produce a "negative pressure" in the entire department. In other words, there should be a greater volume of air vented out than there is coming in. This creates an atmospheric pressure that is undetectable to the workers but helps evacuate any low level radioactive airborne particles from floating in the department and moves them out through the ventilation system.

Patients With Radioactive Implants

There are generally no barriers in rooms of patients who have implants such as radium or cesium. The most important barrier or precaution that can be taken under these circumstances is that the doorway leading to the patient's room be properly posted with appropriate warning signs, and that the nurses and other hospital personnel use the principle of time and distance to their advantage. All activity by workers around patients with implants should be planned before entering the room, so that the exposure to the working personnel is minimized. Equally important as time is that working personnel should keep as much distance from the patient as is practical for proper patient care. (The inverse square law can be used to your advantage.) It is also important to know that the energy of radiation used for implants makes the use of aprons and lead gloves practically worthless, and false securities should not be felt by wearing these items to replace time and distance.

CHARACTERISTICS OF A SAFE ENVIRONMENT FOR PATIENT AND PERSONNEL

Considering radiation protection and MPD specifications, in addition to the effects of background radiation, the amount of radiation given to a patient by the technologist must take on special meaning—since it is the technologist who is responsible primarily for the most significant amount of radiation that a person normally receives. One of the most significant causes of excess radiation exposure to the patients from medical x-rays comes in the form of performing repeat studies. Below are three primary reasons why x-ray studies are repeated:

1. incorrect exposure selection

2. incorrect positioning
3. technical mistakes

Technique Charts versus Patient Dose

When we think of excessive exposure dose we seldom if ever consider the amount of exposure unnecessarily given to the patient whose radiograph might be acceptable, and therefore not repeated for diagnostic purposes, but is simply overexposed, with as much as 30 to 40 percent more radiation than was actually needed to produce the radiograph. Modern x-ray film manufacturers provide us with wide latitude characteristics in their film; this may, in fact, produce a situation by which overexposure is even encouraged: since it is not necessary to calculate too carefully exposure factors, errors do not have an obvious effect on the film. With this in mind, technologists who know they are using an especially wide latitude film might be less precise in calculating their original exposure factors; this could obviously lead to overexposing the patient yet still produce a diagnostic image. If one mentally reviews the number of "dark films" that go through the department in an average busy day, knowing that at least a 30 percent increase in MaS (radiation) is needed to produce a perceptible increase in density, the resulting unnecessary exposure to the patient population over the course of a day is actually much higher than one would expect. If an examination is performed on a young patient that requires three or four abdomen films, each of which were exposed with at least 30 percent more than was necessary, and in addition there is poor shielding of the genital structures, the total amount of unnecessary exposure delivered to the gonads becomes very unsettling (especially when we consider the possible biological effects this may have on the patient's offspring or on subsequent generations). These are the very practical situations that weigh heavily in terms of the technologists' responsibility toward their patients relative to radiation safety. With some basic statistical research one can say that the unnecessary dose of radiation can be traced back to a lack of awareness and poor professional judgment on the part of the technologist.

An attitude held by many technologists is that the use of calibers and following an established technique chart is in some way childish or elementary, if not of inflated importance. It almost seems as though the use of technique charts is an attack on the technologists' pride and technical skill. Whether the basic reason is pride, an unjustified and an inflated perception of one's skill, or just plain laziness, the primary issue is the patient who, in the long run, receives the excessive dose of radiation. For this reason, many departments have resorted to phototiming systems, since the accuracy they produce with practice is greater than manual techniques or pure guess work by technologists.

If the reader learns the sensitivity and awareness of responsibility toward contributing to excessive patient exposure that is one of the primary goals of this chapter, the patients will receive the benefit in a very direct way. With more skill, discipline, and professionalism, repeats and unnecessary overexposure due to improper selection of technique factors, poor positioning, and technique mistakes (such as not taking necklaces off a patient before a chest examination or not double checking to see if a patient might have metal snaps) could be reduced dramatically.

The Role of the Radiologist and Supervising Technologist

The chief radiologist and supervising technologist have considerable responsibility as well for reducing patient dose to the minimum. They are ultimately responsible for establishing the "type" of radiographic images that are obtained in the department in terms of requiring a degree of sharpness, density, and contrast. As you have probably already learned, there is frequently variation between several radiologists regarding an optimal density for a given examination such as a chest. The author has seen chest films that are routinely considerably darker when compared to the same type of examination from another hospital, and it should not be surprising that the variations could cause a routine and unnecessary increase in exposure dose to patients.

There is also a notable variation in the degree of contrast a group of radiologists might like to see, and in many instances this involves the Kv level that has been chosen for a particular exam. The author has also seen unnecessarily high exposure doses delivered to patients for examinations simply because the KvP chosen to obtain a high contrast radiograph was low, causing greater doses than necessary because of the high MaS that were used to compensate. Figure 214 shows two radiographs, one produced with about 40 percent more radiation than was necessary for the patient and the examination. In practice, the radiologist must choose to have radiographs of suitable contrast and density so that the supervising technologist can transcribe his preferences into technique charts with the lowest possible dose to the patients.

The chief radiologist and supervising technologist are responsible for other aspects of patient dose as well. The remainder of these fall in the realm of quality control checks on processors, output of x-ray equipment, and enforcing proper shielding policies. The major responsibilities in a diagnostic department or office for these individuals are listed below:

1. The supervising technologist must be able to transcribe the radiologist's preferences into actual Kv and MaS values onto a technique chart, so that the general population of patients who are radiographed receive the lowest and most realistic dose possible for the examination. Some important factors regarding how this can be done are discussed in detail during Chapters Seven and Eight of this text.
2. The output of all the x-ray equipment must be monitored at least monthly,

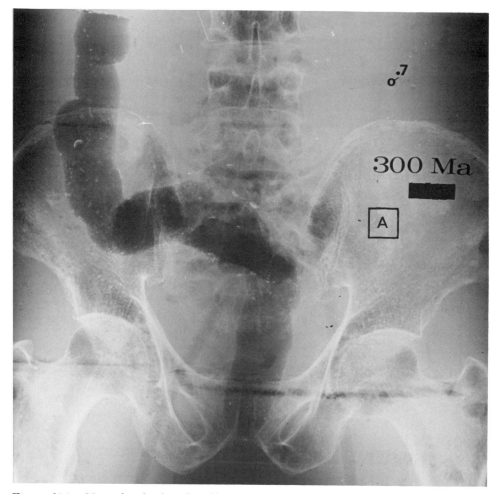

Figure 214.   Note that both pelvis films are "diagnostic," but *B* was produced with 40 percent more radiation using a wide latitude film which provided a diagnostic image, but with excessive exposure.

as prescribed in Chapter Five. Also, the timer and Kv should be calibrated monthly so that the timing is accurate and the Kv selected by the technologist at the control panel is actually producing the proper beam energy for a given exposure.

3. All processors should be monitored at least once a day and preferably twice a day. In addition, it is good practice to set up a policy whereby processors are checked when they have been on standby for a length of time that might affect temperature or other processing parameters. So that complicated examinations such as surgical cholangiograms or special procedure runs can be processed without the necessity or fear of repeats because of some simple function of the processor (such as temperature or

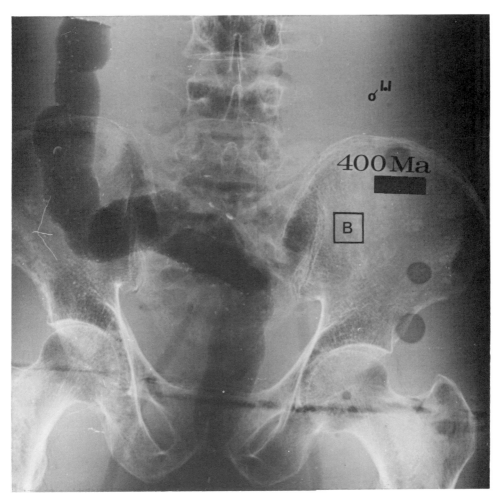

Figure 214B.

$$15 \times \frac{\text{KvP}^2 \times \text{MaS}}{\text{D}^2} = \text{Exposure In Milliroentgens}$$

Exposure A   75 KvP
50 MaS $15 \times \dfrac{5626 \times 50}{10,000}$ = 421.9 mr
100 cm (40″)

Exposure B   86 KvP
25 MaS $15 \times \dfrac{7396 \times 25}{10,000}$ = 277 mr
100 cm

Exposure C   97 KvP
12.5 MaS $15 \times \dfrac{9409 \times 12.5}{10,000}$ = 176 mr
100 cm

All of these exposures would produce similar radiographic densities but with much less exposure when higher KvP techniques are used.

Figure 215. These calculations illustrate the effect high KvP, low MaS has on patient dose compared to the low KvP, high MaS techniques. From NRCP handbook #48.

replenishment rates out of adjustment), it is good practice to run a sensitometric strip before an angio run is processed.

4. The supervising technologist is responsible for reporting any abnormalities or improper functioning to the proper repair companies so that corrective action can be taken promptly.

5. The supervising technologist must be sure that he has documentation that a licensed radiation health physicist has monitored all of the x-ray equipment, lead gloves, and lead aprons at least once a year. In addition, the supervising technologist must be able to document that corrective action was taken promptly after the health physicist's report was received. (The same type of checks as when the equipment was first installed and approved for use are made.)

6. That all of the diagnostic equipment is operating with the proper filtration must be documented.

7. The supervising technologist or the individual technologists themselves can check the films coming from the processor. For quality radiographs, the technologist must have competence, experience, and common sense to evaluate the relative diagnostic value of a radiograph versus its repeatability. Here there is no substitute for experience and general knowledge, since a careful evaluation must be made to determine the diagnostic value that might be gained by taking a repeat film and weighing that against both the additional exposure to the patient and time of the technologist. Radiography, like all medicine, is an art rather than a strict categorical science, and it is no more true at any phase of radiography than it is when checking films for technical quality. These decisions weigh heavily on the overall quality of images going through the department, as well as operating as an effective enforcement or encouragement for more careful and accurate initial selection of exposure factors, and controlled patient dose.

Patient Shielding

Adequate patient shielding must be performed on two separate fronts. The first is simply to reduce the field size so only the area of immediate interest will be involved in the exposure. Second is to provide effective gonadal shielding to both male and female patients, especially children. There are, of course, many schools of thought on how much and for which studies gonadal shielding should be applied. The following is an example of a reasonable and effective approach in establishing a gonadal shielding policy.

Our responsibilities in performing good quality diagnostic films should be second only to keeping the amount of radiation our patients receive to a minimum. With this in mind, we must take the initiative to provide suitable shielding, particularly when pelvic regions are radiographed.

It is difficult, if not impossible, to make categorical prerequisites for shielding because of the many variables involved; however, the statement below should be reviewed and given careful consideration by each technologist and included in daily practice.

Basic Guidelines for X-ray Shielding

1. There must be evidence of some type of gonadal shielding visible on all x-ray films for all male patients of any age who are having examinations such as IVP, abdomen, pelvis, hip, femur, except when special views are ordered for the bladder area.
2. When multiple films are made of female patients of any age with all the same position (such as the hip, abdomen, pelvis, or femur), one view is needed showing the entire area covered by the film; however, the remaining projections must show some evidence of shielding to protect the genital structures.
3. There must be evidence of some type of gonadal shielding on all patients who are having femur and hip studies performed.
4. There should be evidence of collimation visible on the film for all general diagnostic films as permitted by the intent of the examination and the body part radiographed.

The use of rare earth screens in radiography has provided an almost automatic ability to reduce the patient exposure by one-half. When these screens were first introduced to the marketplace the speed they offered was desirable; however, the amount of detail and general quality of the image they actually produced was unsatisfactory. Subsequent improvements in technology of manufacturing these screens have been able to correct their deficiencies in terms of quality; the images these fast screens produce are equal and in some cases better than those obtained with the conventional calcium tungstate crystals. Although these screens typically cost 50 to 100 percent more than the calcium tungstate crystals, with a careful and honest cost appraisal of the benefits they provide—relative to extended tube life, less repeats for motion, the fact that the film used with these screens is generally less expensive than those used for calcium tungstate, not to mention the effect of reducing the exposure to the population of patients by one-half—the initial cost can be justified. In many instances, the additional cost is often saved in two or three years.

Specific Criteria for Effective Gonadal Shielding

Gonadal shielding should be provided when the gonads lie within the primary x-ray field or within close proximity. Figure 216 shows a graph demonstrating the dose levels to various tissues positioned from the center ray to a point 10 cm outside the border of the x-ray field. With "shaped contact" shields placed over male gonads, the exposure to these parts can be reduced by about 95 percent. The reduction of exposure to the female ovaries can be as much as 50 percent if simple contact shielding is used. Shaped contact shielding cannot be used for female patients because the genital structures lie with the pelvic cavity.

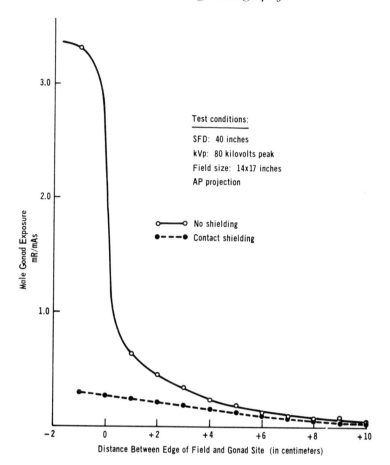

Figure 216.   This graph identifies the degree of dose fall-off when contact shielding is used when the gonads are located at various distances *from the edge of the primary field.*

Types of Gonadal Shielding Available

There are three classifications for gonadal shielding devices.

FLAT CONTACT SHIELDS. This type of contact is a piece of lead-impregnated rubber which is also used as a lead blocker when more than one exposure is made on a single cassette.

SHADOW SHIELDS. Shadow shield devices are typically attached to the collimator of the x-ray tube. They have an extending arm with a metal leaf at the end which can be positioned to block out the primary beam from the gonadal area as shown in Figure 217.

SHAPED CONTACT SHIELDS. These are shown in Figure 217 and are more effective than the two previously mentioned since they can be used easily for obtaining lateral and oblique projections as well as offering some degree of protection during fluoroscopic procedures. These devices, however, can

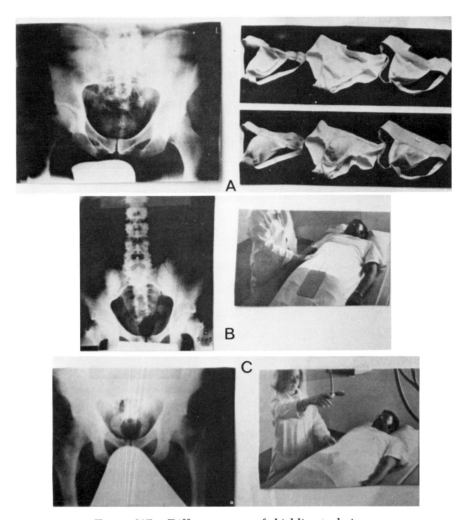

Figure 217.   Different types of shielding techniques.

become awkward for the patient to use and for the department to maintain proper aseptic considerations.

As pointed out earlier in this chapter and shown in Figure 216, shielding devices are effective for areas of the body that are simply within close proximity of the border of the x-ray field. X-ray photons interact with air molecules and produce additional scattered photons that can account for as much as 50 percent of the total gonadal dose. If genital structures are outside the primary x-ray field, they are also exposed by scatter photons resulting from interactions inside the body between the primary beam and body tissue. Enough of these scatter photons travel in all directions through the body to account for a significant exposure to the genital structures, especially in females, even if proper shielding were used. Actually all the statistics, graphs, standard deviations, etc., can be pondered and analyzed, but the essence of protection for the patient is little more than discipline in observing basic shielding policies. The items listed below can serve as a complete and quick review that the technologist, supervising technologist, or radiologist should observe to keep patient exposure dose to a minimum.

1. Total filtration for a general diagnostic should not be less than 2.5 mm of aluminum.
2. On all the equipment without automatic collimation, the diaphragm should be adjusted to the immediate area of diagnostic interest regardless of film size used.
3. Collimation measurements should be taken routinely to assure that the collimator light through which the tube is positioned is aligned with the actual field of primary x-ray beam.
4. Rare earth screens should be employed.
5. There should be reevaluation of image contrast and density characteristics and a tendency toward higher KvPs if at all possible.
6. Provide reliable exposure charts.
7. Monthly checks of the equipment should be performed and documented regarding Kv values, timer accuracy, and Ma output.
8. Processor should be checked at least once a day, with twice a day preferred.
9. Take prompt corrective action when a problem in any of the surveys mentioned above indicates a maladjustment.

Pregnant Patients

Patients who are known to be pregnant require very special care with regard to radiation dosage, especially during the first thirteen weeks (3 months). Radiation doses *to the fetus* of 5 rads or less have shown little or no evidence of leading toward fetal damage, but at approximately 15 rads some evidence of fetal damage can be noted. The whole body MPD for a pregnant woman for the full time of gestation through delivery is 0.5 rem.*

A single AP abdomen exposure using average technique values with a full 14 × 17 film in which the ovaries are part of the primary beam has been calculated to deliver to the ovaries approximately 0.22 rads, taking into account some tissue absorption of the structures immediately over the ovaries themselves. When many such exposures are made for an IVP study, the dose accumulates quickly. If other examinations such as upper and lower GI are done, the dose is ten to twenty times more. With this in mind, the necessity for shielding the reproductive organs in both female and male patients is clear. It is especially important to take reasonable precautions and not radiograph patients in the abdominal structures, especially those who might be pregnant, until a radiologist or the patient's physician is consulted. The kind of film-

---

*For occupational radiation workers.

screen combinations are very important, and the inclusion of a rare earth imaging system is strongly recommended when it is time to purchase new screens.

Female patients of childbearing age should be asked before abdominal films are taken whether there is a possibility the patient could be pregnant. If a reasonable possibility exists in the patient's mind, a radiologist or the patient's referring physician should be contacted. If a physician can not be contacted and it is an elective or nonurgent type of examination, it might be wise for the patient to be rescheduled after a physician can be contacted; most hospitals have emergency room physicians who can be consulted.

In general, there is a difference in philosophy as to which group is primarily responsible for checking patients with regard to their possibility of pregnancy. On one hand, there is a feeling the patient's referring physician—being the patient's primary medical advisor, having the patient's complete record at his disposal, and knowing the potential risk of radiation—should consider his patient accordingly before the order of an x-ray examination is written, especially if he feels the patient could be pregnant at the time. On the other hand, there is a feeling that because the radiologists and the technologists are specialists in their field and know more completely the relative danger of radiation versus the diagnostic value of the examination order by the referring physician, the radiology department should be primarily responsible for screening female patients of childbearing age regarding their possible pregnancy. Regardless of the controversy itself, it is certainly good practice for the technologist to ask a female patient of childbearing age if she could be pregnant before a series of radiographs are taken, especially if that area of the female body is involved in the primary radiation field or within an area of approximately 5 cm from its outer border.

Some hospitals have adopted the ten day rule, which basically means that elective studies are not normally done on female patients who are beyond ten days after menstruation. This rule helps prevent unnecessary exposure to a female patient who might have conceived but is not yet aware of the pregnancy. For non-elective, more urgent, and certainly emergency cases, the ten day rule is abandoned.

Assuring Safety for the Technologist

The various state and federal regulations established for monitoring primary and secondary barriers discussed previously go a long way to protect the technologist from excessive radiation exposure. Surveys regarding female technologists that have been conducted and then followed for many years by various survey groups have shown no evidence whatever of increases in any type of disease or biological effects typically related to chronic x-ray exposure, or any type of genetic problem such as malformation of offspring in subsequent generations. With these points in mind, however, there are still some day-to-day circumstances that the technologist is confronted with that, if not handled prudently, could be a source of excessive radiation. The following four situations are some of the most common that technologists come across in the normal work day.

IMMOBILIZING PATIENT. The various mobilization devices on the market often leave much to be desired under certain circumstances. Emergency and pediatric patients frequently present a very real problem in terms of patient motion, and there always seems to be the need for someone to physically hold the patient or at least stand by in some proximity to the patient during the exposure. It should be obvious that mechanical devices are to be used in all possible circumstances, in lieu of another person. An imaginative and sometimes voluminous amount of adhesive tape, Ace® bandages, and sheets can provide very effective patient restraints. If it is necessary to have a person help immobilize a child or another adult, it should if possible be a "nonradiation worker." Nurses, nurses aides, orderlies, and other nonradiation people should be summoned whenever possible; when they do arrive, make sure they are wearing lead aprons and gloves. Even if these "holders" are wearing lead gloves and aprons they should be carefully placed so that no part of their body is inside the area of the primary beam. See Figure 218.

Exposure factors must be established well before the positioning is complete. The author has seen many situations in which an uncooperative patient is being held while the technologist works frantically in the control panel to conjure up a last minute technique, which often produces an inadequate film diagnostically and causes reexposure of both the patient and the individual holding. Instead of relying on these last minute calculations, the patient should be measured and the exposure factors set on the generators before positioning actually begins. Also, the fastest time possible should be used to reduce to the need for a repeat exposure because of patient motion.

FLUOROSCOPIC EXAMINATIONS. Fluoroscopic examinations provide higher radiation doses to the patient, the technologist, and the radiologist than other common diagnostic examinations. Technologists' daily schedules should be arranged so that there is a regular rotation in and out of the fluoroscopic section so the dose to any one technologist is reduced as much as possible. Still, if proper procedure is followed, the amount of radiation received during these examinations will be well within accepted practical standards. To help assure this, the following points should be observed: (1) The technologist should stay outside of a six foot area of the x-ray beam emitted from the fluoroscope so that the only radiation received will be that of low intensity scatter from the patient. (2) Aprons with a minimum of 0.5 mm lead equivalent must be worn during the fluoroscopic session, and the technologist's hands should always be held either under the apron or behind the technologist's back. (3) If it is necessary at any time to hold or to help position a patient

Exposure Rates from Secondary Radiation to Persons Manually Restraining Patients During Radiographic Procedures

Basis of Measurements. Measurements made on adult trunk phantom. Eye location assumed to be 18 inches above table top, at edge of table. Gonad location taken at table top level, 6 inches from edge. Phantom center 12 inches from edge of table. X-ray unit: 3-phase overhead tube;[3] leakage radiation less than 1 percent of useful beam. Target-film distance: 40 inches. Film size 14 × 17 inches.[4]

Results.

Operating potential (kVp)	Exposure in milliroentgens per 100 milliampere-seconds[3]	
	Exposure at eye location	Exposure at gonad level under 0.5 mm lead apron
60	15	0.08
75	30	0.3
90	40	1.5
110	75	2.0

Use. To determine actual exposure, multiply values in table by appropriate exposure factors, such as those listed in footnote 3.

[3] Recommended conversion factors for other operating conditions

Single phase generator	0.5
Exposure at x mAs	x/100
8 x 10 inch film	0.7

[4] Unpublished data by Christopher Marshall and Gerald Shapiro.

Figure 218. With this chart we can calculate an exposure dose. From NRCP handbook #48.

Example A. Scatter exposure to gonads using 200 MaS at 75 KvP with single phase generator for a 14 × 17 field.

$$\begin{array}{r} 0.3 \text{ rems} \\ \times\ 200 \quad (200 \text{ MaS}) \\ \hline .60 \end{array}$$

$$\begin{array}{r} .60 \\ \times\ .5 \quad (14 \times 17) \\ \hline .3 \end{array} = \text{Mrem total dose to gonads of holder}$$

Example B. Scatter exposure to gonads using 50 MaS at 90 KvP with a three phase generator for an 8 × 10 field.

$$\begin{array}{r} 1.5 \text{ rems} \\ \times\ .5 \quad (50 \text{ MaS}) \\ \hline 0.75 \end{array}$$

$$\begin{array}{r} 0.75 \\ \times\ 0.7 \quad (8 \times 10) \\ \hline 0.52 \end{array} = \text{Mrem total dose to gonads of holder}$$

while the fluoroscope is on, it is mandatory that a technologist wear gloves with a minimum lead equivalent of 0.5 mm, and the hand should be kept as far away as possible from the primary radiation field.

THE CONTROL BOOTH. The control booth is, of course, a primary barrier and regulations require that the x-ray exposure switch must be positioned so that it is impossible for the technologist to initiate an exposure while standing outside of the barrier in the x-ray room itself.

PORTABLE WORK. After fluoroscopy, portable examinations perhaps produce the next greatest accumulative dose to the technologist. Once again lead aprons with a minimum lead equivalent of 0.5 mm should be worn during the exposure, and the technologist should not stand any closer than a six foot distance from the primary irradiated area of the patient. Regulations specify that the minimum target-to-patient distance should be not less than 12 inches, including portable image intensifiers.

For patients with radioactive implants such as cesium or radium, because of their high energies, lead aprons provide no protection. In fact, because of the interactions between lead and radiation, they cause scatter to the technologist. With this in mind, the most effective approach for a technologist is to plan his moves carefully before even entering the room. For example, a technologist should use his experience and expertise wisely in simply observing the patient's size and potential absorption of the primary beam. The technologist should also have calibers in hand and measure the patient as quickly as possible. The technologist should plug the portable x-ray machine in an electrical outlet in the nursing floor corridor and set the proper exposures. It is also helpful for the technologist to place the x-ray cassette in a fresh pillowcase. This at first may seem odd; however, a cassette placed directly against the skin of a patient often sticks to the skin, which makes small adjustments in positioning difficult or impossible and much unnecessary exposure is absorbed by the technologist while negotiating a cassette into its proper place behind the patient. With the cassette inside the pillowcase, it can be adjusted and positioned with relative ease, thus saving considerable time in close proximity to the radiation source. Before the technologist goes into the patient's room, it is a good idea to locate the outlets in the wall so the one furthest from the patient but nearest to the machine can be used. This will reduce the amount of time and exposure one would receive while hunting for an outlet, or unnecessary time in moving furniture. The technologist can even mentally move various pieces of furniture before entering so that the amount of time spent in close proximity to the radium sources can be kept to a minimum.

The film should not be brought into the room and set behind the patient until the portable machine is in place and has been turned on, the tube head positioned at the proper angle, and other things in general made ready for the exposure, since the x-ray film is vulnerable to the radiation coming from the patient's implant sources. If a film is placed too early during the set-up process, scatter from the implant sources will fog the entire film and the examination will be worthless. With regard to the exposure itself, the same exposure is required under these circumstances as would be necessary for a patient without radium sources, since the density on the film from the implants does not contribute to the "diagnostic" information. There is some fogging of the film no matter what precautions are taken; however, since it is caused mostly by scatter, the use of a grid would noticeably reduce the fogging. From Figure 219 we can determine that a technologist doing a portable examination on a patient with 100 mg radium and requiring 10 minutes with the patient at an average distance of 5 feet would have a total exposure of 2.0 millirems.

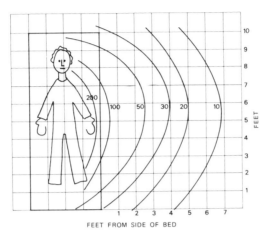

FEET FROM SIDE OF BED

Figure 219.   Approximate exposure rates emitted by a patient who has 100 mg of radium, or 300 mCi of gold, or 300 mCi of iodine. Note that although only one side is shown, the exposure pattern surrounds the patient. If the patient's bed is near a wall of beaverboard or other light material, the radiation penetrates the wall with little reduction, and the exposure pattern continues beyond the wall. From NRCP handbook #48.

FLUOROSCOPY. As mentioned earlier, fluoroscopic examinations contribute a greater amount of radiation exposure to the patient than other diagnostic studies. For this reason, special state and federal regulations, which are listed below, have been established.

1. The exposure rate measured at the fluoroscopy tabletop from the fluoroscopy tube must not at any time exceed 10 r/min. With the high efficiency image-intensifying systems now available, good fluoroscopy on average-sized people is possible with a tabletop dose rate as little as 5 r/min. It should not be necessary to use more than 3 milliamperes of current during a fluoroscopic session. Calculating the maximum exposure allowable, an average patient could receive an entering skin dose in a 5 minute fluoroscopy session of 50 r.

2. The minimum distance for the fluoroscopy tube to the tabletop is 15 inches, with 18 to 20 highly preferred.

3. Total filtration for fluoroscopy tubes should be not less than 2.5 mm of aluminum equivalent.

4. The lead aprons and gloves worn during a fluoroscopic session must be at least 0.5 mm of lead equivalent, and various pairs of lead gloves and individual aprons should be numbered for identification purposes and

checked by a radiation physicist (or a knowledgeable and responsible individual) twice a year.

5. The radiologist should use intermittent fluoroscopic technique during the fluoroscopic session; this type of "off-on" procedure effectively reduces fluoroscopic time and exposure dose to the patient.

6. The fluoroscopist should keep the fluoroscopic field and spot film field size controlled so that only the structures under immediate interest are exposed.

7. The fluoroscopic Kv used should be as high as possible without sacrificing the diagnostic quality of the image. A typical and acceptable fluoroscopic Kv level for upper and lower G.I. work should be not less than 100 KvP; for extremity work such as arthrography of the shoulder and the knee, the minimum kilovoltage used should be 80 KvP.

Using appropriately high KvP values results in an unfortunate trade between reduced patient dose and the image quality seen on the T.V. monitor. As KvP increases too much for a given body part, the T.V. image tends to become grainy and snowy because the internal circuitry of the imaging system compensates by reducing Ma. This reduction in Ma cuts back the number of remnant photons that strike the image intensifier and produces a type of quantum noise that we see on the T.V. monitor as snow or grain. High KvP fluoroscopic technique requires low Ma values which carry insufficient quantities of primary radiation through the patient to the image intensifying tube; the T.V. system responds by artificially increasing the brightness. However, with an insufficient *number* of photons striking the phosphorus coating of the intensifying T.V. image, contrast drops substantially. Although higher KvP techniques are better for patient dose, they cause a noticeable reduction in the quality of the fluoroscopic image.

8. Precautions must be taken to absorb lateral scatter from the patient during fluoroscopy. A lead-impregnated rubber apron is typically installed to hang between the space of the image intensifier and the tabletop, thereby absorbing lateral scatter from the patient during fluoroscopy.

9. The x-ray equipment must provide adequate protection from scatter radiation resulting from interactions between the primary fluoroscopic beam, the Bucky, and the tabletop, and escaping toward the radiologist and technologist through the open space under the table used for the Bucky travel. X-ray equipment manufacturers provide this protection, often by building a metal gate-type arrangement which folds down to the side of the patient when not in use. During the fluoroscopic session, this metal gate is raised upward in a vertical position and suitably absorbs the radiation coming from between the tabletop and the Bucky.

Who Should Wear Radiation Monitoring Devices

Frequently, circumstances develop that raise the question of whom should be monitored. Radiology employed escorts, physicians, nurses, and anesthesiologists assigned to cases with x-ray control are often concerned and want to know how much radiation they are actually receiving and if they should be wearing some type of monitoring device regularly. The standards established by the NCRP (National Council on Radiation Protection) say that "anyone for whom there is a possibility of exposure of more than 1/4 MPD for a radiation worker must be classified as a *radiation worker* and, therefore, regularly wear a radiation monitoring device as a radiation worker. A simple way to verify the dose that such individuals receive is to have an individual or group in question wear a film badge or dosimeter under supervision of a radiation safety officer for a few months until a sample reading that is representive of absorbed dose is obtained. If it is 1/4 MPD or more, the radiation worker's monitoring must be a permanent situation of those individuals. In other words a radiation worker is an individual for whom there is a possibility of receiving 1.25 rem per year or 0.75 rem in a thirteen week period.

The Radiation Health Officer

Radiation safety officers can be employed on either a full-time or part-time basis, depending on the size of the facility. If the facility employs a part-time radiation officer he should maintain an "on call" agreement with the facility to handle or direct any urgent situations involving radiation safety. Below is a brief listing of the primary duties of the Radiation Safety (Health) Officer or Radiation Protection Supervisor. In general, the overall responsibility is to evaluate radiation areas and procedures to assure all radiation work is being performed without undue exposure to anyone.

1. Check protective barriers for the effectiveness of primary and secondary barriers.
2. Advise which areas are designated as controlled areas and noncontrolled areas.
3. Advise as to which people are designated as radiation workers.
4. Be responsible for monitoring the radiation history records of all people classified as radiation workers.
5. Supervise the procedures for wearing monitoring devices by individuals.
6. Make sure the dose to general public does not exceed the maximum allowed dose for people in noncontrolled areas.
7. Inspect all ionizing radiation equipment and sources according to regulations established by the NCRP.

After any x-ray equipment has been completely installed and is ready for normal use, state regulations commonly require that the machine be checked by a licensed radiation health physicist for the items listed below:

1. Calibration of KvP
2. Calibration of Ma
3. Calibration of timing devices
4. Proper filtration

5. Radiation leakage from collimator
6. Alignment of light field from the collimator
7. Proper field size produced by collimator shutters
8. F.F.D. markers correctly located

Other checks are often for specific types of equipment such as tomographic and fluoroscopic units.

MINIMUM PERMANENT FILTRATION IN PRIMARY X-RAY BEAMS

Operating KvP	*Minimum Total Filtration*
Below 50 KvP	0.5 mm Al
50-70 KvP	1.5 mm Al
Above 70 KvP	2.5 mm Al

WARNING SIGNS INDICATING RADIATION HAZARDS

Various state and federal regulations require specific warning signs, tags, and labels to be posted on patient's charts, in control areas, and outside of the patient's rooms. Figures 220 through 223 show samples of these various posting signs and their specific applications.

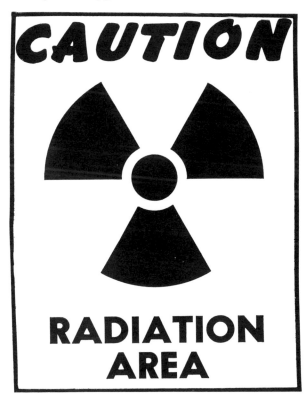

Figure 220.   Red on yellow label designating an area in which there is the same degree of radioactivity.

Figure 222.   Red on yellow label attached to the front of the patient's chart; removed when source is taken out.

Figure 221. This multicolor tag is attached to containers ready for shipping. The activity level of its contents is signified by the number of red bars shown, ranging from 1 to 111 with 111 being high activity.

_____HOSPITAL

PATIENT'S NAME_____ UNIT NUMBER_____

CAUTION
RADIOACTIVE MATERIAL

TEMPORARY IMPLANT

Radionuclide_____mCi_____

Inserted_____
                (DATE)
Initial Exposure Rate at 1 Meter_____ mR/h

(SIGNATURE)_____

To Be Removed_____
                    (DATE)

INSTRUCTIONS:
  Patient must remain in hospital until implant is removed.
  When implant is removed, "Radioactivity Precautions
  Tags" may also be removed.
  For further information call Radiation Protection Office
  (Ext_____)
  In case of emergency, the telephone operator has a call
  list for use when the Radiation Protection Office is
  not open.

Date_____ Signature_____
                    RADIATION PROTECTION SUPERVISOR

Label for a patient's chart: temporary implant.

_____HOSPITAL

PATIENT'S NAME_____ UNIT NUMBER_____

CAUTION

PATIENT CONTAINS RADIOACTIVE MATERIAL

DO NOT REMOVE THIS LABEL UNTIL:
1) Radioactive material is removed from patient, or
2) Removal is authorized by Radiation
   Protection Supervisor (Ext_____).
VISITORS MUST CHECK WITH NURSING STATION
        BEFORE GOING TO PATIENT.

Date_____ Signature_____
                    RADIATION PROTECTION SUPERVISOR

Institutional storage region and container labels.

Figure 223.   Red on yellow label designating a storage area or container for radioactive materials.

CHAPTER FOURTEEN

RADIOGRAPHIC TUBES

WHEN ROENTGEN FIRST discovered x-radiation in 1895, he was actually experimenting with high voltage conductivity through a vacuum. The electric tube or diode he had been working with produced a new form of energy, a serendipitous discovery. Later this form of energy was identified as being a member of the electromagnetic spectrum and was called x-rays, x representing the unknown.

The tube that was used by Roentgen was a cold cathode tube with a partial vacuum inside the glass envelope. In fact, Roentgen depended on the air molecules remaining inside the envelope to produce x-rays.

When a current was passed through this tube from cathode to anode, the air became ionized. The ions were driven toward the anode at a very high rate of speed. When the electrons struck the glass envelope, x-rays were produced. The energy of the x-rays produced by Roentgen was barely sufficient to pass through 7 centimeters of tissue.

The partial vacuum tubes used by Roentgen had many limitations in medical diagnosis, not the least of which was inconsistency of output. It was also very difficult to control the amount of air remaining in those early tubes after some period of use, because the rate at which outside air seeped back into the glass envelope varied. This lack of consistency of air supply in the tube was certainly important, because the amount of radiation produced depended heavily on the amount of air there was inside the glass envelope to become ionized.

The tubes used by Roentgen were known as cold cathode tubes. The supply of electrons in those tubes that were to take on kinetic energy was obtained from the air molecules inside the glass envelope. A positive and negative charge was applied across the tube but a filament was not used; thus, these early tubes that operated with a partial vacuum were also called cold cathode tubes.

About the year 1910, the American physicist William D. Coolidge developed a glass tube that could be evacuated to a very high percentage. So effective was this new vacuum process that it became possible to heat the filament substantially without it burning. The Coolidge tube was so named in recognition of its discoverer; it took its place as a major breakthrough in technology and become the immediate forerunner of the modern x-ray tube.

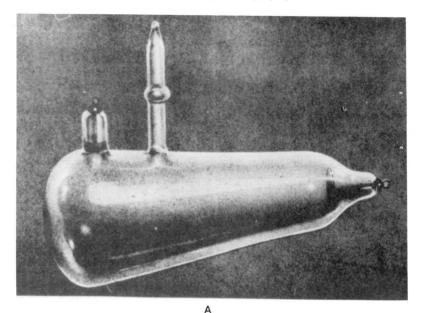

A

Coolidge

B

Figure 224.   *A.* The Crooks tube shown here is one of many types. *B.* The Coolidge tube shown is undamaged and still useable.

The significant breakthrough this tube offered was that the amount of tube current could be controlled precisely by regulating the temperature of the filament through incandescence rather than being dependent on a random quantity of air inside the glass envelope, as was the case with Roentgen's tube.

By the late 1920s and early 1930s the exposure values were very much limited, with maximum heat capacity at about 75,000 heat units. The resulting x-ray beam produced by the Coolidge tube was a significantly harder x-ray beam as well. The primary reason is that the tube current was striking a hard tungsten metal causing a much greater deceleration of electrons compared to what occurred earlier in the Crooks tube. Also by this time more advanced methods were in use that accelerated the tube current speed to the anode at a much higher rate.

THE MODERN X-RAY TUBE

Modern x-ray tubes are capable of operating at 150 KvP and 1,500 Ma, values even Roentgen might not have foreseen. Special x-ray tubes are now manufactured that can store over one million heat units when compared to a maximum of 250,000 heat units only ten years ago, and general diagnostic exposures of 400 Ma, 120 KvP, at 1.5 seconds are commonplace today. Equally important are the advances in reducing a focal spot size to a point where a 1.2 mm focal spot is considered large by today's standards, with the typical small focal spot at approximately 0.6 mm. Special purpose x-ray tubes, which will be given more attention later, make possible Ceni radiography for heart catheterization and serial filming for other major vascular work, to keep pace with the modern medical treatments.

The main problem of controlling heat that Roentgen and succeeding scientists faced is the foremost problem of the x-ray tube manufacturers today. Today's exposure values generate temperatures at the anode as high as 1,200°C and occasionally even higher. With today's exposure the *average* temperature imposed on the anode is approximately 800°C. A great deal of research and development has gone into the manufacturing of today's x-ray tubes to produce an anode that can endure this type of heat on a routine basis and at the same time continue to produce high quality radiographs that are obtainable over a normal life expectancy of two to four years. Yet despite the modern day developments, today's x-ray tubes still remain relatively inefficient devices. The emphasis placed on heat generation may serve as a clue to its low efficiency operation, since most of the kinetic energy carried by the individual electrons of the space charge are converted to heat rather than x-rays. In fact, generally less than 0.2 percent of the total interactions taking place at the target emerge from the tube in the form of usable x-radiation.

Components of the X-ray Tube

A quick look at Figure 225 will help to identify each of the major components of the x-ray tube and will provide some basis for understanding how they work in conjunction with each other.

Our discussion with this section will begin with the glass envelope itself. The basic purpose of the glass envelope is to provide an environment by which the cathode (containing a hot filament) and anode can operate in an almost perfect vacuum condition. What seems, at first glance, to be a rather unremarkable component is very misleading. For example, it is important that the glass envelope be particularly resistant to heat since the heat generated at the anode must be transmitted via convection to the insulating oil

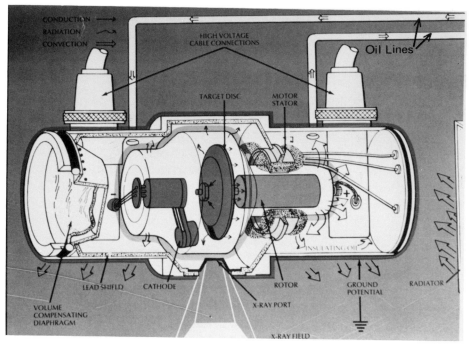

Figure 225. A diagram of a typical radiographic tube. Courtesy of The Machlette Corp.

outside and surrounding the glass envelope. The temperature of the anode can get as high as 1,200°C; such a temperature could melt a piece of common steel material almost instantaneously.

Because of the almost perfect vacuum inside the glass envelope, there exists a tremendous negative pressure inside, which exerts a great deal of atmospheric force on the glass. In fact, this pressure is approximately 10^{-6} torr (millimeters of mercury).

The glass envelope is given its final shaping by skilled master glass blowers. The glass envelope is actually made of a Pyrex® material and is approximately 0.10 inch thick overall; it is somewhat thinner where it forms the window through which the primary beam passes. This helps reduce the amount of x-ray absorption by the glass. The glass window provides approximately 0.5 mm of aluminum filtration inherent.

After the glass envelope is formed by the glass blower, the other components are assembled and the air is then evacuated slowly over a period of many hours at gradually increasing pressures with special vacuum air pumps. Also, the various components in the x-ray tube are heated (baked) to high temperatures and allowed to cool, only to be reheated and cooled several times over. This process forces any gas (air) molecules present within the metal parts, i.e. filament anode and focusing cup, to the surface and out of the tube entirely by the pumps. The baking or degassing process as it is called continues for several days until evacuation of air molecules is complete. At this point, the vacuum in the x-ray tube must be so complete that special precautions are taken during the later assembly stages because if a tiny lint particle the size of a grain of sand was left inside it could supply sufficient numbers of air molecules to destroy the integrity of the vacuum and make the tube defective.

The Cathode Assembly

The cathode encompasses the focusing cup and the filament, and has a negative electrical charge.

The focusing cup serves both as a supporting device for the filament and an aiming mechanism for the space charge. It is important to control the size and the direction of the space charge as much as possible so the electrons released by the filaments, through a process known as thermionic emission, will strike the anode at a precise spot. Thermionic emission with respect to the filament of a x-ray tube is when the filament material is heated so the centrifugal energy of the circling electrons is raised to a point where they cannot be contained by the usual magnetic pull of the nucleus. At some critical point as the filament is heated, the electron's centrifugal force goes beyond the pulling strength of the nucleus and is released from the material. In this way many electrons are released from the tungsten filament and accumulate in a cloud formation in front of the filament. This cloud or ball of suspended electrons is known as the space charge.

The nature of electrons at this point is to move away from each other, since they are of similar electric charge. If this, however, were allowed to occur the results would be very chaotic and undesirable: when the high voltage was applied across the tube (when the exposure is made), the electrons would be in disarray and would bombard all parts of the anode without any concentrated effect (Fig. 226). In this event, x-rays would be produced across the entire surface of the anode in addition to the target area and the resulting sharpness would be totally unacceptable, since the focal spot would be the entire surface of the anode. The x-ray output through the collimator would be greatly reduced also because the x-ray beam would not be concentrated.

The importance of confining the space charge is extremely vital and is accomplished by the focusing cup. Figure 227 illustrates how this occurs. The placement of the filament within the focusing cup is extremely important during the manufacturing stage. If the filament was set further inside the cup, the space charge would be confined to a smaller spot on the anode during the

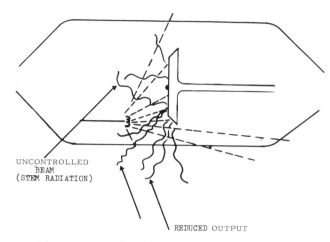

Figure 226.   The effect of a nonfocused tube.

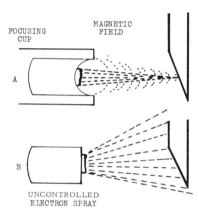

Figure 227.   We can see here how the tube current is aimed at the target to form the focal spot.

exposure than if the filament were placed further out of the confines of the cup. Thus, when the filament is placed properly only the intended target area will be bombarded by the bolus of electrons, namely that area known as the focal spot or target of the anode.

Most diagnostic tubes contain two filament sizes, which in turn produce large and small focal spot sizes at the anode. The filaments are actually a continuation of the filament circuit and held in place by their tungsten wires.

As we will see later, the diameter of the filament is extremely important because as the diameter changes with use it causes greater resistance to the filament circuit with time. As the filament diameter decreases, it produces more resistance to the filament current and its temperature, and thus the number of electrons it produces to form the space charge increases. This, of course, affects directly the amount of x-rays that will be produced during the exposure. Through the process of thermionic emission its diameter reduces or evaporates in size over many exposures and causes important changes in x-ray output. During each exposure of the normal life cycles, the filament evaporates as its electrons are continually emitted. This evaporation continues and, barring any other tube problems, would result in the filament becoming so thin that it simply breaks, rendering the x-ray useless. In fact, it is generally known that one of the primary reasons for tube failure is the filament breaking as a result of this evaporation process. If the diameter of the filament is reduced by only 10 percent, the filament will in all likelihood not be able to withstand the temperature and will break, since the filament is only from 0.005 to 0.010 inch when new.

Filament Current and the Resulting Milliamperage

The filament of a common light bulb is heated by current supplied to a point where it maintains incandescence. The light you see emitted by the filament of the x-ray tube is an electromagnetic energy caused by the energy of the accelerated electrons ionizing as a result of the filament heating process.

In an x-ray tube, the heat of the filament is produced by a rather low current of approximately 3 to 5 amps at 10 to 12 volts. This current originates in the primary x-ray circuit. The filament circuit can be seen in Figure 228; how the filament circuit actually works in conjunction with the other components of the main x-ray circuit can be seen. As simple as the filament may appear, it is extremely important to x-ray production because it controls and

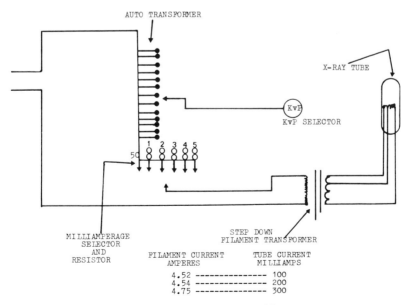

Figure 228. Basic diagrams of the filament circuit.

regulates the amount of tube current (milliamperage) that will bombard the target. The contact points at the primary side of the filament circuit are actually the point at which the technologist selects the appropriate Ma stations for the exposure when adjusting the Ma control knob.

As explained earlier (in Chapter Five) and shown here in Figure 228, the Ma stations are actually an array of points that produce different values of electrical resistance. If the technologist makes an exposure at 400 Ma, and for the next exposure selects a 200 Ma station, the resistance to the filament circuit will increase, allowing less current to flow. Thus the 200 Ma station would cause more resistance in the filament circuit, allowing less current to travel to the filament. With less current to the filament, its temperature will be reduced to a point where the electron emission and space charge will be equal to 200 Ma for the second exposure. Only slight changes in the filament circuit produce disproportionate effects on the temperature of the filament and ultimately the amount of tube current produced. With this in mind, one can surmise that when the technologist selects Ma stations from 400 to 200 Ma the resistance in the filament circuit would not be reduced by one-half. Figure 228 will demonstrate the actual filament values versus the relative tube current produced for a given x-ray unit currently in use.

The Booster Circuit

An addition to the circuitry so far discussed is the booster circuit, which serves to prolong the life of the x-ray tube. Since the heat required to produce a suitable space charge is quite substantial, especially for today's high Ma exposures, a rapid evaporation process of the x-ray filament could occur if the filament remained at its maximum temperature while the x-ray machine was turned on but not actually being used for making x-ray exposures. If this temperature was maintained, the filament would evaporate to a breaking point in a very short period of time. To guard against this happening, the filament operates in two stages. When the x-ray machine is in a so-called dormant state between examinations, the filament is maintained at a very low temperature. However, when the technologist is about to make an exposure the rotor or ready button is pressed. This serves two purposes: first, the anode begins to turn at the proper speed for an exposure, and second, the booster circuit raises the temperature of the filament to the appropriate value so that the correct amount of space charge can be produced. In this fashion, the filament temperature is raised to its optimal point only for a brief amount of time, thus holding evaporation to a minimum. There is a slight delay just after the rotor or ready button is pressed and before the exposure is possible. This brief amount of time is usually not greater than one or two seconds and allows sufficient time for the booster circuit to take over and raise the filament temperature and for the anode to reach its optimal spinning speed.

THE ANODE

As was mentioned earlier in this chapter, the amount of heat generated is a primary problem in tube life, and this is true of the anode especially. The heat generated in an average diagnostic tube can be as much as 1,200°C with normal operating temperatures ranging from 700°C to 1,000°C. The anode is the portion of the x-ray tube that has a positive electrical charge. The major components of the anode are seen in Figure 225. They include the anode disc (which contains the focal track) and the shaft of the anode (which connects the anode disc to the stator and induction motor). It is necessary, for reasons that will be discussed later, that the anode disc spin at a relatively high rate of speed. In fact, the anode rotates at approximately 3,300 revolutions per minute under normal circumstances. The shaft is hollow to minimize the amount of heat that can be carried from the disc to the bearing in the rotor. This is important because overheating the rotor bearing will ultimately increase friction and reduce the amount of revolutions per minute the anode will turn during the exposure. This, of course, would prevent the heat generated by the exposure from being evenly distributed along the focal track and this would lead to overheating and significant pitting of the anode. Figure 229 shows the difference between the total area bombarded on the rotating anode compared to the stationary anode. Actually, the rotor part of the x-ray tube is analogous to a simple electric motor. The rotor relates to the armature and the induction coils, which are located outside the glass envelope surrounding the area of the rotor. When electric current is applied to the induction coils (as the technologist presses the rotor button) the magnetic field generated by the coils begins to turn the stator. The speed at which the rotor turns is dependent

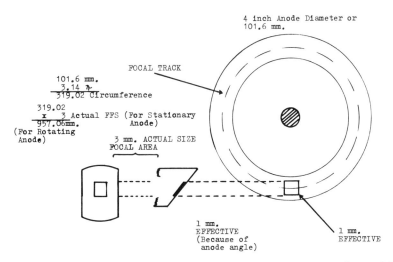

Figure 229. A stationary and rotating anode showing comparison of *actual* focal spot sizes. Note the great difference in actual focal spot struck by the space charge.

to a large extent on the electrical phasing of the induction coils; if calculated, the normal speed of the anode would be about 3,300 revolutions per minute. It would be a mistake, however, to think this speed is absolute. Fluctuations throughout the electrical circuit and wear of the rotor bearings cause slight variations in anode speed.

The Anode Disc

The anode disc must serve three specific purposes. First, it must provide immediate deceleration of the electrons of the tube current. Second, the anode disc must conduct current through to the shaft, the rotor, and eventually through connecting wires to the rectifier and other parts of the secondary circuit to complete a closed circuit. Third, the anode via convection must emit tremendous volumes of heat through the vacuum of the x-ray tube to the oil surrounding the glass insert. The heat is then, via conduction, transferred from the oil to the metal housing and ultimately radiates to the room itself.

The primary material used in making the anode disc is tungsten. In recent years, a material known as rhenium has been used to coat the focal track of the anode as well as the entire disc. Rhenium serves to improve greatly the efficiency of heat dissipation to the oil, and thus provides greater heat-loading capabilities to the tube. One of the most striking advances in tube loading is the mega heat unit tubes currently used in CT scanning machines. Tungsten is a primary material because of its resistance to heat. The melting point of tungsten is very high (3,370°C), but extreme heat still imposes limits on exposures. Tungsten also has a high atomic number, which serves to effectively stop or decelerate electrons, causing the dramatic energy conversion process from kinetic energy to x-ray photons.

The size of the anode varies somewhat between manufacturers and for the tube's intended function. As tubes are required to perform more demanding functions, such as in special procedures work, manufacturers have resorted to using anodes with greater disc diameters. Typical anode discs for general diagnostic tubes are approximately three to four inches in diameter. Greater anode diameters are available, up to five inches. These larger 5 inch anode tubes provide two benefits: (1) the focal track substantially increases in circumference as the diameter increases, which allows the tube to better distribute the heat from any given exposure through the anode, especially when very short exposures are used; (2) the larger overall size of the anode can physically handle higher heat loading produced by the exposures required for special purposes such as arteriogram examinations. In general, the larger discs allow the technologist to use overall greater exposure values and quicker series of exposures (i.e. for angiography work), and make the use of fractional focal spot sizes for such examinations more practical.

The Target Area (Focal Spot)

Considering that the rotating anode disc has no defined area from where x-rays can be emitted but rather a track that circumscribes the face of the disc, one is reminded of the importance of the focusing cup both in establishing focal spot size and in its effect on radiographic sharpness.

The electron cloud conforms in size and strikes the anode disc at precisely the right location. Figure 225 shows that the anode is trimmed or beveled. The degree of angle of the beveled portion of the disc for routine purposes is usually between 17 to 20 degrees, depending upon the manufacturer. This angle provides a variance between the actual focal spot size (the area bombarded by the electrons during the exposure) and the effective focal spot size (the size of the beam as it leaves the target toward the x-ray film).

The relationship between the actual focal spot size and the effective focal spot size is described as the line focus principle and is illustrated in Figure 230. In short, the line focus principle acts to the technologist's favor by providing a relatively large area of the anode that is struck by the electrons, but because of the angle of the anode, the beam is projected as a smaller focal spot size toward the film; this has a very positive effect on the geometry of the beam producing greater sharpness. The concept of the line focus principle is discussed in detail in Chapter Six.

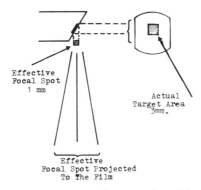

Effective
Focal Spot
1 mm

Actual
Target Area
3mm.

Effective
Focal Spot Projected
To The Film

Figure 230. The line focus principle is produced by the angled anode.

The angle at which the bevel is made on the anode disc in Figure 231 can range from 10 degrees for special purpose tubes to approximately 17 degrees for portable and general diagnostic tubes. In general, as the angle of the anode decreases toward 10 degrees, the tube can withstand greater heat loads than the same type of tube with an anode at 17 degrees, because the actual focal spot can afford to be a little larger without affecting the effective focal spot size (because of the line focus principle). Figure 231 also shows how decreased anode angles, such as 10 degrees, can improve the line focus principle significantly. The drawback to a low anode angle is that the anode heel effect is increased substantially so that the field covered by the radiation decreases to the point where the entire surface of a 14 × 17 inch film is not exposed at a distance of 40 inches F.F.D. Thus, as the angle decreases and tube load increases the coverage area is limited. If it were not for the limited coverage ability of the low angled anodes, they would be used more extensively in general diagnostic tubes.

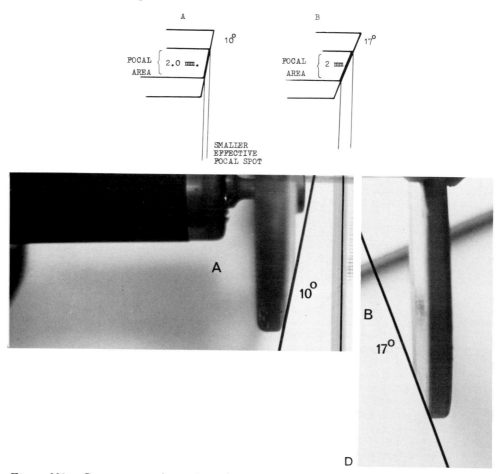

Figure 231. Comparison of actual anodes with different angles. Notice how the angle of an anode can affect the effective focal spot size. You will note that when using a 10 degree angle one can afford to have a large focal area producing the same effective focal spot. This would serve to distribute the heat over a wider area thereby increasing tube rating.

Anode Surface Heat

Steel normally used in construction work of buildings melts at a point of 1,000°C; in contrast, the possible operating anode temperature is between 700°C and 1,200°C. Figure 232 illustrates the effects of heat on the anode. The anodes shown here have been involved in various circumstances where this intense heat caused immediate damage to the entire anode disc.

Figure 232. *A*, The result of a heavy exposure on a cold anode. *B*, This anode shows minimal wear; note the smooth focal track. *C*, Anode with melted spots. *D*, Moderately used anode. *E*, A new anode.

High Speed Anodes

The heat generated along the focal track must be dispersed throughout the focal track as evenly as possible. For example, if an anode is rotating at a standard speed of approximately 3,300 revolutions per minute, an exposure taken at 1/20 second is long enough to allow for 2.75 revolutions; under these circumstances, heat would be adequately spread throughout the entire focal track and ultimately the entire anode (8.25 revolutions per minute with a high speed anode).

Many modern tubes are available with the flexibility of two anode speeds in the same tube. When the technologist selects a high intensity "quick timed" exposure, additional circuitry automatically takes over and increases the voltage of the induction wires, which drives the anode at a considerably faster speed of about 10,000 revolutions per minute. When more normal, less stressful heat loads are used for diagnostic exposures, the normal current runs through the induction wires and the anode turns at the conventional 3,300 revolutions per minute. These dual speed tubes, if properly used, increase the longevity considerably.

Before reviewing additional applications for high speed anodes, it must be pointed out that high speed anodes have one important disadvantage. Friction generates heat; even with the highly sophisticated metal alloys and lubricants used in tube bearing assemblies, the faster the anode turns, the more friction and heat is produced on the bearings. The anode bearings are similar in operation and somewhat analogous to the bearings in the wheel of an automobile. As long as the wheel bearings are lubricated and are in good condition, the wheel will turn freely along the highway. However, without proper lubrication or if foreign materials erode the smooth surface of the ball bearing, friction increases and it becomes more difficult for the wheel to turn freely. When the tube operates at a high rotating speed, there is additional heat generated, thus wear on the bearing increases and there is a great potential for reduced life expectancy. In brief, there is usually considerably more wear on the bearings and thus a shorter life span than what would normally be experienced with a standard speed rotation.

Complicating the problem of bearing wear is that conventional grease-type lubricants cannot be used, since air molecules trapped in the grease would escape into the vacuum of the x-ray tube and cause a gasy tube, rendering it worthless. Because of this problem, the composition of the bearings must be engineered and manufactured in such a way to provide smooth rotating motion of the anode without the aid of grease-type lubricants.

Emphasis is placed later in this chapter on operating the tube in such a way as to provide the longest possible life span. The greatest reason for this is related to the cost of new tubes, in addition to the labor cost for replacement. An average general radiographic diagnostic tube will cost several thousand dollars, not including labor for installation. This alone suggests that much care should be given in selecting the proper tube for the examination or the type of patient load that is expected. It is far less expensive initially, both in base cost and in down time, to purchase a moderately more expensive tube that will provide more reliable and consistent service over the years.

The anode must turn at a fairly consistent rate of speed. If the anode were to slow down during the exposure the portion of the anode directly bombarded by the electron cloud would experience increased heat. If this continued, irreversible damage to the anode would result and the tube would become unuseable. Also, the anode disc must rotate without any type of wobble. If the anode disc wobbled during the exposure, radiographic unsharpness would certainly result, since it would have the effect of projecting a larger focal spot toward the film. Nearly all anodes experience some degree of wobble, even though manufacturers take every conceivable precaution to design the tube to spin evenly. The effect of noticeable unsharpness on the film due to wobble is, in part, a function of the focal spot size. The smaller the focal spot size, the more critical the anode wobble becomes. As the anode wobbles, it produces a quivering pulsing kind of motion, which in effect serves to produce a kind of pseudo or ghost target area that is larger than the actual focal spot size. Figure 233 shows schematically how wobble could effect radiographic sharpness. Once excessive wobble becomes evident, a sequence of events usually develops with irrevocable deterioration to the tube. Even a slight amount beyond normal tolerances, combined with the centrifugal force generated by 3,300 rpm, tears at the stability of the bearing and would damage the tube in a relatively short period of time.

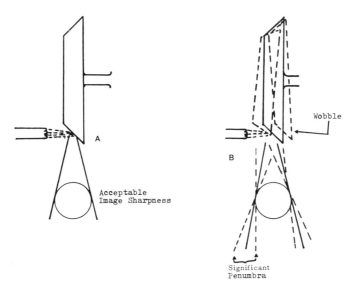

Figure 233.   Demonstrates the importance of a smooth turning anode.

The additional friction generated during this process can be noticed by a shortened amount of coast time of the anode. Coast time refers to the amount of time it takes the anode to lose its inertia and come to rest after the technologist releases the exposure button. As the anode spins slower and slower, the heat on the focal track with each exposure builds greatly; Figure 234 shows a very common result. Here we can see an anode with melted spots where the pulses of tube current struck the anode while turning slowly.

One might be suspicious of a wobbly anode when noticing (1) consistent unsharpness (when all other factors such as screen contact and Bucky motion have been satisfactorily ruled out); (2) increasing noise or vibration coming from the tube when the anode is turning at its top speed; and (3) short coast time of an anode. Many tubes have a mechanism that works as a braking system to stop the anode within 20 to 30 seconds of the exposure. Tubes without this mechanism will spin for 15 to 20 minutes if the bearings are in good condition.

Figure 234.   The melted spots shown were caused by a slow turning anode.

Size of the Focal Spot

The size of the focal spot is primarily dependent on the effectiveness of the focusing cup (in aiming the tube current) and the angle of the anode.

Recent advances have brought us to the development and use of microfocus tubes. With this tube, a 45 degree hollow stationary anode is used. The anode is hollowed out so that oil can be circulated through just behind the target area to obtain very efficient target cooling. The effective focal spot is in the area of 0.1 mm and yields superb geometric resolution.

Even with the hollowed anode target, heat is still a major problem: maximum KvP and Ma used are impractical for routine work of even small sized patients. The primary role might be in mammography and in magnification work of extremities.

Measuring the Effective Focal Size

We have discussed focal spot size frequently throughout this chapter and will now present some information on how focal spots are actually measured. When focal spots are quoted in literature and are spoken of in general, it is with reference to the effective focal spot size rather than the actual focal spot size. The focal spot sizes vary noticeably and are not always exactly the same size quoted in the literature but will fall into an accepted and established range. An organization known as the National Electrical Manufacturers Association (NEMA) sets acceptable tolerances, which tube manufacturers follow. For example, a tube purchased and presumed to have a requested focal spot of 0.6 might in fact have as much as 0.8 or 0.9 mm.

The issue of focal spot size becomes more difficult when the topic of how focal spot sizes are actually measured is considered. The so-called pinhole camera will produce slightly different measurement than when an alternate method of measuring focal spot size known as the star pattern is used. Manufacturing x-ray tubes is not an assembly line operation, but rather they are handmade assembled units. As the craftsmen work in assembling the filament and focusing cup structures, along with the other components, some tolerance is forgiveable, if not necessary. Hand making an x-ray tube with no allowable tolerances would increase the cost of each tube as much as three or four times. Also, in most exposure conditions there are many variable circumstances that often weigh more heavily in determining diagnostic quality than the slight variations with the focal spot size of each tube.

The two accepted methods for measuring a focal spot size are the star pattern and the pinhole camera, which can be seen in any quality control catalog. Generally, the star pattern is used for evaluating focal spot sizes smaller than 0.3 mm and a pinhole camera is used for evaluating focal spot sizes greater than 0.3 mm.

THE X-RAY TUBE HOUSING

In comparison to the x-ray tube, the complexity of function and operation of the tube housing seems rather plain and uninteresting. However, much of the battle in handling the heat generated by each exposure is ultimately given to the tube housing. As was implied earlier, if the intense heat on the anode is not transferred quickly, the focal track would severely pit and even melt. One of the keys to maintaining a healthy anode over a long period of time lies in its ability to radiate the heat away from the anode disc to the surrounding oil and ultimately on to the housing, which in turn submits its heat to the air of the exposure room. Thus, the housing must act as a kind of radiator to keep the process of heat transfer active.

Figure 225 shows the various structures of the tube housing. In it we can see that the insert is bathed in oil, which provides two very important functions. The first is to provide electrical insulation, keeping in mind that with the extremely high voltages used, the electrical pressure is extreme and that if the electrons jumped beyond their intended path directly to the metal housing (arcing) rather than from cathode to anode, the tube would be immediately ruined since the glass envelope would be shattered. Such arcing to the metal housing would certainly occur if it were not for the oil acting as an electrical insulator.

There are approximately 2.5 quarts of oil in the tube housing acting as a buffer between the tube insert and the metal housing; so important is the quality of electrical insulation that if any bubbles or air pockets are present in the oil during the exposure some degree of arcing could result through the air pocket.

The second function of the oil is to act as a heat absorber by drawing the heat from the anode and transferring it to the tube housing as quickly as possible. In this process the oil temperature can get quite hot, as much as 180°F. Also, as the oil heats, it tends to carbonize and become much less effective as an electrical insulator, increasing the possibility of arcing.

The tube housing must act as a radiation barrier with shielding qualities strictly defined by federal and state regulatory agencies. The amount of radiation leakage from the tube housing varies with the energy of the photons (KvP); however, at no time should a diagnostic tube emit more than 1.0 R/hr measured at any point one meter distance from the tube housing.

Finally the tube housing offers effective protection from the anode breaking away from the shaft and causing damage, because the housing of the x-ray tube is fitted with a steel liner capable of keeping an anode that breaks from its shaft within the housing and safe from the technologist and the patient, even when operating at 10,000 revolutions per minute.

High Tension Cables

The secondary side of the high tension transformer carries very high voltage, ranging from 35,000 to 150,000 volts, and relatively low amperage. Special cables must be provided to handle safely these extremely high electrical pressures from the secondary of the high tension transformer and rectifier to the x-ray tube. There is, in fact, pressure caused by electricity. An example of this involves a situation experienced by a serviceman who had worked with the author. While replacing an x-ray tube, one of the high tension cables had not been properly fastened to the coupling at the tube housing. When the first test exposure was made, the electrical pressure generated within the high tension cable blew the end of the cable away from the housing and sent it wheeling through the room in much the same fashion as water pressure would throw a garden hose around on the lawn if left unattended. In fact, the amount of electrical pressure in the high tension cable is enough to cause injury if someone happened to be struck during such an episode. Figure 235 shows a cross section of a common pair of high tension cables for the anode and cathode side of the x-ray tube.

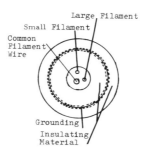

Figure 235.   Cross section of a typical cathode H.T. cable. The anode cable is the same in appearance and construction except it has only two line conducting wires.

MAMMOGRAPHIC AND OTHER SPECIAL PURPOSE TUBES

The information covered up to this point has been for the most part about the general diagnostic tube. We will now review other types of tubes that are used to serve special functions in radiography.

The first of these is the mammographic tube. These are of two different types. One type of mammographic tube is powered by a conventional generator with special wiring added to provide low KvP outputs. The tube is very much like the general radiographic tube except that it may have a 0.3 mm focal spot. Even though breast tissue is relatively easy for x-rays to penetrate, its subject contrast is very low; the technologist must use kilovoltages in the range of 35 to 45 depending upon whether conventional screen or xeroradiographic imaging is used. With such a low KvP setting, very few photons actually penetrate the breast, and this must be compensated for by using up

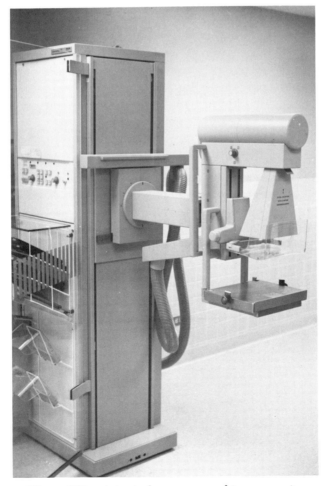

Figure 236.   A typical mammographic x-ray unit.

to 600 MaS depending on the image system. A typical exposure for caudalo-cranial breast examination for xeroradiography would be 40 KvP at 300 Ma at 0.7 seconds as compared to perhaps more conventional exposure of another part of the body, such as the shoulder, of 70 KvP, 100 Ma, at 1/8 second. These tubes must be able to handle high heat unit accumulations.

The second type of mammographic tube is made especially for mammographic equipment, such as the one pictured in Figure 236. These tubes have a molybdenum anode. Molybdenum is used because it produces a very soft x-ray beam compared to the conventional tungsten-rhenium target material used for general purpose tubes. In fact, the resulting x-ray beam from a molybdenum target is so soft that it would be almost totally absorbed by the standard Pyrex® glass material made for conventional x-ray tube windows. With this in mind, the special mammographic tubes can be made with a special beryllium window. Beryllium is a metal that has a very low absorption rate and is for all practical purposes transparent to this soft beam. As might be expected, the amount and type of external filtration applied to the beam must be very carefully designed. The material used for the added filtration is made of 0.25 mm and is approximately equivalent to 0.5 mm of aluminum in thickness. These mammographic tubes can be purchased with a stationary or a rotating anode. There are disadvantages with the beryllium window because if it is not used with care, the patient could receive an excessive dose of soft photons.

Magnification Tubes

Magnification tubes tend to be physically larger in overall size because of the oversized anode. Large anodes are required to better handle the concentrated heat loads that are generated at the very small focal spot sizes mandatory for magnification work (0.3 to 0.1 mm). Most of these tubes have an anode that is 5 inches in diameter and weighs approximately 4 pounds. The increased diameter produces a significantly longer focal track, which provides better exposure and heat distribution.

In even more demanding cases (such as diagnostic heart catheterization procedures), special tube housings are designed with a flexible plastic plumbing system as shown schematically in Figure 225, which allows oil to be pumped from the tube housing through a heating exchanger (which helps to cool the oil substantially) and then back to the tube housing.

You will recall that the normal heat dissipation path originates along the focal track, then spreads through the entire surface of the anode, at which time heat is radiated via convection to the surrounding oil in the tube housing; from the oil, heat is conducted to the metal housing and from there into the room's atmosphere. The technologist should keep in mind that this route of dissipation is most effective or efficient when there is a great variance in temperature from the anode to the oil. As the anode temperature drives the oil temperature up, the amount of heat that can be pulled from the anode to the oil decreases substantially. With the special oil circulation system mentioned above, a maximum amount of heat can be pulled from the anode to the cooled oil, resulting in substantially greater tube heat loading capacities. The tube rating charts shown in Figure 241 indicate the tube housing cooling' rate is approximately twice that of non-oil-cooled x-ray tubes.

Although conventional x-ray tubes equipped with high speed anodes are so commonplace today it seems hardly appropriate to make special mention of it here, these tubes do offer special loading characteristics and are specially suited for chest and angiogram work where 500 to 1,000 Ma stations are used routinely. Equally important, the exposure times are usually very short, in the range of 1/40 second or less. Under these conditions, an anode rotating at a high rate of speed is important to distribute these high intensity short exposures along the focal track as evenly as possible. The high speed rotating anode is well suited for practically any routine examination where high intensity exposures are required in combination with very short exposure times. The technologist should keep in mind, however, that the heat can be adequately distributed during an exposure only if the anode rotates a minimum of two times per exposure. A general diagnostic tube rotating at 3,300 rpm will make only 55 revolutions per second, and approximately 1.4 revolutions at 1/40 second as compared to 2.8 revolutions with a high speed rotating anode.

Grid Pulsed Tubes

There are certain special applications where it is essential to produce x-rays only when the electrical phase is at its peak. A variety of energy levels are produced when the entire electrical phase is used to make an ordinary x-ray exposure. However, for certain studies when a high speed Ceni camera is used and the burst of photon emissions from the x-ray anode must be in an exact synchronization with each frame of the Ceni camera moving at 60 frames a second, the firing of the x-ray tube must be extremely short and precise so that each exposure will be coordinated with each frame of the high speed camera.

When doing certain special procedure work, it is especially desirable to produce the optimum contrast possible on angiogram films, and improved contrast can be accomplished if the photons striking the film are all of equal energy.

In order to produce the proper synchronization of the x-ray exposures with the high speed camera and to produce photons of relatively equal value during each exposure blast from the anode, a special subcircuit is built into the x-ray tube and secondary circuit that controls the timing of each exposure. Figure 237 illustrates how the circuitry is arranged. X-ray tubes that are con-

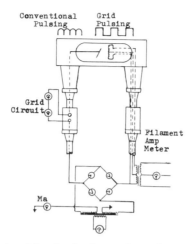

Figure 237.   A grid pulsed tube with modern wiring scheme.

trolled with this kind of circuitry are known as grid pulsed tubes, because originally they were manufactured with a metal grid placed between the cathode and anode. When the technologist pressed the exposure button, the secondary circuit was closed and energized the subcircuit operating the grid in the x-ray tube. The grid in turn would take on a strong negative charge, greater than the potential across the entire x-ray tube that was established with the KvP. The negative charge of the grid was strong enough to oppose the electrons of the tube current for a brief moment until the electrical phase reached its maximum point, established by the KvP at that moment of peak electrical phase when the electrons in the space charge would overcome the opposing grid field and move in precise coordination with each other to strike the anode, all at the same potential. This entire process, of course, happens fast enough so that up to 60 exposures per second can be made.

This "delay" and "blast" action of the space charge in up-to-date grid-controlled tubes is accomplished by producing an electrical bias between the anode and the cathode via a special circuit without the use of the metal grid.

PROPER HANDLING OF X-RAY TUBES

It seems more than appropriate to discuss in some detail what constitutes abuse in handling x-ray tubes, from the viewpoint of exposure values used in handling x-ray tubes.

Using Proper Exposure Values

The first and perhaps most obvious need is to observe the written tube loading and tube cooling specifications of the x-ray tube being used. For most general radiographic work, the technologist will not need to refer to the tube rating charts frequently since the average exposure requirements are usually well within the loading capabilities; however, there are some instances where the patient size or exposure conditions have been underestimated. A common examination such as a complete lumbar spine or the overhead filming for lower G.I. examination can generate sufficient heat by the end of the examination to make the anode cherry red. With this in mind, even with these common examinations there is sufficient heat generated that the tube rating chart should be consulted, especially when very heavy patients are radiographed. Evidence of this intense heat can be demonstrated by touching the tube housing, which will often feel very hot. Although one exposure made at the tube limit with a relatively new x-ray tube would probably not cause any immediate damage, it could cause fatigue or stress points in the anode that would with subsequent exposures shorten the normal lifespan of the tube. The greatest protection from this is to refer to the tube rating chart whenever there is a question about the tube loading. In these instances, the x-ray tube responds similarly to the human body, in that an early injury may not show an immediate effect but could impair the person's health as time advances more than would occur without the previous injury.

Fortunately, many of today's x-ray tubes have safety circuits built into the generators to prevent the technologist from making an exposure that is beyond the capacity of the x-ray tube. This special circuitry is located in the primary circuit of the generator and is wired in such a way that if any combination of Ma, time, or KvP are selected that would produce an exposure beyond the heat loading capabilities of the tube, a relay is energized to prevent the exposure. An important point for the technologist to remember is that the special circuitry does not detect the prevailing temperature of the x-ray tube at any one time but senses only the load on a tube for the pending exposure. Under these circumstances, an otherwise safe exposure made with a tube that has already been heated to the upper limits will elevate the heat units in the tube beyond the capacity of the x-ray tube and could cause immediate damage. Thus, this safety circuitry often does not completely protect the tube from damage, so the primary responsibility lies with the technologist using the equipment.

Even though a single exposure is safe and does not exceed this safety circuitry, it may still cause damage to the tube. Using exposures that "routinely" drive the temperature of the anode into the upper third of the tube heating capacity will greatly shorten tube life. This situation often occurs when a tube was improperly selected and is of a lower heat loading capacity than normal work situations require.

In one incident known to the author, the technologist was assigned to radiograph an obese patient for an I.V.P. examination. The scout film exposure barely made an impression on the film, and the technologist reasoned that perhaps double-exposing the film with two quick exposures would be the answer to the problem. He proceeded to double and then triple the exposure; and in so doing, the anode disc became so heated that the amount of heat dissipated to the oil drove the oil temperature to a point where it began to cook, creating a "pressure cooker" effect which, in turn, caused an explosion of the tube housing. The x-ray tube housing burst in the process, as did the glass insert, and scalding oil sprayed throughout the room and onto the patient. Recently the author examined the remaining parts of a glass insert that had been shattered as a result of a similar instance at another hospital.

Today most tube housings are equipped with a kind of bellows assembly that compresses when the oil begins to expand as a result of excessive heat coming from the anode disc. As the bellows is pushed back to a critical point by the heated and expanding oil, it trips a relay circuit which prevents the technologist from making any additional exposures. See Figure 225.

The second type of abuse by overloading the tube is when moderate to high output exposures are made on a cold anode. Under these conditions the heat generated along the focal track causes a sudden and nonuniform expansion of the tungsten metal, which can lead directly to a cracked anode. These cracks usually are of very small hairline type fractures; however, as shown in Figure 232A, the anode can actually crack. This can occur from only one exposure on a cold anode.

Another indication of abuse with regard to overexposure results when the x-ray tube is used frequently in the upper third limit of its rating capacities. These circumstances cause a certain roughening (pitting) of the focal track, which has two very noticeable results: (1) the x-ray output can typically be reduced by at least 20 percent and (2) the roughened anode projects the x-ray photons in a geometric pattern that results in unsharpness on the radiographic image.

The reduced x-ray output can be compensated for by a serviceman, who is called to recalibrate the Ma stations. He does this by boosting the filament current so that the tube will produce the required x-ray output obtained when the tube is operating properly. The boosted tube current unfortunately sets up a cycle of events which causes even greater strain and heat loading on the x-ray anode, which in turn causes additional roughening and an accelerated breakdown of the x-ray tube's anode.

Physical Abuse

Another pattern of x-ray tube abuse comes in the form of physical rough handling of the tube housing. Figure 238 gives some indication as to how fragile the glass insert can be. As mentioned earlier, the anode disc alone can weigh up to 4 pounds; with the addition of the weight of the stator, which is connected to the anode, there is a considerable mass. Figure 238 shows the method by which manufacturers bind the glass envelope to the stator and anode assembly. To get some concept of just how difficult this is, obtain an anode and stator assembly as shown here that has been discarded and attempt to hold it in your hand as the glass envelope must do. Once attempted, the technologist can begin to appreciate how damaging a hard bang or knock of the tube housing against another structure can be. This kind of physical abuse can easily result in small cracks or stress points in the fragile glass insert which, as in the anode, may not cause immediate damage but in the long run may contribute to cracks sufficient to allow air to seep into the vacuum of the tube.

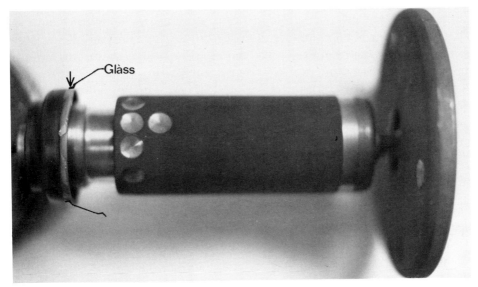

Figure 238. Banging or jerking the tube housing can damage an x-ray tube.

Damage to the tube can occur even if it is not struck against another object, however. A technologist can damage the tube by jerking or swinging the tube from one position to another during the set up for an exposure such as in cross-table hip examination or especially in portable work.

Because the early high speed rotating anodes turned at a speed of approximately 10,000 revolutions per minute, a gyroscopic effect developed which made the anode very vulnerable with respect to physical shock. Some instances were reported where the technologist twisted or jerked the tube housing suddenly while the anode was turning at high speed and against the gyroscopic effect of the anode disc; the anode tore from its shaft and right through the exterior tube housing into the room. Manufacturers of modern high speed anode tubes have been able to design and build tubes where these circumstances are no longer a problem. The overall weight or mass of the anode has been somewhat reduced, and the aluminum housing material is lined with a steel shielding insert sufficient to retain an anode if spun from its shaft.

The technologist should develop a healthy respect for the fragile nature of the glass, and the almost unimaginable heat that is generated at the anode with each exposure. In addition, the technologist should follow a prudent course of action with every exposure, especially under conditions where there might be excessive stress on the anode from heat. The following six suggestions will help guide the technologist in choosing safe and realistic exposures for the x-ray tubes that, if followed, will go a long way to extend tube life.

1. The cold anode should be warmed each morning by using the following exposure techniques: make two exposures using 200 Ma, 80 KvP, and 1/2 seconds. The rule of thumb is the warm-up exposure should be 1/10 the heat units of what would be produced by the first diagnostic exposure.

2. Be aware that a routine examination such as a basic lumbar spine series or a series of overhead films after a lower G.I. examination normally generates enough heat to drive the disc to a point of incandescence: when radiographing even moderately heavy patients consult frequently the tube rating charts.

3. Keep in mind that jerking or banging the x-ray tube will damage the glass insert.

4. Use exposures that utilize optimal beam energy (high KvP) for ample penetration of the body part so that the MaS is not increased unnecessarily to compensate for the poor penetration.

5. The technologist who is responsible for choosing the equipment should be sure that a tube is obtained that can handle the patient load and the type of examination conditions routinely required in the exposure room.

6. To make an exposure immediately after the ready light or the click signals the filament has reached its proper temperature and the anode has reached its optimal rotating speed is important for two reasons: first, delays in making the exposure after the tube is ready causes excessive filament evaporation to take place; second, delays in making the exposure after the ready signal requires the anode to spin at its top speed unnecessarily, which causes the bearings to wear rapidly.

Selection of Exposure Values and Rectification

There are some final points regarding tube damage reduction that should be made at this time. The first is that the x-ray tube operates much more efficiently at high Kv setting than at low Kv setting. An easy example of this can be made by comparing the following two exposures that would ultimately produce a very similar radiograph yielding similar information: an exposure made at 60 KvP at 200 MaS would produce approximately 12,000 heat units, but an exposure yielding similar radiographic results and density would result if made at 92 KvP at 25 MaS and would produce only 2,340 heat units. The reason is that higher Kv exposures give x-ray photons individually more energy and so fewer photons are needed to pass through the body and produce a given density. Thus equal amounts of work, so to speak, can be done with fewer photons and the tube operates at an overall lower temperature, because there is less energy conversion with fewer electrons needed in the tube current.

The one disadvantage in using high Kv values is that visibility of detail might suffer somewhat, due to the relatively low subject contrast that is characteristic of high KvP exposures. Thus a pragmatic and rational choice should be made when exposure charts are established with phototimed equipment as well as manually operated x-ray equipment, weighing the general effects of the reduced contrast against improved tube life.

The second point worthy of attention is that of selection of actual time and Ma values to be used once the MaS has been established. Here again conflicting circumstances arise, and one must be careful to balance the ideal with the practical. The ideal situation for extending tube life would be to use exposures of moderate to long times, say from one-quarter to one second in length. The reason, of course, is to allow the heat generated at the focal spot to be distributed as evenly as possible over the entire surface of the focal track. This more even heating of the anode track would effectively reduce the development of tiny hairline cracks in the anode. The cracks are caused by overheating and are a prelude to anode pitting. Using longer exposure times reduces substantially the tremendous heat variations along the focal track that cause cracking of the anode as the metal attempts to expand under the force of uneven anode temperatures.

The conflict is that the patient motion would surely be a problem, resulting in poor radiographic sharpness from long exposure times. Once again, exposure times and Ma values for a given MaS should be given prudent consideration and if possible held at not less than one-tenth of a second for routine exposures. This would allow a minimum of 5.5 revolutions of the anode and provide acceptable heat distribution of the exposure along the entire focal track when the anode is turning at the standard speed.

Rectification: Single versus Three Phase

The type of rectification has an important bearing on expanding tube life as well. Figure 239 shows tube rating for a similar tube with single and three-phase equipment. That heat is the tube's greatest enemy also applies to rectification. As you know, the basic goal is to provide as much x-radiation as possible with the least amount of heat on the anode. With this in mind, one should consider the electric current phase as it passes through the x-ray tube, as illustrated in Figure 240. As is discussed in Chapter Seven, three-phase equipment can provide relatively constant potential during the exposure.

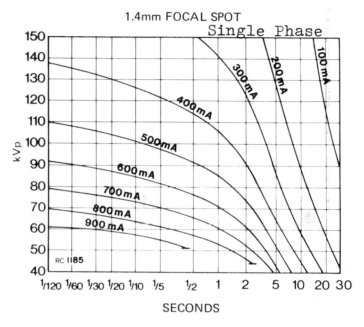

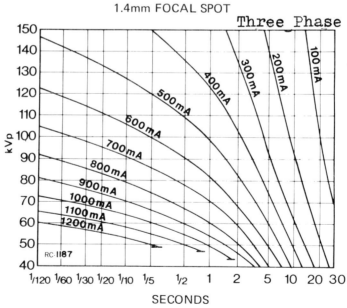

Figure 239. Notice the overall higher loading capabilities when used with three-phase equipment. *Courtesy of Machlette Corp.*

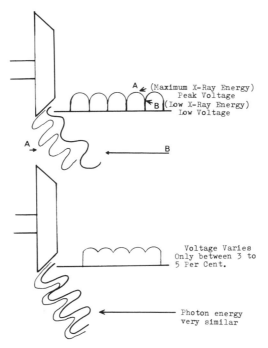

Figure 240. Variances in wave form of the tube current, comparing single phase current with three-phase current.

In most instances, useful x-rays are not emitted until the electrical potential raises to approximately 80 percent of the peak voltage (KvP) set by the technologist; however, heat on the anode is generated throughout the entire electrical phase. For example, if the technologist sets the KvP at 100 for a given exposure, relatively few useful x-ray photons will be emitted until the voltage rises and the electrical phase moves from zero to approximately 80 percent of its peak, or 80 KvP; however, the anode is experiencing heat while it is bombarded by electrons at the low voltage tube current. This means that although useful x-rays are not being produced, the anode is being bombarded continually by low potential tube current and generates little except heat on the anode surface. This disadvantage is reduced noticeably with three-phase rectification, especially when a twelve-pulse system is in operation. With three-phase rectification, very few if any low voltage electrons strike the anode as shown in Figure 240. Here the tube voltage is maintained at a level fairly consistent with the KvP selected by the technologist, allowing virtually no heat at the anode that is not accompanied by useful x-radiation. In addition, the more consistent voltages produced with three-phase equipment yield an average of approximately 35 percent more radiation than an equal exposure using single-phase full-wave rectification, so the MaS can be reduced proportionately relieving the anode of heat.

Normal Life Cycle of the X-ray Tube

Even the tube that has not been abused can not last forever. However, a few moderately sized x-ray tubes have produced good results for as much as ten years in a moderately busy exposure room.

Actually there are many possible progressions or avenues of aging an x-ray tube would go through when it is used under proper conditions. These will be discussed now with no attempt to identify them in order of probability, since variations in normal use are great.

One typical pattern is that over time the diameter of the filament through the process of evaporation (resulting from incandescence) decreases to a point at which it simply breaks, as does a filament in the common light bulb. This, of course, opens the filament circuit and the tube will not operate.

In the interim, however, it is possible to use the tube until the replacement is available by adjusting the techniques so that the second filament in the tube can be used. This is especially easy with phototiming, since the technologist need only to select an Ma station that utilizes the remaining functional filament size.

Through the process of normal filament evaporation, the x-ray output could actually increase. As the filament becomes thinner it offers more resistance to the filament circuit and thus generates more heat. Because the filament gets hotter it produces a larger space charge and a higher tube current results. Only slight variations in filament circuit and minimal differences in temperature cause disproportionate increases in output.

The increased x-ray output, however, is often somewhat moderated by the filtering effect of deposits of tungsten evaporants that accumulate on the window of the glass tube. These accumulations absorb very effectively some of the radiation. In older tubes, the evaporated filament is the dominating factor in terms of affect on net output of the tube. Under these conditions, tungsten evaporants from the anode accumulate at the window of the x-ray tube. Tungsten is, of course, a very dense material and thus has high heat absorption properties. As the high Ma exposures continue to be made over time, excessive anode heat occurs and the deposits of tungsten from the anode become greater. This in turn absorbs increasing quantities of the existing beam and causes a reduction in output; soon a serviceman is called upon because the machine is "shooting light." He typically boosts the current in the filament circuit to compensate for the absorption of the x-ray beam by the anode tungsten evaporant; however, the tube must now work harder to produce the same output it did before the tungsten accumulated. In the process, the exposure to the anode is greater and its focal track becomes increasingly heated and pitted. The increased pitting reduces the exposure output even more and, along with possible reduced sharpness on radiographs, the tube quickly becomes unuseable.

the exposure to the anode is greater and its focal track becomes increasingly heated and pitted. The increased pitting reduces the exposure output even more and, along with possible reduced sharpness on radiographs, the tube quickly becomes unuseable.

The second possibility is when the bearings of the x-ray tube begin to fatigue. This was mentioned earlier and leads directly to increased friction on the bearings, which causes even more heat on the bearings. This cycle continues and the anode begins to spin at slower and slower speeds, which reduce the degree to which the heat is spread along the focal track. An anode spinning at 3,300 revolutions per minute will only "turn a new face" approximately 1.4 revolutions, which certainly does not provide a great deal of heat distribution along the focal track. Eventually the anode slows to a point where the heat from the exposure becomes concentrated on a smaller length of the focal track, causing a high degree of pitting.

In more advanced stages of bearing damage, the bearings could actually "freeze" and the anode will not turn at all. When this occurs, small or large melted welds develop on the focal track, as shown in Figure 234.

The third and final condition that tubes may go through involves the accumulation of tungsten on the glass but with a slight variation. Under optimal conditions, the two points of electrical potential (positive or negative) should only be the anode and cathode; however, with the metal tungsten accumulations on the window a third point of electrical charge may develop. If the tube continues to operate under these conditions, it is very possible that some of the electrons in the space charge will jump toward the tungsten accumulations and impact against the glass envelope, causing it to fracture. At this point even if the result is only a small hairline crack, the tremendous negative pressure within the glass envelope would draw the insulating oil inside the envelope and render the tube useless. The author has seen a number of instances where this has occurred, in older tubes and in new tubes that have been used routinely at the upper limits of the heat loading capacity.

Once again the technologist should keep in mind that there is no single circumscribed route by which tubes eventually break down; however, the different situations described above provide some realistic examples of why tubes break down.

Tube Cooling Chart

There is a distinct difference between the tube rating chart and the tube cooling chart. The tube rating chart is thoroughly discussed in Chapter Five, but we will now review in some detail the tube cooling chart.

It has been emphasized throughout this chapter that heat generated at the anode must be disbursed as quickly as possible. Among the factors that affect the heat are the kind of exposure used, the size of the anode, the type of rectification in use, the quality of oil inside the tube housing, and whether or not any special devices are used (such as fans or recirculating oil systems). All these factors along with others that have a more subtle effect on tube cooling are calculated by the manufacturer and converted into a graph for easy reference. Each different type of tube manufactured has its own tube cooling chart. Figure 241 is a typical example of a tube cooling chart. It might be helpful to obtain various types of cooling charts (with the tubes equipped with fans and circulating oil systems) from an x-ray service company or a tube manufacturer so that you can evaluate how the cooling characteristics change with various tubes. It should always be kept in mind that even a "safe" exposure can cause considerable damage to an x-ray tube if the tube has already accumulated heat units close to the upper limits from previous exposures. Under these circumstances, the tube cooling chart must be consulted before any further exposures are made no matter how safe they may seem individually.

There are some devices now available for purchase, known as tube heat calculators, that operate from a microcomputer system. These little devices, approximately the size of a cigar box, are mounted on top of the generator and provide a continual digital read-out of the heat still in the tube housing.

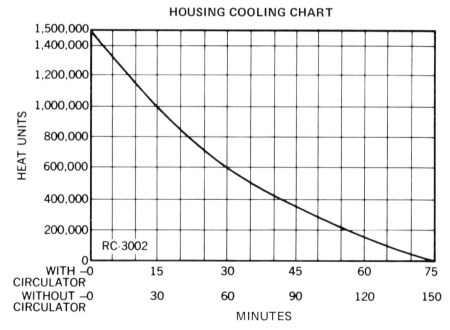

Figure 241.   A typical tube cooling chart. This chart indicates 15 minutes is needed for the tube to cool from 1 million H.U. to 600,000 H.U. with special cooling. Without special oil cooling, 30 minutes is needed to cool from 1 million H.U. to 600,000 H.U. *Courtesy of Machlette Corp.*

These devices are also capable, if wired into the x-ray control panel, of preventing any additional exposures if the heat in the tube housing has reached a certain point. Often this "shut-off" point is set to occur when the tube reaches 80 percent of the total heat capacity.

Saturation Point

The x-ray tube must operate at saturation current because this allows the technologist to independently control both the quantity of the x-ray beam and its quality. Without any additional circuitry, an x-ray tube will operate at saturation current, but when kilovoltage rises to a certain point the condition of saturation diminishes. Up to a certain KvP level the supply of electrons at the filament is sufficient so when they are attracted to the anode the supply from the filament does not outweigh the demand from the anode. If the supply of electrons available in the space charge remains abundant by comparison to the demand of those pulled by the KvP during the exposure, the situation of saturation exits. Under these conditions the technologist can increase the KvP without effecting any changes whatever in the quantity of current, thus maintaining the advantageous independent relationship between Ma and KvP. However, unless aided by a special circuit as mentioned above, if KvP is raised to a point where additional electrons are drawn from the cathode, then the selected Ma will be affected. Under this nonsaturated condition, increases in KvP will pull additional electrons from the filament itself in addition to those in the space charge, and these "extra" electrons would cause an increase in the Ma. Increases in KvP would also cause an increase in Ma.

With the inclusion of a space charged compensator, a special circuit is used to automatically adjust the filament current downward as the KvP is elevated by the technologist. This compensates for the extra filament electrons that will be drawn by the high KvP setting. In this way the space charge compensator makes just enough adjustment in the filament circuit to keep the net Ma striking the anode at the proper level regardless of the KvP.

Selected Ma	Selected KvP	Automatic changes in filament circuit via space charge compensator
200	100	4.51
200	80	4.54
200	60	4.58

Selection of X-ray Tubes

The proper selection of an x-ray tube is to a large extent governed by common sense plus some specific facts that have been already discussed.

Below are various considerations that should go into tube selection.

I. The overall work load should be determined, i.e. major type of work being performed:
   A. General overhead radiographic work.
   B. Fluoroscopy.
   C. Chest work.
   D. Emergency work.
   E. Extremity work.
   F. Special procedures.
   G. Portables.
   H. Magnification.

II. Overall volume of work (number of exposures considering the degree of rest time the tube has for cooling between each exposure):
   A. Moderate to heavy work load with relatively no time between exposures for the tube to cool.
   B. Moderate to heavy exposures allowing some time between exposures for tube cooling.
   C. Light to moderate work load.

III. Selection of technique and types of exposures:
   A. Moderate to high Kv with relatively low MaS values.
   B. Moderate to low Kv with relatively high MaS values.
   C. Moderate exposure times.
   D. Very short exposure times.
   E. Field size to be covered.

IV. Review power ratings of the generator and rectification:
   A. KW ratings.
   B. Consider auxillary cooling equipment.

The subject of tube selection could conceivably become very involved and embellished with charts of all types to evaluate each set of exposure circumstances versus the many tube characteristics and options that are available. However, this is not within the purpose or goal of this text and we will instead discuss the practical points of the topic.

The items listed in the first group suggest there are, in fact, tube characteristics available for many different purposes. For example, a general purpose tube usually does not need the help of a high speed anode, nor would it likely need a large 5 inch anode. Even more unlikely would one want to purchase an insert with an unusually small focal spot size or have it equipped with special accessory cooling devices such as recirculative oil or air fans.

In short, one might do well to use a standard sized 3 or 4 inch anode tube with a 0.6 mm small focal spot and 1.2 mm large focal spot size. A common rule of thumb is that an average exposure room sees about 25 patients per day, or makes about 37 examinations per day, which requires approximately 148 exposures. General purpose tubes should have a KW rating of about 50 and have a heat loading capacity of approximately 250,000 to 300,000 heat units.

The average sampling of work sent through such a general radiographic room will usually give the tube sufficient time between each exposure to cool, and even more time between patients, so fans mounted on tube housings are usually not required. However, the work flow may be kept at a fast pace with long, complicated studies where many exposures are made over relatively dense parts of the anatomy, such as might occur in a busy emergency area. In this case, an x-ray tube is required to make exposures almost continuously for a 16 hour day for examinations such as lower lumbar spine, hip, pelvis, and obstructive series, which usually create a great deal of tube heating.

Under these more demanding conditions, although a single exposure would be well within the heat units applied to the anode, the total accumulated heat units to the tube is likely to stay at a high level. A high speed anode tube would be of relatively little value, since the primary problem is to transmit the heat accumulation to the oil and to rid the anode of those high temperatures as quickly as possible (not just disburse the heat generated from a single exposure over the full length of the focal track as evenly as possible). In this example, a tube with a minimum of 300,000 heat units should be considered; perhaps, in addition, the housing should be fitted with a fan to dissipate the heat on the exterior of the tube housing into the room so more heat can be transferred from the anode.

A fluoroscopy tube has a similar problem regarding accumulated heat, since on a normal schedule a fluoroscopic tube will seldom go below one-half its rated heat capacity before more fluoroscopic or spot films are made, driving the overall accumulated temperature up again. The fluoroscopy tube must handle dual problems. First, many exposures are made at a relatively short time (because of the high Kv techniques often used) so the problem of distributing a single exposure equally along the entire anode track is common. Second, the accumulated heat units stay fairly high because of the fluoroscopy function. The anode spins normally at approximately 3,300 rpm, so it will rotate 0.45 times during a 1/120 second exposure. Here a high speed tube is justifiable, as well as one with a relatively high capacity of heat units (between 300,000 and 400,000), and a cooling fan could be put to good use as well. It is also wise to consider a tube with a relatively low target angle, thereby taking advantage of the line focus principle to keep the effective focal spot relatively small without decreasing the amount of actual target area bombarded by the tube current. The focal spot size for these tubes are usually 0.6 and 1.2 mm.

Chest work is perhaps the easiest situation to calculate tube specification. First, the body part is usually not very dense, so single exposure tube ratings are not a significant factor. Second, the Kv setting is relatively constant for the majority of exposures and it is often in the upper Kv range (from 120 to 140) with a relatively low MaS, allowing the tube to operate at lower temperatures. The overall amount of heat units produced when compared to a fluoroscopic tube or a busy emergency room tube is low. The primary difficulty with chest work is that very short exposures are used, and this makes the problem of disbursing heat along the entire length of the focal track crucial. Thus it is advisable to have a high speed anode tube to spread the heat and also to have one of the larger 5 inch diameter anodes to handle the short instantaneous exposure burst as effectively as possible. (The circumference of a 5 inch anode is 398.7 mm compared to 239.2 mm of a 3 inch anode.) Because of the long focal distance, the geometry of the beam is quite good; therefore, it is not necessary to consider the use of a tube with a focal spot size much less than 0.5 to 0.6 mm plus a large anode with a large focal spot 1.0 to 1.2 mm to obtain good quality chest radiographs. It is wise to have a tube rated at not less than 300,000 heat units. A high volume patient load is common in dedicated chest rooms, and this could produce a moderately high anode temperature accumulation.

Extremity work, as you would expect, is quite simple and tube selection is relatively uncomplicated. Here the focal spot should be of primary interest, since quite frequently fine bony architectural lines are scrupulously evaluated. If the primary purpose of a given exposure room is to perform bone work, a fractional focal spot with a low tube angle (perhaps 10 degrees) would be a good choice. Radiographs of the femur, tibia, and fibula might pose some problems since a 10 degree angle might not provide sufficient field coverage at 40 inches. In this event the F.F.D. would have to be increased to perhaps 48 inches.

Special procedure radiography imposes heavy demands on the tube and even a well-selected tube might not last more than 18 months. The exposures on a serial run alone would put most tubes used for general radiographic purposes at a great disadvantage. First, most serial runs are made with relatively low Kv and high MaS selections to enhance radiographic contrast, which places an added strain on tube heating. Second, the average angio run for a renal arteriogram is 2 or 3 per second for 3 seconds; followed by 1 film per second for 3 seconds; followed by 1 film per second for 2 seconds. This, without figuring cooling time, would produce approximately 60,000 heat units. Even though there is some cooling continuously between angio exposures, it is (for practical purposes) usually a point to take into consideration. To add to the problem of accumulated anode heat, the requirement for high resolution must be satisfied to such a degree that fine vessels can be visualized routinely, and this makes the use of the smallest focal spot possible highly desirable. This combination of demands becomes even more complicated when magnification technique is used, because fractional focal spots come into play. For departments that use magnification technique routinely, tube selection is very important: oil-cooled circulating units deserve consideration.

For arteriograms without routine magnification, a focal spot of 0.6 would be acceptable and still remain reasonably safe for handling heavy heat loading demands on an arteriogram run. These tubes should have a 5 inch anode to increase heat storage capability and better distribution of focal track heat for short exposures. They should be equipped with a high speed rotor, and an air cooling fan over the tube housing for more effective cooling characteristics. Even oil-circulating equipment could be justified to help dissipate excessive heat from the anode. Since anode heat is discharged most effectively when the oil and tube housing are kept relatively cool, a fan is often mounted to blow air over the housing, thereby keeping its temperature as low as possible.

Portable work is primarily a problem of single exposure loading and physical abuse to the tube. It never ceases to amaze the author how inconsiderate and ill-mannered technologists can be when handling portable equipment. It seems the general dislike associated with doing a portable examination is childishly transferred to abuse of the portable equipment itself. As a result, decisions to purchase less expensive equipment are sometimes made so that the loss of abused equipment would not be so great. This, however, backlashes to the technologist, since usually the less expensive equipment is more difficult to handle and offers fewer advantages in terms of exposure capacity.

The author has frequently seen portable machines no more than six months old with unforgiveable dents, scratches, and assorted broken parts on the control panel as well as stripped locks on the tube mounting, indicating the extent of abuse to the equipment.

Portable units purchased twenty years ago were equipped with simple stationary anodes and were very limited on their Kv and Ma output values, but modern portable generators require rotating anodes and are capable of delivering much greater exposures to the tube. Thus, selection of portable tubes for contemporary portable equipment warrants somewhat more consideration. Very often, however, the type of tube options that are available for any given portable unit are relatively few and in some instances only one type of tube is available for a given portable unit. The tube life of a portable x-ray tube can be increased substantially by simply handling the tube, as well as the unit itself, with the respect it deserves.

PRODUCTION OF THE X-RAY BEAM

The topic of x-ray production could easily require many pages of information; however, the concepts presented here will provide sufficient insight that the technologist will be able to understand what affects the production and varieties of x-rays that are formed to make up the primary beam.

The x-ray tube itself operates as a rather inefficient x-ray convertor. As it changes kinetic energy into x-ray energy, about 99.8 percent of the kinetic energy available (carried to the anode by the electrons) is wasted and given off in a form of a low energy heat, which in effect causes tremendous problems in operating an effective x-ray tube.

X-ray production is possible by providing for three crucial conditions. The first is to produce a supply of electrons; second, to give the electrons a high level of kinetic energy; and third, to abruptly dispose of the kinetic energy in order to force its transformation into another energy form known as x-radiation. The kinetic energy that is created through electron acceleration causes the electrons to reach approximately one-half the speed of light in the short few centimeter distance between cathode and anode, and of course is controlled by the kilovoltage.

Characteristics and the Nature of X-ray Photons

A discussion of x-ray production would not be complete without a brief description of the nature of x-rays. However, even a lengthy and detailed explanation of what x-rays are leaves the student at an immediate disadvantage, since x-ray energies seem to contradict some of the basic laws of physics. The first and by far most striking characteristic is that x-ray energy has no mass and yet it can interact and affect atoms of living tissue as though it had mass. X-rays are perceived to be tiny packets of energy in the electromagnetic spectrum that vibrate in wavelengths of from 0.1 Å to 0.25 Å and travel the speed of light with no electrical charge, occupying a portion of the magnetic spectrum as shown in Figure 242.

Earlier in this chapter numerous points were discussed about how the working components of the x-ray tube operate both separately and in conjunction with each other. Despite its lack of efficiency, a very high level of technology was necessary to provide this extremely difficult and complex task of producing x-ray energies so that high quality radiographs are obtain-

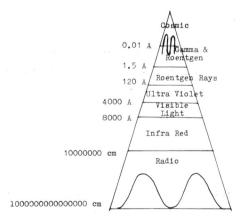

Figure 242.   Electromagnetic spectrum.

able routinely. We must consider what occurs at precisely the moment in time when the kinetic energy is transformed into x-rays. There are actually two types of x-ray energies that emerge to produce what we know as a primary beam. The first is called Bremsstrahlung, or braking radiation, and the second type is characteristic, or general, radiation.

Bremsstrahlung (Braking Radiation)

Bremsstrahlung, or Brems radiation, is produced when the speed of an electron in the tube current is slowed substantially by the positive attraction of the target atoms (target atoms being those comprising the tungsten anode). As the speed (kinetic energy) of the tube current electron is reduced, a proportional amount of energy must be transferred, which happens to be an electromagnetic frequency in the range of 0.1 to 0.25 Å. Thus the kinetic energy lost by the slowing speed of the space charge electron is instantly transformed into x-ray energy. Braking radiation gets its name because the positive attraction from the target nuclei provides a slowing or braking effect to the fast moving incident electron and results in physically slowing down.

The braking radiation photons produced have varying energies and so they are not uniform in penetrating power. The primary reasons for this have been previously discussed in the text. First, the tube electrons do not always travel at a consistent speed (see Fig. 240). It is dependent on the rise and fall of the pulsing that is characteristic of electrical phases. An electron striking the target when the electrical phase is at a level close to the peak of the potential (KvP) set by the technologist will have a substantially greater amount of kinetic energy than an electron striking the target material during a point less than the peak voltage value. In this instance, the amount of kinetic energy available to be transferred is substantially less, and thus the energy of the resulting x-rays is also diminished. This is also illustrated in Figure 240.

The second reason for energy variances between Brems photons in the primary beam is the relative proximity that the incident electron moves with respect to the nucleus of the target material. If the incident electron's path (Fig. 243) brings it close to the nucleus of the target material, the deceleration will be greater and will result in a higher energy transformation (into more powerful or energetic x-ray photons). If, on the other hand, the path taken by an incident electron happens to take a path further from the nucleus of the target material, the positive pull from the target nucleus will not be as great and so the deceleration process will not be as dramatic; the conversion process from kinetic energy to x-ray energy will be somewhat diminished, leaving the resulting photons with less energy and penetrating power.

The third circumstance that contributes to the variations in Brems radiation occurs when the accelerated electron actually strikes the target nucleus. In this instance, the total amount of kinetic energy is converted to x-ray

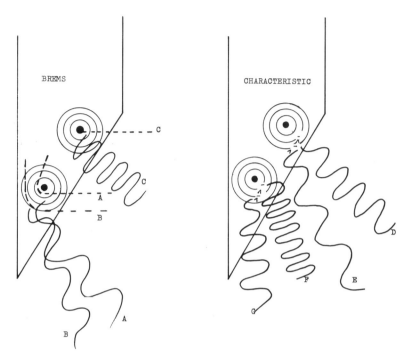

Figure 243. *C*, the most energetic and penetrating photons via interaction between the space charge electron and the target nucleus producing the most dramatic level of braking radiation. *A*, the next highest energy of a photon occurs when a space charge electron's speed diminishes more, slowing by very strong positive attraction from the target atoms nucleus, as it passes close by the nucleus. *B*, the least energetic braking photon occurs by virtue of the less dramatic deceleration as the electron passed slightly further from the target nucleus, and thus feels less of the positive pull by the nucleus. Photons *D thorugh G* indicate various types of characteristic photon energies produced by electrons jumping from one orbit to another, discharging appropriate levels of radiation according to the atomic number of the element and the orbits involved.

energy, yielding an x-ray photon with substantially higher energy than any of the circumstances mentioned above, as all the kinetic is transferred.

The fourth possibility is when the accelerated electrons collide with each other, resulting in a relatively low energy conversion compared to those mentioned above, seldom reaching the film.

The technologist uses different KvP levels for various exposures. The overall amount of energy produced that results in Bremsstrahlung radiation varies substantially with the amount of kinetic energy initially given to the electrons. In general, as the KvP increases, the amount of kinetic energy given to the accelerated electrons increases, and their slowing or braking effect causes a somewhat greater conversion and transformation into higher x-ray energies. Conversely, as the technologist uses somewhat lower KvP levels for an exposure, the average kinetic energy originally given to these incident electrons in the tube current provides less of a conversion process, and the general overall energy level of the resulting Brems photons is reduced by comparison.

Characteristic Radiation

The second type of x-ray contained in the primary beam is called characteristic radiation. Characteristic radiation results from energies given off as electrons move from one orbit to another to fill existing voids created by a collision between the target atom and incident electrons. The resulting x-ray emissions are "characteristic" of the atomic number of the atom involved, along with the relative orbits that are involved in the ionization process. The atom draws an electron from an outer neighboring orbit to the vacancy and in so doing a discharge of a certain amount of energy by the electron is necessary to make it comfortable in its new orbit environment. There is now a new void to be filled in the outer orbit caused by the transplanted electron, and so the vacancy results in a similar way in a discharge of energy "characteristic" to the orbits involved. Although their emissions as electrons are transplanted from one orbit to another, photons emitted as useful primary radiation originate only from the K and L shell of the target atom. Figure 243 shows both Bremsstrahlung and characteristic radiation patterns.

The composition of primary photons emitted as characteristic radiation is about 10 percent when 80 to 150 KvP exposures are made, and at lower than 80 KvP, the energies of the characteristic radiation do not provide sufficient penetration to pass through the glass tube. Thus in any one exposure a variety of x-ray energies are produced, ranging from hard to soft photons. How these photon energies work to produce a given radiographic effect is discussed in Chapter Seven.

Other Significant Factors Affecting X-ray Production

Figure 244 shows the distribution of energies of x-ray photons emitted by a single exposure. The maximum energy of x-ray photons is dependent on the voltage of the tube current. In fact, one can say the maximum energy possible for any photon emitted is directly related to the maximum kinetic energy of any given electron moving from filament to anode. The distribution of these energies, shown in Figure 245, is the result of the varying tube voltage caused by the sinusoidal form of full-wave rectified current, as in the specific circumstances just discussed. The distribution of photons in a given x-ray beam also changes with different kilovoltages, since the maximum amount of photon energy is related to the kinetic energy generated in the tube current electrons by the KvP.

Figures 244 and 245 also show a few spikes representing the emissions in characteristic radiation that occur when the tube current electrons collide and remove electrons from the K shell of the target atom. Characteristic radiation

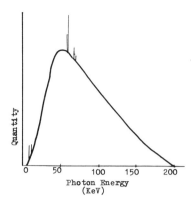

Figure 244. Demonstrates the distribution of photon energies given off by a typical anode target.

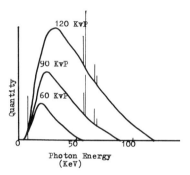

Figure 245. Shows how the minimum and maximum Kev energies of the x-ray photons change with KvP.

is associated with L shell and are not shown, since their significance to the radiographic image and to the dose to the patient is nil. Characteristic radiation (K shell) gradually diminishes in quantity as the KvP increases over 150, until at 300 KvP it disappears entirely.

The type of target material does not noticeably affect the *energy* of the photons that are emitted, only the *amount* of photons. This is primarily since x-rays are produced by stopping kinetic energy associated with the tube current electrons, and this can be accomplished by using a number of materials besides tungsten. In the earlier Crooks tubes, the glass envelope served as the point of sudden deceleration of the electrons. In other words, the maximum photon energy emitted is equal to the maximum kinetic energy of the individual electrons at the moment they strike target material; however, the efficiency of the conversion of a massive number of electrons with associated kinetic energy into heat and x-ray energy is dependent on the atomic number of the target material.

Figure 246 shows two energy distribution curves of photons coming from an x-ray beam that were produced by an equal exposure value but with different target materials used. You will note that the spread of Brems energy is the same for both target materials. The only noticeable difference is the overall increase in the amount (quantity) of photons produced by the different target materials.

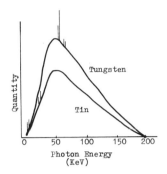

Figure 246. Notice the minimum and maximum distribution of photon energies in Kev with tin or tungsten targets, with the only difference being in the (efficiency) quantity of photons produced at any given Kev.

If lead were used as target material, the amount of photons produced would be even greater than with tungsten. Thus, materials with a higher atomic number produce more efficient conversion of the kinetic energy of the electron into x-ray energy (efficient conversion equals more photons but less heat). A lead target, because of its relatively low melting point, makes a very poor choice for target material. Although target material does noticeably affect the maximum and minimum extremes of energy distribution of the beam for Brems radiation emissions, it also affects the energy of the characteristic radiation from the K shell, since by definition characteristic energy changes with binding energy between the nucleus and the K shell.

The wavelengths of the most energetic photon for a given exposure can be calculated using the following formula:

$$\text{Minimum (\AA)} = \frac{12.4}{\text{applied KvP}}$$

tube current voltage	minimum wavelength that will be produced
60 KvP	0.206 Å
80 KvP	0.155 Å
100 KvP	0.124 Å
120 KvP	0.103 Å
150 KvP	0.082 Å

This formula is applicable to both single and three-phase equipment since both units can produce comparable *peak* tube current voltages. Using the formula, one can easily calculate the wavelength of x-ray photons that would be produced for any given KvP.

Exposure Values versus Radiation Production

It is very beneficial to be able to calculate an actual R value that any given exposure would yield. A quick estimate, for example, could be made as to how much a skull x-ray, chest x-ray, lumbar spine, etc., would produce if the actual exposure values were known. The formula shown below can be used for this purpose.

$$\text{Exposure} = 15 \times \frac{\text{Kv}^2 \times \text{MaS}}{\text{D}^2 \text{ (in cm)}}$$
$$\text{(in mr)}$$

For values of 80 KvP, 300 Ma, 1/4 sec, 40 inch (100 cm) F.F.D.:

$$\text{Exposure} = 15 \times \frac{80^2 \times 75}{100^2} = 15 \times \frac{480,000}{10,000} \times 15 = 720 \text{ mr}$$

The exposure value above should be used for estimates only. The actual output of radiation, as expressed here in milliroentgens, is dependent on a number of variables such as the age of the tube, fluctuations in current of the x-ray equipment circuits, filtration, rectification. It is helpful, however, in making quick and accurate calculations when comparing radiation delivered to the table (for skin dose to the patient) using high Kv, low MaS exposures, and vice versa. For example, all other things being equal and to produce the same radiographic density as the technique shown above, we can compare the milliroentgens generated by the exposure using 80 Kv with an exposure using 92 KvP. It should be pointed out that the MaS in both incidences was adjusted correctly, so the density between these two radiographs is relatively constant.

$$\text{Exposure} = 15 \times \frac{92^2 \times 37.5}{100^2} = 15 \times \frac{317,400}{10,000} \times 15 = 476.1 \text{ mr}$$

Comparison: exposure of 80 KvP and 75 MaS yields 720 mr
          exposure of 92 KvP and 37.5 MaS yields 476.1 mr

PROPERTIES OF X-RADIATION

As you are probably already aware, x-radiation is a part of a continuum of electromagnetic frequencies that is shown in Figure 242. With this continuum in mind, the various types of energy shown in Figure 242 have very similar properties and characteristics. The major factors differentiating one energy from another in this spectrum are frequency and wavelength. For example, at the lowest energy end of the spectrum one wavelength of radio waves can be as much as 400,000,000 (4×10^8) cm, while the most extreme short wavelength would measure approximately 0.000,000,000,003 (3×10^{-11}) cm. The following is a brief summation of the properties of x-radiation.

1. X-ray photons travel at the same speed as light (186,000 miles per second when measured in a vacuum).
2. X-ray photons are very directional; they do not change direction as they

travel through air unless they interact with matter and produce secondary or scatter photons.

3. X-ray photons have sufficient energy to ionize both solid and gaseous substances. (They can remove orbiting electrons from atoms.) They can thus cause certain luminescent materials to emit light, as in an intensifying screen.

4. X-rays do not respond as does light to standard optical glass lenses, and therefore cannot be focused as lower energy light photons can.

5. Because of their higher energy ionizing characteristic, they can produce latent images on photographic film.

6. They do cause biological and chemical changes when interacting with human or any other type of living tissue.

7. Upon interacting with other attenuating substances, x-radiation produces "byproduct" types of emissions known as secondary and scatter photons.

8. Because of their high energy level, x-ray photons can penetrate solid substances.

9. Because x-radiation is a massless (having no mass) energy form, x-ray photons cannot take a positive or negative electrical charge and are therefore neutral.

10. Because x-ray photons are electrically neutral, they are not affected nor do they affect magnetic fields through which they might be passing.

11. Photons having all the characteristics mentioned in this section can be seen or observed over a wide range of frequencies, with the useful diagnostic range being about 0.1 to 0.25 angstrom (an angstrom is equal to 10^{-8} cm, or $1/100,000,000$ cm).

CHAPTER FIFTEEN

THE X-RAY CIRCUIT

ALTHOUGH THE BASIC x-ray circuit has been discussed in terms of its specific parts and how they relate to kilovoltage, milliamperage, and time as well as how these components work together to produce a radiographic image with required quality, it is important—for the sake of continuity and to point out other pertinent information—to discuss the basic x-ray circuit as a whole.

THE INITIAL SUPPLY OF CURRENT

We shall begin this discussion with the supply of incoming current. The electric company is responsible for supplying the facility, whether it is a one room office or a fifty room department, with 60 cycle alternating current. This current enters the building from a designated transformer outside of the building and leads through special hospital or office circuit lines to the specific x-ray equipment it is to energize. The path of incoming wire is often very tortuous as it travels through the hospital to the x-ray department, but eventually it appears at a main x-ray breaker box, which is usually mounted on a wall close to the x-ray control panel. The wall breaker box contains a series of circuit breakers and, of course, the main disconnect switch, which extends out from the wall box as a small metal arm or is in the form of a single large circuit breaker located inside the door panel. Wires are then connected from this junction box to the x-ray control panel as shown in Figure 247. The wall breaker or junction box serves two purposes: (1) it provides a safe hookup between the supply current coming from the outside and the main wires going directly to the x-ray control panel; (2) the circuit breakers or fuses provide a safety point since they will "trip" and open the circuit to the x-ray control panel if an electrical overload or short occurs at some point in the circuitry before entering the wall box. If unchecked, such a current surge from the outside could cause damage to the x-ray generator itself (the x-ray control panel).

Although there are some variations and special requirements for incoming current, most general diagnostic equipment requires an incoming line that will carry 220 volts at 100 amperes. Since the current supplied by the electrical company is 110 volts, as shown in Figure 247, two separate 110 volt lines are connected with a common ground to produce the desired 220 volts supplying the control panel. This junction of the two 110 voltage lines occurs in the wall

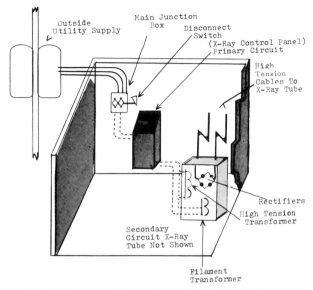

Figure 247. A simplified drawing of the relationship between the incoming current from the power company to the wall junction box. One can in addition see the actual room layout with respect to the various components. You will note that the components to the primary circuit are contained in the x-ray generator, and those to the secondary circuit are located in the high tension transformer and the x-ray tube.

box. It is important to restate that the 220 volt line requirements do change with various types of equipment; however, this is the common hook-up used currently.

THE PRIMARY CIRCUIT

The X-ray Control Panel

There is a great variety in outward appearance of the x-ray control panels (sometimes called generators) from one manufacturer to another. This is because manufacturers have varying philosophies as to how their control knobs and meters should be wired into the basic system, just as automobile manufacturers have different approaches to how their automobiles are to be designed and assembled. Although these x-ray control panels vary in appearance and in detailed function, they are basically similar and provide the same end result. We will now proceed with a more detailed discussion of the function of each item commonly seen in the x-ray control panel. See Figure 248 which shows basic x-ray circuitry.

The Prereading Voltmeter

Earlier x-ray control panels had an assortment of meters and control knobs that seldom appear in more modern generators. See Figure 249. The

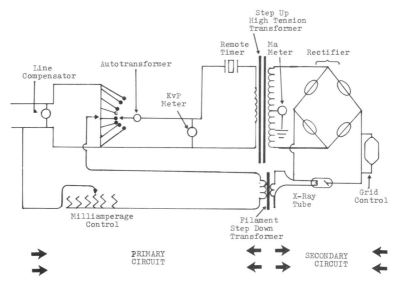

Figure 248.   The basic x-ray circuit.

x-ray control panel has two types of devices. Basically one group consists of monitoring devices, shown here in the form of meters; they simply react to the amount of voltage or amperage going through a particular part of the circuit. The second group are the actual control knobs and buttons which cause the changes in voltage and amperage that go through various parts of the circuit.

As you can see in Figure 248 the prereading voltmeter is connected in parallel after the autotransformer and samples voltage going into the high tension transformer. It is called a prereading voltmeter because it samples the primary voltage before the high tension transformer which ranges from about 60 to 250 volts. Any amount of voltage at the x-ray control panel will eventually be boosted or stepped up by a high tension transformer, so the actual voltage going through the x-ray tube is considerably higher, ranging from 45,000 volts to 150,000 volts. The prereading voltmeter is wired into the primary circuit after the autotransformer, but the meter itself is located on the x-ray control panel so the technologist can view it and verify the level of KvP going through the x-ray tube. This meter is calibrated to "preread" the peak voltage that will ultimately be produced at the time of the exposure in the x-ray tube. It would not be practical to directly connect the prereading voltmeter to the actual current after the high tension transformer because of the extreme voltage, which can go as high as 150,000 volts during the exposure; this would present a great problem both electrically and from a safety point of view for the operator. Some manufacturers find it more practical in design and in manufacturing to eliminate this meter altogether and simply provide a

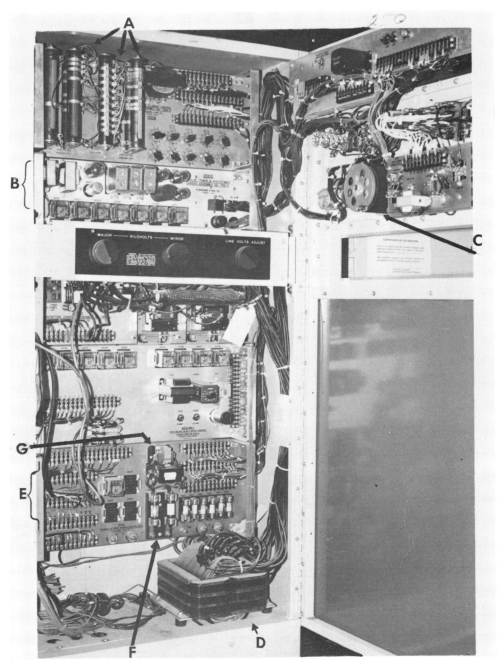

Figure 249. The inside of a modern generator. (A) Resistors for controlling Ma. (B) Timing circuit. (C) Timer control. (D) Autotransformer. (E) Accessory panel to supply tables, locks, etc. (F) Main fuses for incoming 220 volt current. (G) Remote timing contactors.

series of push buttons or Kv knobs individually marked with appropriate increments that when selected by the technologist adjust kilovoltage. It should be pointed out that in any system, the Kv controls, whether they are buttons or knobs, are directly connected to the autotransformer, as shown in Figure 250. Here we see an opened x-ray control panel with Kv controls going directly to the autotransformer.

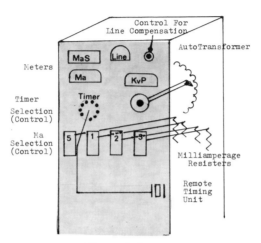

Figure 250.

The autotransformer is a variable transformer, meaning that it can either boost (step up) the voltage or decrease (step down) the voltage. A schematic of an autotransformer is shown in the basic x-ray circuit illustration (Fig. 248). The KvP knobs move an arm which makes contact at various points of the windings, which serve to increase or decrease the voltage sent to the high tension transformer.

The Voltage Compensator

Once the technologist selects a given set of exposure factors, in terms of kilovoltage and milliamperage, they will likely yield the desired exposure. However, if the incoming current from the utility company through the hospital or office building lines fluctuates more than about 10 percent, the output of the x-ray tube will be noticeably affected and will result in either decreased or increased radiographic density. These incoming line voltage fluctuations are common, and in certain hospitals they do, in fact, fluctuate as much as 20 percent.

With this in mind, equipment manufacturers install a type of compensating mechanism to correct these changes in incoming voltage. This is accomplished in one of two ways. The first is to install a line voltage compensator knob and a line voltage meter, which is usually wired before the autotransformer as shown in the x-ray schematic drawing. The meter and the control knobs are located on the generator but are wired before the autotransformer. Figure 250 shows these devices, and it is important that technologists use them to their best advantage. For example, before the technologist begins to set the exposure factors, the line voltage meter should be viewed to see if there are, in fact, any incoming line fluctuations.

These meters usually are equipped with a calibration point, and the dial in the meter should be adjusted so that it is aligned with that calibration point. The incoming line also carries current to the filament circuit. Small fluctuations in the filament circuit cause disproportionately great changes in the amount of tube current (Ma) that is produced, so it is doubly important that the line compensator be monitored frequently by the technologists before exposures are made.

The second method used by some manufacturers to stabilize incoming current is to install an automatic current regulator with no visible control knobs or meters present on the x-ray control panel. With this system, there are various types of current regulators electrically installed inside the circuitry which monitor and then control automatically the incoming voltage in an effort to keep it stable.

The Milliamperage Meter and Control

As you know, electricity contains two major components which when multiplied together produce a total amount of electrical energy called watts. We have just seen how voltage is initially set by the technologist at the autotransformer; however, the amount of amperage must be monitored and controlled as well. This is accomplished by a Ma meter and a Ma control system, both of which are located on the x-ray control panel.

In the basic electrical schematic we have identified three major circuits, the primary, the secondary, and the filament circuit; each of these has important characteristics. For the sake of discussion, however, the filament circuit is divided, as shown in Figure 248, into a primary and secondary side as well. The milliampere meter (Ma meter) is located on the x-ray control panel but is actually wired directly and reads the current moving through the center of the secondary of the high tension transformer. The high tension transformer is located in the large steel oil-filled tank usually placed on the floor in the exposure room separate from the table and from the x-ray control panel. The control knob for the milliamperage and the meter appears on the x-ray control panel, with the actual mechanism used to change the Ma located within the control panel itself and thus a part of the primary filament circuit as shown on the electrical schematic drawing. Although it will be given more attention later in this chapter, the purpose of the Ma control is to vary the amount of current moving through the filament circuit, which in turn controls the temperature of the x-ray filament in the tube.

The X-ray Timer

The x-ray timer contactors are remotely wired into the primary circuit, with the actual mechanism used to control the time located in the x-ray control panel. One of the important characteristics of the primary circuit is that it carries a very high amperage and relatively low voltage as compared to the secondary circuit, which has very high voltage and low amperage. Actually the primary circuit will carry more than 100 amperes. Because of the high amperes, it is safer if the actual contactors or shut-off mechanism is not operating in close proximity to the technologist, especially in older machines that use mechanical contactors. These contactors permit current to flow through the x-ray secondary circuit when in the closed position; however, when the timer acts to terminate the exposure, these mechanical contacts abruptly separate causing a certain arcing of the current as it tries to maintain its continuity from one point to the other. A similar circumstance occurs when you pull an electrical plug from a wall receptacle while the electrical device is working at a maximum strength. This arcing is dangerous to the operator and must be placed in a more remote position than at the face of the x-ray control panel. These contactors (remote timing unit) in more modern equipment have been replaced with solid state circuit boards.

With older generators these electrical contacts are usually located at the bottom and back of the x-ray control panel and are remotely operated by the actual timing knob or buttons of the x-ray control panel. There are a variety of different timing systems: synchronous, electronic, and impulse timers. These various timers have been used over the years and do not warrant a detailed discussion; basically the synchronous timer is operated by a synchronous motor and is accurate only to approximately 1/20 second. These timers work in harmony with the standard 60 cycle electrical current, as a motor on a common electric clock at home. The electronic timer is slightly more accurate, but the impulse timer can control the time exposures accurately down to 1/120 second.

Today's timing systems are all solid state and have special circuit boards that control the timing without any type of mechanical operation. With changes in technology, it is becoming more common that exposures of less than 1/120 second are necessary; in fact, exposures in milliseconds (1/1000 second) are becoming common. For example, if high milliamperage techniques are used with rare-earth imaging systems, the exposure time for a medium to small patient's chest examination could very well be 1/250 second or shorter for pediatric work. The most critical timing devices are necessary with cardiac catheterization work, arteriography, and digital subtraction fluoroscopy. Such timing systems must be able to stop the current in the primary circuit at any point in the electrical phase and not merely at a convenient point, when the voltage is at zero. In Figure 251 we see a typical single and three-phase current pattern and the termination points for each.

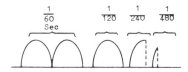

Figure 251. This illustration gives an idea of how much of the electrical phase is used at various exposure times. One can imagine this "forced extinction" of the exposure requires special circuitry, considering that up to 150,000 volts is moving these pulses of tube current.

Forced Extinction Circuitry

Since there is no convenient stopping point relative to the electrical phase with exposures less than 1/120 second, special circuitry is needed to stop the current at any point through its electrical phase. This special circuitry is known as forced extinction and works as a supplemental circuit to apply special electrical power to the timing circuit to instantly stop the flow of current at any point in the electrical phase.

Phototiming Systems

Phototiming or automatic timing systems are becoming more and more widely used and offer some distinct advantages. These advantages have already been given appropriate attention in Chapter Five and warrant only brief additional discussion here. A quick review of the electrical schematic of the basic x-ray circuit will indicate where these automatic timers are actually wired into the primary circuit.

Circuit Breakers

A circuit breaker is a device that "trips" when a surge of current passes through it. It is in effect a modernized fuse. A circuit breaker is wired into the primary circuit just before or after the autotransformer, depending upon the manufacturer. When a fuse experiences a surge of current, the excess heat generated by the "extra" current is sufficient to melt the conducting wire running between the end points of the fuse. When the conducting wire melts, it separates and opens the circuit. The circuit breaker is an updated version of the x-ray fuse and is more convenient to operate. There is always a circuit breaker of some design located in the x-ray control panel. This device protects the technologists from electrical hazards and protects the equipment from sudden surges in current which can cause internal damage. For example, if a wire becomes loose in the x-ray control panel it could cause a short and an associated surge of current; the circuit breaker would sense this current increase and trip to protect the remaining equipment from electrical overload. Of course, the x-ray control panel is grounded, and any such short or electrical malfunction will be grounded immediately, keeping the technologist safe from electrical shock.

Usually if the circuit breaker trips it can be reset by the technologist. However, if it trips a second or third time no further attempts should be made to work the piece of equipment until the serviceman can evaluate the problem, since repeated attempts to make an exposure under these conditions could cause electrical damage to the equipment.

INTRODUCTION TO THE SECONDARY CIRCUIT

It is important to discuss how the current flows through the rectifying system, the high tension transformer, and the x-ray tube. While the exposure switch is in the open position, there is no electrical activity in the secondary side of the x-ray circuit. However, when the exposure switch is closed, three distinctly different parts of the circuit begin to operate in harmony. The filament current begins to flow through the x-ray tube filament creating enough heat to produce a thermionic emission and adequate space charge for the desired Ma. Concurrently with this, the high tension transformer creates a situation where the electrical potential between the cathode and anode of the x-ray tube is increased to a very high degree; however, no "tube current" is flowing until the exposure is closed. When this occurs, the electrical current moves through the tube in spurts according to the pulses (or phases) flowing through the rectifiers. With each pulse of current (in concert with each electrical phase) a bolus of electrons (the space charge) streak toward the anode at approximately one-half the speed of light and collide with the target material, causing a phenomenon that transfers the kinetic energy into x-ray and heat energy.

The current continues to flow as each succession of space charges strike the anode as long as the exposure switch is closed. Even though a great deal of kinetic energy is lost at the target material, the electrons continue to flow through the anode, through the shaft of the anode and the anode wires to the rectifying system—the secondary side of the H.T. transformer and the Ma meter to complete the flow of current through the secondary circuit. As the current through the secondary circuit flows, the milliampere meter, which is wired to the secondary of the high tension transformer, begins to read the milliamperage, as seen on the control panel.

High Tension Transformer

A vital component of the basic x-ray circuit is the high tension transformer. In order to produce x-rays, a high degree of kinetic energy must accompany the electrons of the space charge. The kinetic energy is given to each electron by this high voltage transformer producing from 45,000 to

150,000 volts between the cathode and anode of the x-ray tube. The high tension transformer is a step-up transformer with the ratio of approximately 1:500. This ratio means that for every one volt of current entering the primary side of the transformer 500 volts leaves the secondary side.

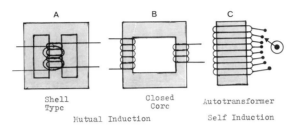

Figure 252. The basic construction of three common transformers. *A* has its primary and secondary coils wound around the same area yet they are insulated from each other. *B* has fewer coils on the secondary side, making it a step-down type. *C* is an autotransformer with numerous taps.

With the assumption that the reader is already somewhat familiar with the basic construction and operation of transformers, only a brief review will be presented here. A total amount of electrical power (watts) moving through a circuit can be obtained by multiplying the amperage by the voltage. Current must, however, be altered somewhat for any given subcircuit within the main x-ray circuit so that each individual component can be supplied with the appropriate ratio of amps and voltage for proper operation. These alterations in current through the various subcircuits are produced by capacitors (which have the ability to store and suddenly release specified amounts of current) and resistors (which take away a specific amount of current from a circuit by converting electrical energy into heat).

Another of these current-altering devices is the transformer, which is used to increase or decrease voltage. Whenever a transformer is used to change voltage, the amperage is also changed since the transformer is not an energy producer and therefore cannot simply produce more voltage. The total amount of energy in watts coming from the secondary side of the transformer is approximately the same as what entered. For example, if a step-up transformer increases the voltage, the amperage is reduced, and conversely if the step-down transformer decreases voltage, amperage increases so that the net amount of watts leaving the transformer is very similar to the wattage that entered the primary side of the transformer. With regard to the entire x-ray circuit, the primary circuit is said to be a low voltage high amperage circuit, and the secondary circuit is said to be a high voltage low amperage circuit: the initial incoming current at the control panel is approximately 220 volts at 100 amps but changes after the high tension transformer to 40,000 to 150,000 volts with an amperage ranging from 0.05 (50 Ma) to 1.5 amps (1,500 Ma).

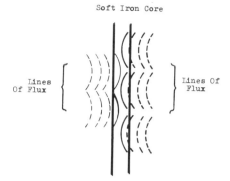

Figure 253. Illustrates the magnetic flux from the A.C. current crossing the iron core. The magnetic domains change direction as they are pushed and pulled by the external flux lines caused by the A.C. current.

There are basically three types of transformers: the autotransformer already mentioned, the step-down transformer, and the step-up transformer. The basic concept is that the rate of voltage change (between the primary and the secondary side) is controlled by the number of coils that are placed around the magnetic core. Transformers can only operate with alternating current, because when the electron flow changes from positive to negative phase, a pulsating field is generated around the magnet. This pulsating field causes the domains of the iron core to change in concert with alternating phases of the A.C. current. In turn, these changing domains of the core produce a magnetic flux that cuts coils of wire in the secondary side with current in these coils. See Figure 253. It is in fact the relative number of coils on each side of the transformer (primary or secondary side) that determines to what extent the voltage will be changed. The number of coils needed on each side of the transformer can be calculated once the voltage needed is determined. If, for example, the voltage is to be increased, we know that more coils have to be placed on the secondary side, but if the voltage is to be reduced, fewer coils must be placed on the secondary side compared to those of the primary.

Transformer Energy Loose

It was stated earlier that the wattage leaving the secondary side of a transformer is similar to the total power entering the primary side. In reality, whenever energy is changed or goes through some type of adjustment, there is always some loss of energy toward heat. This is also the case with transformers since a certain amount of energy is lost as the result of three circumstances that occur within the transformer simultaneously. The first, hystersis, occurs as the magnetic domains in the iron core change rapidly as they adapt to each positive and negative flux of the alternating current. This fast-changing flux (120 times per second for single phase and 360 times per second for three phase) sets circumstances whereby the magnetic domains in the iron core almost "run into themselves" as they shift from one direction to another. During this realignment, no current can be induced to secondary coils, causing a reduction in electric flow, but some heat is generated in the core. The second circumstance that contributes to heat and loss of current in a transformer is the production of eddie currents, which are small quantities of electrical current that are a kind of "byproduct" or "secondary current." These low energy currents in effect oppose the general flow of current moving through the coils. Eddie currents are in some ways similar to small swirls or ripples of water that seem to oppose the general flow of water around twigs or stones that lay in the path; in a transformer they are caused by the pulsing flux lines. The third circumstance which contributes to heat of the transformer is the conductive wires themselves. As we know there is always some amount of heat produced by current flowing in any type of electrical circuit. The resistance (ohms) is caused by the wires themselves and it is especially important with transformers because of the great length of wire needed to make up the windings on the primary and secondary side. Altogether the electrical loss caused by these three major circumstances will produce about a 10 percent decrease in power. The amount of electrical current entering the primary side of the transformer in watts will equal the amount of current leaving the secondary side of the transformer minus approximately 10 percent for transformer heat loss.

Transformer ratio identifies the relative increase or decrease in the secondary side of the transformer, compared to the primary side. The high tension transformer of the x-ray circuit usually has a ratio of 1:500. Since this ratio is fixed, the voltage fed to the primary side of the high tension transformer will be adjusted by the technologist by using an autotransformer. For example, if the technologist adjusts the KvP selector to a position that will allow 90 volts to come through the high tension transformer, the voltage will be stepped up 500 times and will produce 45,000 volts (45 KvP). However, the scale on the prereading voltmeter is calibrated to read what the voltage will be as it moves through the secondary circuit and x-ray tube.

The Autotransformer

The autotransformer is known as a self-induction transformer as compared to a mutual induction transformer such as in the step-up or step-down type. Figure 252C illustrates an autotransformer. Here there is only one set of coils, but they contain "taps." If contact is made with any one of these points, the flow of current is precluded at that point from the remaining coils of the transformer, and thus, the amount of current and flux lines cutting the iron core is reduced. This, of course, limits the amount of voltage that will be increased. As you can imagine by looking at this schematic, the more coils that are involved in carrying current along the core, the higher the voltage that will be produced.

The Milliampere Meter and the Milliampere-seconds Meter

It has already been pointed out that the current going through the secondary circuit has a very low amperage, so low in fact that it is referred to in milliamperes. The term *milli* equals 1/1,000, and the tube current is measured in milliamperage. When the technologist selects 200 Ma on the control panel, a specified amount of the tube current will ultimately be produced by virtue of the temperature of the tube filament. The only amperage available to the secondary circuit and the rectifying system is what is produced by the filament temperature and the resulting space charge. The current going through the secondary is actually selected by the technologist on the control panel. For example, an exposure using 80 KvP and 400 Ma would actually produce 80,000 volts of current through the x-ray tube and rectifiers, and 400 Ma would yield 0.2 amps of current through the x-ray tube and rectifying system.

One should be careful to distinguish between the current traveling through the filament circuit and the tube current. Filament circuit current is rated in amperes, usually 3 to 5 amperes in value, whereas the tube current is measured in milliamperes, usually ranging from 50 to 1,500 milliamperes.

The purpose of the Ma meter is to measure the actual amount of current flowing through the secondary circuit and the rectifiers, even though the meter itself is located on the x-ray control panel. The Ma meter is important to the technologist since it supplies important information leading to the quality of the resulting radiographic exposure. For this reason we will itemize three specific circumstances the Ma meter can indicate to the technologist while the exposure is on. Also since the amount of current flowing through the tube fluctuates many times a second, the Ma meter registers the average amount of the tube current. In this way the Ma meter registers the average Ma during a given exposure.

1. *Verifies actual exposure:* since the only current going through the x-ray tube is tube current and this is eventually what the Ma meter reads, when the technologist presses the exposure button a zero reading on the Ma meter will instantly inform the technologist there was no exposure.
2. *Verifies the milliamperage actually produced:* because of the wiring architecture noted in the schematic of the basic x-ray circuit, if the technologist selects 400 Ma at perhaps 1/5 a second, the Ma meter should read 400 when the exposure is on. If the Ma meter indicates a maximum of 200 Ma on its scale, the technologist knows instantly the radiograph will be underexposed because only one-half of the intended quantity of tube current was produced. On the other hand, it is possible for the Ma meter to register in excess of the selected Ma, which will of course indicate to the technologist that the resulting radiograph will be overexposed.

3. *Indicates a gasy tube:* it is very important that the vacuum of the x-ray tube is maintained to its utmost degree. For a number of reasons, the x-ray tube can become gasy (a leakage of air molecules into the glass envelope). In these circumstances the abundance of air molecules will give up some of their electrons and be added to those electrons in the space charge, and will in effect add to the tube current in a degree depending upon the amount of air that has seeped into the glass envelope. If the technologist makes an exposure and the Ma meter "spikes" to the far side of the meter, it usually indicates that an x-ray tube is gasy because the ionized air produced many more electrons moving toward the anode. The Ma needle spikes because it is attempting to read the tremendously increased volume of tube current that is traveling from filament to anode. When this situation is noted on the Ma meter, the technologist should discontinue the examination and call a serviceman since the x-ray machine is unuseable and probably the x-ray tube has to be replaced.

The MaS Meter

The Ma meter reads the quantity of electrons moving through the tube during any given exposure, providing the exposure time is long enough for the meter to register to the maximum amperage accurately. In other words, it takes a brief moment of time for the needle to respond to the total flow of current through the x-ray tube and to stabilize. The amount of time needed for this to occur is generally considered to be about 1/10 second. For shorter exposures another meter must be used to measure tube current, and it is referred to as the MaS meter or milliampere-seconds meter. The MaS meter as the name implies measures the combination of tube current and exposure time and integrates these factors to produce what the total MaS should have been. The MaS meter is often the same meter as the Ma meter, however, the integrated figures are printed on a second scale of values. In this way the Ma meter is also calibrated for MaS values and when that scale is used it becomes an MaS meter.

THE FILAMENT CIRCUIT

We have already discussed the three criteria for producing x-rays. A supply of electrons is needed, a high degree of kinetic energy is necessary to accelerate the electrons, and the electrons experience sudden deceleration. We have also just seen how the circuit produces the required kinetic energy by virtue of the high tension transformer and how the timing circuit controls the duration of the exposure itself.

To understand how the filament circuit actually controls the tube current, it would be helpful to think of the filament circuit as an independent circuit operating in conjunction with the basic x-ray circuit as discussed so far. The filament circuit is, of course, itself divided into a primary and secondary side as you can see on the basic x-ray drawing (Fig. 248). The filament circuit supplies the x-ray tube with the required amount of electrons to be bombarded against the target material. As you can see from the large schematic, the basic filament circuit is relatively simple, and extends from the primary side all the way to the filament of the x-ray tube. Thus the filament of the x-ray tube is actually a continuation of the filament circuit.

The current values moving through this circuit usually range between three and five amps in 4 to 12 volts; thus, in relative values, the filament circuit carries a low amount of electrical current. We will now discuss the various components separately.

The Filament Ammeter

Although this device is not used in newer pieces of equipment, when it did appear it was used to read the amperage moving through the primary side of the filament circuit. It was pointed out in Chapter Fourteen that amperage flowing in the filament circuit is not equal in value to the tube current (Ma); therefore, the filament ammeter should not be confused with the Ma meter. When used, the ammeter was wired to the primary side of the filament circuit and located on the x-ray control panel.

Filament Current Control (Ma Selector)

This device is made in the form of a series of resistors, one for each Ma station. The actual configuration was discussed in Chapter Fourteen and little additional information is needed beyond what was said there. When the technologist chooses a given Ma and presses the appropriate button, a specific resistor comes into play which changes the amount of current flowing in the filament circuit and in this manner regulates the temperature of the filament in the x-ray tube which, of course, controls the size of the space charge and the tube current. The filament control (Ma selector) also appears on the x-ray control panel in the form of a knob instead of push buttons, In any event, the filament control indirectly regulates the tube's filament temperature by the array of various resistors it comes in contact with, as shown in the basic electrical schematic.

It would not be correct to state categorically that the Ma controls the size of the focal spot. However, it is true that a supplemental circuit is included in the Ma selector control so that when the technologist turns the Ma beyond a certain point (usually at 200 Ma and above) it switches current to the large filament of the x-ray tube automatically. With this arrangement one can correctly say that the Ma station has some indirect effect on the focal spot size since there are approximately 5 to 10 Ma stations on some of the larger x-ray control panels and only two filament sizes; however, changes in Ma are not generally considered to affect focal spot size.

Filament Step-Down Transformer

The purpose of the filament step-down transformer is to step down the voltage to the secondary filament circuit and ultimately to the x-ray tube so that the proper temperatures can be created at the filament of the x-ray tube. The actual voltage going to the primary of the step-down filament transformer is approximately 3 to 5, and ranges from 10 amps to 16 volts as it comes out of the secondary side of the filament step-down transformer. The filament step-down transformer is located in the large lower steel tank remotely located from the table and the x-ray control panel.

Filament Current Stabilizer

As we have discussed earlier in this chapter, the incoming lines to the wall box from the utility company do experience some current fluctuations and thus alter the voltage in the autotransformer and high tension transformer. The filament current also experiences incoming current fluctuations, and it is especially important for the filament circuit to remain stable during small changes in the filament current. With this in mind, a very effective filament current stabilizer is wired into the primary side of the filament circuit and is designed to hold the filament current to very close tolerances. Please refer to the basic x-ray circuit drawing. The device that performs this task in modern equipment is a solid state circuit, taking the place of the obsolete choke coil.

RECTIFYING SYSTEM

The current moving from the secondary of the high tension transformer must be an alternating type as shown in Figure 254. This illustration shows wave forms of five alternating and direct single phase and three-phase current:

1. self-rectified (single phase)
2. half-wave rectification (single phase)
3. full-wave rectification (single phase)
4. six valve rectification (three phase)
5. twelve valve rectification (three phase)

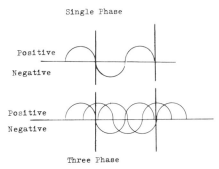

Figure 254. Single and three-phase wave forms.

How Valve Tubes Work

Rectifying tubes used until recent years were actually thermionic diode tubes that had both cathode and anode elements. Recently, however, this type of tube has given way to solid state rectifying "sticks" usually made of a silicone type material. Many pieces of equipment are still equipped with the more conventional valve tube system and a brief explanation of how they work will be presented at this point. Figure 255 shows the construction elements of the conventional thermionic diode rectifying tube. As you can see, the filaments built into a valve tube are noticeably different in shape and in thickness when compared with those of the x-ray tube. The anode is also strikingly different from the ones used for x-ray tubes, but their shape allows the electrons to arch easily from the filament to the anode and continue through the remainder of the rectifying system and the secondary circuit. This is important since holding back or resisting current flowing through these valve tubes could seriously affect the overall efficiency of the x-ray tube. With this in mind, the cathodes are made to fit inside the cylindrical anode and are very much elongated so as many electrons as possible can jump from the filament to the anode side of these diodes. Figure 256 shows how a pair of valve tubes can be wired into the secondary circuit containing single phased current to provide both full-wave and half-wave rectification.

Once the space charge has developed and high voltage is applied across the cathode and anode of the rectifying tube, current begins to flow. Since the anode does not produce a space charge of its own, it is impossible under normal circumstances for current to reverse its direction and move backwards from the anode during the negative phase of alternating current. For this reason, no electrons are available to flow and the negative is said to have been suppressed. There is, however, a great deal of heat impressed upon the anode during its negative phase even though no current flows.

Self-Rectification

With the earliest types of x-ray equipment, the x-ray tube was wired directly from the high tension transformer without any valve tubes present. This arrangement led to a considerable amount of heat generated on the anode of the x-ray tube, and when higher exposure values were used it was possible that the anode could become so heated that it could produce its own

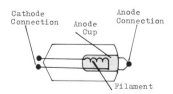

Figure 255.　Schematic of a diode rectifier tube.

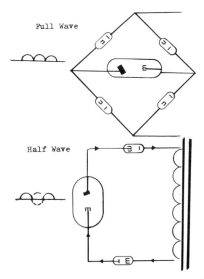

Figure 256.   Half- and full-wave rectifying schemes for single phase.

electron space charge. When the alternating current switched from positive to its negative phase, the small space charge of electrons at the face of the anode would travel toward the filament and cause it and the tube instant damage. In addition to this circumstance of tube damage, it was a very inefficient system since there were no x-rays produced during one-half of the electrical cycle, the negative phase.

Half-Wave Rectification

The first step toward advancement beyond the self-rectified system was half-wave rectification. In this method two valve tubes were placed between the high tension transformer and the x-ray tube. The primary advantage of this setup was that the x-ray tube could be spared the excessive heat brought on by the negative phase impressing on the anode directly because the valve tubes were placed in such a way after the high tension transformer to absorb the negative phase before the x-ray tube. This system basically allowed the technologist to choose slightly higher Ma values than with the self-rectified system without fear of imminent x-ray tube damage from inverse voltage. However, the problem of inefficiency in x-ray production continued, in that only one-half of the electrical current was being used to produce x-rays.

Full-Wave Rectification

The second and final advance in rectification came with the addition of two additional valve tubes placed between the high tension transformer and the x-ray tube. The illustration in Figure 256 shows how these valve tubes are arranged and plots the general flow of current through the rectifying system to the x-ray tube. This system was developed many years ago and because of its effective operating characteristics has been relatively unchanged since its inception. With full-wave rectification the negative phase is not suppressed as it had been with the self- and half-wave rectification, but rather converted to a positive phase which can drive the x-ray tube with more than twice the efficiency of self- or half-wave rectification.

Of course the anode in the valve tube is constructed in such a way that it can easily tolerate the heat generated by the negative phase of the alternating current without any reasonable chance of allowing the condition known as inverse voltage. With this four valve arrangement the negative phase of current can be utilized by the x-ray tube to produce x-rays and therefore in a sense full-wave rectification converts the negative phase to a positive phase.

The Spin Top Test

Actually the phases of electrical current moving through the x-ray tube as shown in previous illustrations makes for a convenient method to check the accuracy of the x-ray timing circuit and Figure 257 shows how this is done. Here we see the sequential pulses of current through the x-ray tube as seen by the spinning top. A spinning top (which is usually made of a brass disc in which a small hole is drilled) is spun by the technologist before the test exposure is made. As the spinning top turns rapidly, the test exposure is made and the hole passes over different parts of the film with each moment of exposure time. With the knowledge that x-ray photons are emitted from the tube in spurts which are actually in sequence with the electrical phases going from cathode to anode, it can easily be imagined that each pulse of photons from the x-ray tube will find the hole in the spinning disc at a different location over the film. Since the disc is opaque to x-rays the only exposure to the film is through the tiny hole.

Since the bursts of x-ray photons are synchronous with each phase of electrical current we can deduce that 120 such bursts will be produced during a one second exposure for single phase current. If the one second exposure was made in this manner, so many dots would be produced on the film that they would superimpose each other and in fact produce a single circle on the film. However if a shorter exposure time was used, such as one tenth of a second, only twelve pulses of current and x-rays would be produced by the tube, thus twelve dots showing on the film. The table shown below will show the correct number of dots that should be seen on an x-ray film with corresponding exposure times. Usually the longest practical time that can be tested with the spin top is 1/10 second, since the twelve dots will usually span the entire circumference of the circle with any additional dots beginning to superimpose, thus making it difficult to differentiate one from the other for an accurate count. However, it is practical to check the x-ray timer with the

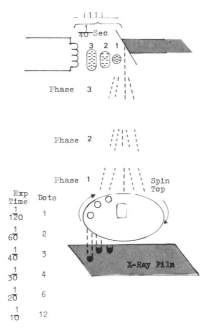

Figure 257. Spin top test. This test is useful only when single phase equipment is used at 1/10 second or shorter.

spinning top during 1/120 a second which would, of course, produce only one dot on the film.

Dots for Spin Top Test

Exposure time	Dots showing on film
1/120	1
1/60	2
1/40	3
1/30	4
1/20	6
1/10	12

In addition to checking the exposure time, the rectification system can be evaluated by the appearance and density of the dots as well. For example, if only one-half of the number of dots appear on the film that you would expect with the table above, there is a reasonable chance that a defective valve tube exists in the rectifying system because one defective tube will break the continuity of that particular phase of current.

In general, the density of the dots on the x-ray film are fairly similar; if they vary it is often an early indication of trouble. If there is a noticeable alternating density from one dot to the next (one dot being darker than the other) it is a reasonably good indication that a valve tube is not totally defective but is not able to conduct properly the full electrical current for the phase of current it is assigned to conduct through the circuit.

Disadvantages of Spin Top Test

There are two disadvantages to using the spin top method for checking the timer. First, as already discussed an exposure time greater than 1/10 second makes so many dots it is very difficult if not impossible to distinguish visually separate dots and thus get a correct count. Second, it cannot be used for three phase equipment and this will become evident as we proceed with our discussion on three phase current.

Three Phase Rectification

Fortunately or unfortunately, it seems someone is always working on a better method of doing things. The opportunity for new invention with x-ray circuitry can be seen upon careful inspection of the wave form of a fully rectified three phase circuit as shown in Figure 258. From this it is apparent that there are times when no voltage whatever is flowing from cathode to anode, and therefore no photons are being produced. In addition, with single phase fully rectified current—since the voltage is zero at the beginning of the electrical phase and rises quickly during the phase to its peak and then reduces back to zero at the end of the phase—there is a wide fluctuation of tube potential during any given exposure. In fact, there are many instances within 1/10 second exposure when this fluctuating voltage causes a wide distribution of x-ray photon energies with great variances in penetrating power. For example, if an exposure was at 1/120 second only one phase of current would be utilized for the entire exposure; however, the voltage of the tube current would have fluctuated from zero to the maximum selected KvP, which might be 90, and back to 0 at the end of the exposure. Thus during this time a wide distribution of photon energies would be produced yielding a very heterogenic x-ray beam. At the same time, the patient would be exposed to radiations during much of this phase that are relatively low energy and that expose the skin but have no direct effect on the x-ray image. This fluctuation in tube current voltage also tends to increase subject contrast, because although 90 Kv might have been chosen as the peak voltage for an exposure there are many photons that strike the patient and pass to the film with energies equivalent to 50, 60, 70, 80, and 90 KvP.

These low Kv values would have no practical effect on the diagnostic image but do contribute to low energy photon skin dose. They also add heat units to the anode without producing useable x-ray photons. The idea, therefore, is to produce a wave form of tube current that will not drop dramatically towards zero, but rather stay as close as possible to the peak voltage that was selected by the technologist, which leads us directly to a discussion of three phase x-ray equipment.

To begin we must look at the incoming current to the x-ray room and junction box before the x-ray control panel. Figure 254 compares the wave forms of three phase and single phase current. It will also be helpful to recall the manner in which current is actually produced at the power station, and to keep in mind that the number of phases per second (cycles of current per second) is controlled by how many times the communicator cuts across the magnetic field per second in the generators at the electric power plant. By coordinating the timing of these generators it is possible to produce a flow current through a conductor that has 360 phases per second as opposed to 120 phases per second with single phase.

The basic electrical circuitry of three phase equipment is the same; however, considerable modifications are necessary for three phase equipment, as can be seen on actual detailed electrical drawings from the manufacturer. In any event, three phase current must also be rectified since it is alternating current, and a typical wiring scheme for this is shown in Figure 258. The rectification process for three phase equipment can be accomplished by using either 6 or 12 valve tubes. It is not important to memorize how these two systems are wired in the rectifying circuit of three phase equipment, but it is important to understand the difference between 6 and 12 valve rectification, as well as some of the general characteristics associated with three phase equipment.

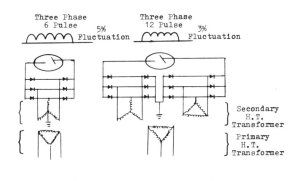

Figure 258. Three phase rectification using 6 and 23 pulsed systems.

Characteristics of Three Phase Exposures

It has already been pointed out that the efficiency of single phase equipment is less than three phase equipment, because the tube current voltage is allowed to drop to zero at the beginning and the end of each individual phase. This translates to approximately a 30 percent reduction of output for single phase generators compared to three phase generators. With this in mind, if equal exposures were made on two equally calibrated pieces of equipment, the radiograph produced by the single phase equipment would yield only approximately 70 percent of the density obtained with the three-phase equipment machine.

Radiographic contrast decreases when three phase equipment is used, since the voltage of the tube current drops only about 5 percent with 6 valve tube rectification and 3 percent with 12 valve tube rectification. See Figure 259. With this in mind, the average energy of x-ray photons penetrating the patient is considerably higher with three phase compared to single phase. One can say, in fact, that this circumstance partially accounts for the distribution of photon energies, whereas three phase equipment is considerably more homogenic. Therefore, the average energy of the beam from three phase equipment is considerably higher when compared with three phase.

Patient Doses and Three Phase Equipment

Patient dose is also affected by three phase equipment. The formula that was given in the previous chapter to calculate milliroentgens per exposure can be useful for comparison here. You will recall that two different exposures were made, each yielding the same radiographic density, one with 80 KvP, and the second with 92 KvP and less MaS. From the simple formula mentioned, we were able to calculate that the 80 KvP exposure produced 720 mr and the 92 KvP exposure only 476.1 mr. The reason, of course, is that high KvP exposures are more efficient, since the photons individually average greater energy; fewer are needed (less MaS is needed to produce the desired density). By reviewing Figure 259, which compares single full-wave rectification with 12 and 6 valve three phase rectification, it is easy to see that the average effective KvP for three phase equipment per exposure is higher compared to single phase equipment, and this more efficient beam reduces patient dose.

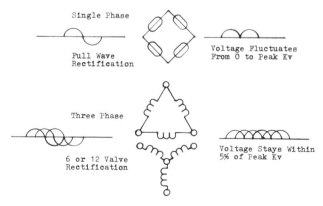

Figure 259.   Compares the tube current pulses for single- and three-phase rectification.

X-ray Tube and Three Phase Equipment

No special tube design is necessary to operate with three phase equipment. However, there is a noticeable advantage in tube rating used with three phase equipment compared with single phase, as shown in Figure 239. Here we can see how the tube rating for the same tube changes depending on its use with single and three phase equipment. It should be pointed out, however, that the tube cooling characteristics are the same regardless of the equipment used, since that is dependent on characteristics of the x-ray tube itself and its ability to rid itself of the anode heat.

In summary, three phase equipment offers some tangible advantages in that it produces a greater amount of radiation per exposure with less patient dose but with some loss in radiologic contrast. The disadvantage is primarily in cost. An average three phase 600 Ma generator will cost about twice as much as a 600 Ma single phase generator.

SUMMARY

There has been a deliberate attempt to explain the operation of individual components of the basic x-ray circuit as well as the overall operation in as concise terms as possible. In fact, the operation of the x-ray circuit as shown in this chapter is quite simple and can be explained with relative ease providing the basic characteristics of the primary circuit, the secondary circuit, and the filament circuit are kept separate from each other. These are, indeed, individual circuits and for purposes of our conversation operate somewhat independently from each other with totally different functions.

The basic function of the primary circuit is to control the voltage that is delivered to the high tension transformer and to regulate the exposure time. The primary function of the filament circuit is to regulate the temperature of the filament of the x-ray tube so that the required or desired Ma can be obtained during the exposure. The function of the secondary circuit is to increase the voltage delivered to it by about 500 times so that the electrons in the space charge of the x-ray tube can be accelerated to a point where x-ray production occurs, as well as converting the alternating current to direct current for maximizing x-ray tube efficiency.

CHAPTER SIXTEEN

T.V. CAMERAS, IMAGE INTENSIFICATION, AND DIGITAL SUBTRACTION

THE EXCITEMENT AND disbelief that ran through the community of physicians and physicists as Roentgen discussed and presented his discoveries regarding the production of x-radiation, and the almost rampant activity to duplicate his experiments and produce similar effect, are unparalleled in medical history. Since that time progress in radiography has been marked with numerous cycles of more modest discoveries and achievements by physicists and radiologists who had been attracted to the medical x-ray field in conjunction with a developing lineage of talented and dedicated technologists. A very interesting book published by Charles C Thomas, Publisher, entitled *The Trail of the Invisible Light* is a virtual encyclopedia of radiologists, physicists, and technologists who contributed to the development of radiology over the years, beginning with Roentgen's discovery to recent times.

Using Roentgen's discovery of x-rays as a standard of measurement, the field of radiology had settled into a refined diagnostic medium contributing enormously to the general standard of practicing medicine. This gradual development in techniques and procedures in the field of radiology, however, came to an abrupt end during the mid-1960s when the first computed tomography (CT) scanning unit was tested and developed for mass production. Its availability on the medical marketplace soon changed many methods by which physicians of other medical disciplines practiced. The primary impetus to the advancement of the CT scanner was the modern computer. Once this marriage in equipment between the x-ray machine and computer was tested, radiology was off on a new roller coaster of development and discovery. Papers were written about CT, probably with enthusiasm equal to those who first experimented and explained Roentgen's discoveries and suggested related medical applications.

As the use of CT scanning equipment has become commonplace in today's modern radiology arsenal of diagnostic techniques, a new and equally startling tool known as nuclear magnetic resonance (NMR) has become available for use as a promising diagnostic tool for the future. As we look into the future, we can easily suggest the threshold to "space age" radiology techniques is near as we view the capabilities of nuclear magnetic resonance.

Between the development of CT and NMR, we have been jolted once again with the development of digital subtraction fluoroscopy.

These three modalities of the present and near future are ushering in a renaissance in radiology diagnostic techniques that is indeed uncommon in medical history. The primary aim of this chapter is to review in some detail the concept, equipment components, and application for digital subtraction fluoroscopy, which is (of the three) most likely to become a part of everyday procedure used by the staff radiologic technologist in the very near future.

MANUAL SUBTRACTION TECHNIQUE

Technologists are already aware of, and in most instances familiar with, the manual technique called subtraction using standard radiographic film. The basic principle involved here is that complimentary densities of opposite polarity (negative and positive) will cancel each other out and in effect become invisible if the two films are sandwiched together and exposed on a third film with the use of visible light. Some further explanation is certainly in order at this point, and will help greatly in understanding the basic concept of digital subtraction fluoroscopic equipment, and a quick look at Figure 260 will help. In reviewing this illustration we are looking at the various steps to manual subtraction technique.

It is important to understand the basic differences between a positive and negative image. We have arbitrarily determined that what our eyes show us as we look at a landscape, another person, or whatever we are viewing is a positive image. The images that are reflecting the most light appear as light or white to us, and the areas that are reflecting the least light appear to our eyes as dark or black areas. The image we see on a standard black and white T.V.

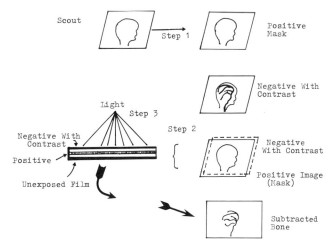

Figure 260.   By using special film and film copier, a mask (positive image) is made from the scout film. The mask is then sandwiched with a negative x-ray image showing opacified vessels. This sandwich is placed into the copying machine and exposed to a special subtraction film; it is then processed and reveals only the opacified vessels.

screen is a positive image since the white and dark areas appearing on the screen are as we would normally expect to see them in real life. For example, the sails of a ship are white compared to the tones of the dark ocean water. We are all familiar with the appearance of a black and white negative film. A negative image of the same ocean scene just mentioned would show the sails to be dark and the water to be white, and so we refer to this as a "reversed" image, or a negative image.

The radiographic images we produce daily are negative images; however, by using various techniques as are described in numerous publications,* the image can easily be reversed into a positive. See Figure 261. An interesting and vital point to appreciate is that the structures seen in these two positive and negative images have complimentary densities. In other words, the structure of a specific part of anatomy in the negative image is as dark as the same area is white on the positive image. In our example of the ocean scene, the density of the sails would be equal to a value of one, and the negative image would have the same numerical value except it would be shown on the negative density scale. Likewise, the density or tone of the ocean would have the same density value on the positive scale as it would on the negative tone scale. Thus we have complimentary numerical tone values but opposite positive and negative images.

The proof of this concept is that if a positive x-ray image and a negative x-ray image of exactly the same anatomical part were placed together very carefully so the structures shown on both images were perfectly superimposed and held together and viewed against a luminator light, the complimentary tones on each film would cancel out and the resulting image would be virtually invisible. See Figure 262. A final step in manual subtraction technique can be performed so that a third film is made which will capture the subtracted image, and this is accomplished in Figure 260. The sandwich of the positive and negative x-ray images is placed in a duplicating machine which is especially designed for image subtraction. In addition to this, an unexposed x-ray film is placed on the sandwich. The duplicating machine is then turned on so that the light passes through the sandwich first and exposes the third film. The white light passes through the sandwich of the two images and the images are transposed to the unexposed film. When the newly exposed film is processed, it will produce a subtracted permanent image of what we saw earlier by holding the negative and positive images to the light of the illuminator.

To this point, the positive and negative images we are subtracting contain identical information so all the structures shown in each image will cancel out on the final film. The next point to appreciate in this process of manual

*Kodak "Subtraction Technic Guide" M3-141.

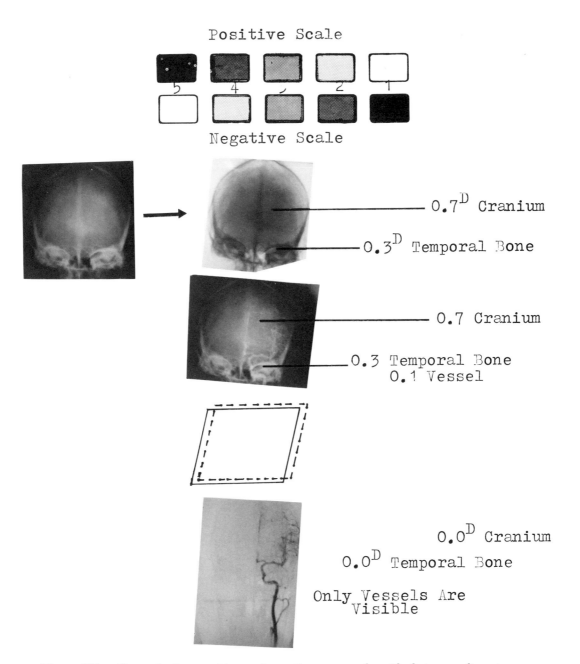

Figure 261. Shows both a positive and negative gray scale with their complimentary numerical values. Also shown is the concept of number density subtraction that is used to produce the final subtracted image. Note there are no complimentary densities on the positive image for the opacified vessels.

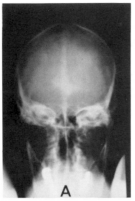
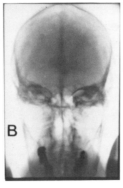
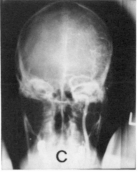
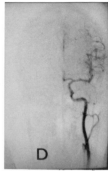

Figure 262.

subtraction is that any structure not "mated" with an opposite structure will not cancel and will therefore appear by itself on the final "subtracted" image. Figure 261 illustrates this point well. We see in this illustration all the complimentary densities have canceled out. However, the density on one of the films is not complimentary and will show with distinct detail on the third film.

The cerebral arteriogram, for example, as shown in Figure 262, shows poor visibility of the internal carotid artery where the opacified vessels are similar in density and superimposed over the dense petrous bone. The image shown in Figure 262D is the final subtracted image, which demonstrates improved visibility of the structures resulting from the complimentary positive and negative densities of the petrous bone canceling or subtracting each other, leaving only the densities associated with the opacified vessel. Thus the advantage in using the subtraction technique is to improve the visibility of opacified vessels where they lay over bone or other body structures that have similar x-ray absorption to the vessels.

MAJOR ASPECTS OF DIGITAL SUBTRACTION FLUOROSCOPY

The system or technique known as digital subtraction fluoroscopy encompasses three distinctly different aspects. The subtraction aspect of this new imaging modality has just been explained with the concept of positive and negative images. The second aspect is of at least equal importance and certainly greater complexity: the remnant beam contains a much greater amount of diagnostic information than had previously been expected but was not fully appreciated since the recording medium used, i.e. screen-film combinations, was not sensitive enough to record these details. An analogy of an individual whose hearing is impaired can be used here. Such an individual could be in an environment with sounds of many different frequencies and intensities, however, the person's reception or recording media (the ears) are insensitive to these signals and therefore they do not exist to the person. Similarly, most film-screen systems require at least a 25 percent variance in MaS before noticeable density changes are made.

The third concept is that the physiology of a healthy human eye makes it unable to distinguish radiographic densities with normal viewing conditions that vary less than 15 percent. Even if the screen-film recording media could capture all the information available in the remnant beam, the human eye could not recognize anatomical structures whose portraying densities happen to vary by 15 percent or less on the radiographic image. To our eyes those structures could not be distinguished from each other and would therefore appear as one whole part. Some benefit could be derived if these minute differences of diagnostic information contained in the remnant beam could be manipulated somewhat and "artificially enhanced" so that our eyes could take advantage of that valuable information. As we will see later, this can be done with the help of a computer.

BASIC COMPONENTS COMPRISING A DIGITAL SUBTRACTION SYSTEM

Fluoroscopic T.V. System

A basic component used in digital subtraction work is the fluoroscopic T.V. system. The remnant beam containing the initial diagnostic information is produced by utilizing the fluoroscopic tube and T.V. image system, as opposed to the overhead tube, thus the term digital subtraction fluoroscopy. We will now discuss in some detail how the entire system works together to form the fluoroscopic T.V. image and this links up to the computer system and the x-ray generator to produce the final images. Figure 263 shows the entire system in block style.

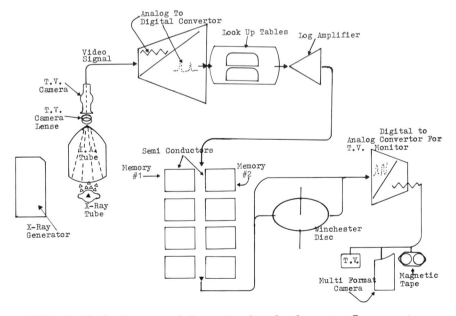

Figure 263. A block diagram of the entire digital subtraction fluoroscopic process. Memory 1 receives the digitized images as a positive, and memory 2 as a negative. The positive and negative images are subtracted to produce the final image.

The Image Amplifier

The image amplifier was originally used for fluoroscopy in the late 1950s and early 1960s. Its primary purpose was to intensify (increase) the brightness of the fluoroscopic image, which was up to that time very low in intensity making it difficult to visualize small detailed structures. Fluoroscopic spot devices today actually look very similar to those in the early 1960s. The main differences with the spot devices that were used before the image intensifier tube and those used today was how the image was viewed, not in how the fluoroscopic spot films were exposed.

The fluoroscopic image on the older units was a flat, lead-impregnated glass plate that lay over a coating of phosphorescent crystals, most commonly made of zinc-cadmium-sulfide material. The fluoroscopic x-ray tube under the table projected a beam through the tabletop and the patient, and the remnant beam was projected onto the fluoroescent screen. The fluorescent screen was illuminated in response to the varying remnant intensities of light that struck its surface, and the radiologist viewed this fluorescent image directly through the safety of the lead-impregnated glass. The quality of the image was very low compared to what we know today, because the brightness of the image was inadequate for easily observing or detecting small anatomical structures in the body. This technique was extremely inconvenient as well, since the examination had to be performed in almost total darkness so that the radiologist could visualize the image emitted by the fluorescent screen. In addition, because the examination was performed in a dark room, the acuity of the radiologist's vision as he worked in the darkened room had to be maintained by wearing dark red goggles any time he stepped out; otherwise, his lack of "night vision" would seriously affect his ability to see anatomical structures on the dimly lit fluorescent screen. The red goggles kept his eyes protected from the normal room lighting and kept his eyes adapted to night vision, or low light vision.

As he began the fluoroscopic examination using those old fluoroscopic units, the room lights were turned out, the goggles would be taken away, and his eyes would be ready or accommodated for the dimly lit images on the fluoroscopic screen. Even with light adaption, the information on the screen was very poor in comparison to what can be visualized using modern T.V. imaging systems. The author can remember seeing more than one radiologist driving to the hospital wearing the red goggles because he did not want to waste the fifteen or twenty minutes that were needed after putting on the goggles to adapt his eyes to night vision and be ready for fluoroscopy.

At any rate the fluoroscopic screen image was improved enormously by a remarkable device known as the image amplifier, and Figure 264 shows a diagram of its basic component parts.

As its name implies, the image amplifier tube amplifies the intensity and the contrast emitted by the fluoroscopic screen, and if designed properly can intensify this image by about 1,000 times more than the original intensity seen on the fluoroscope itself. As you know, modern x-ray fluoroscopic devices contain a cone-shaped structure mounted over the spot film device. In this manner the under-table fluoroscopic tube emits a beam that passes through the patient and projects with varying intensities to the layer of phosphorescent crystals. The image tube is mounted directly on top of the spot device (from which the conventional zinc-cadmium-sulfide screen was removed entirely), is contained within the cone-shaped structure, receives directly the image via the remnant beam through its own input phosphorus, and intensifies the beam, as mentioned earlier, approximately 1,000 times.

Although many other factors contribute to this intensifying factor, the primary concept is similar to experiments done in school science classes that can be duplicated with a simple magnifying glass. If a magnifying glass is placed over a piece of paper or wood and positioned a few inches away from the material so that the sunlight coming through the glass is focused or concentrated on the paper or wood, the temperature from the sun's rays transmitted through the magnifying glass is concentrated and therefore increased in intensity many times. The intensity of the sunlight, with its accompanying heat, that can be generated by this minification is easily sufficient to cause the paper or piece of wood to break into flames. Thus the heat of the sun was intensified greatly, simply by concentrating its energy over a very small confined area. In the same manner the input side of the image amplifying tube which first receives the remnant fluoroscopic image usually measures from 6 to 14 inches in diameter, depending on the type of system used, and concentrates the light to an area approximately 3/4 inch in diameter. Thus the existing brightness of the image on the input side is concentrated to this small area.

The Visible and the Electronic Image

We can see from Figure 264 how this concept works with the construction of the image amplifier (I.A.) tube. The bottom or input phosphor of the I.A. tube is coated with a very fine layer of luminescent material, usually made of cesium iodide, which takes the place of the phosphorescent materials that produced the fluoroscopic images. We will now briefly discuss how the image intensifier operates to produce the increase in image intensity.

As the remnant x-ray photons strike the input phosphor of the I.A. tube, they fluoresce, as do the conventional intensifier screens used in cassettes. The phosphor layer is in a concave fashion and produces a visible image via the phosphor illumination. Immediately after the input phosphor layer is a very thin metal plate coated with a chemical that liberates electrons. This material is commonly known as a photoelectric material, and this produces electrons at any one minute point of its surface in exact proportion to the amount of light that was emitted from the corresponding point of the phosphor coating.

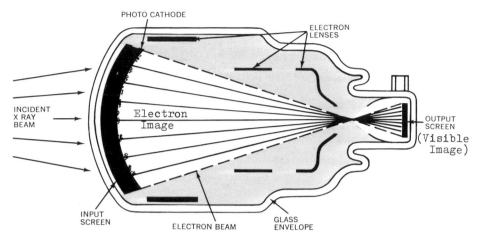

Figure 264. A simple drawing of the entire image amplifier tube. *Courtesy of* General Electric.

Thus, the accumulation of electrons on the metal plate is totally representative in its distribution to the visible image on the fluorescent layer of the input phosphor. Thus the visible image is converted to an electronic image.

The image amplifying tube can correctly be thought of as a diode tube with cathode and anode elements. A high voltage circuit connects these two electrodes and when this high voltage is applied (25,000 volts) between the two electrodes, the electron charges produced by the photon conductive plate are driven from the input side of the amplifying tube to the output side, which is called the anode. As was mentioned earlier, a vital characteristic of this tube is the concept of image minification. In other words, the field of liberated electrons is reduced from approximately 6 or 14 inch diameter (depending on the type of image tube used) to a 3/4 inch diameter on the output phosphor. In addition, one can imagine that the electron image made up of a distribution of free electrons must be perfectly focused, which means the spacial relationship of these electrons to each other has to be perfectly maintained during the minification process. This is accomplished by the placement of an additional circuit with complimentary electrodes positioned at strategic locations around the periphery of the image amplifier tube that produce a strong charged field along the inside walls of the I.A. glass envelope. Once energized, these elements produce an electron field which serves to "focus" the electron image as it moves from the input side to the output side of the image amplifier tube using precisely the same principle as the focusing cup in an ordinary x-ray tube.

Although the design, engineering, and manufacturing of image intensifying tubes is considerably more complex and tedious, this simple functional description of how they operate will serve well in describing the various components of a functional modern digital subtraction fluoroscopic system.

T.V. Lens System

The T.V. camera, as you might expect, does not receive the visible image directly from the output phosphor of the I.A. tube. The T.V. camera contains an optic lens very similar to those in movie or photographic cameras. As you know, a lens is used to adjust the alignment of light rays entering the camera so that the light photons are focused properly on the surface of the film. The degree of bending and focusing of these light rays is dependent on the focal length between the lens and the plane of the film or recording media such as the target surface of the T.V. camera. Thus, optic lensing systems work on the same principle of bending light rays as you have seen or observed when using a prism, except they are more carefully ground and polished. These lensing systems are expensive and must be mounted by experienced and skillful service people so that the image existing in the I.A. tube is properly transmitted and focused to the target of the T.V. camera so sharpness will be assured. Any loss of image sharpness because of the lensing system will be irretrievable regardless of the quality of the remaining electrical components and the computer systems that are connected after.

The T.V. Pickup Camera

At this point the signal has been converted back to a visible image by the intensifier tube's output phosphor because the electrons moving through the I.A. tube are focused to strike a piece of luminescent material which emits light commensurate with the varying intensities of the electrons.

The function of the T.V. pickup camera is to scan and transmit the visible image so that it can be relayed to a T.V. monitor and to a computer so that the picture information can be digitized. To understand how this is accomplished, we must take a few moments to review the operating characteristics of a T.V. pickup tube itself. Figure 265B is a photograph of an actual plumbicon camera tube. It measures approximately 1 inch in diameter and about 6.5 inches in length.

Major Components and Operation of the T.V. Camera

Figure 265A shows a schematic drawing illustrating the primary parts of a typical vidicon camera tube. We will begin with the face plate, which is a highly polished piece of glass that also helps form the outer surface of its evacuated glass envelope. As you might expect, this glass plate must be exceptionally well polished (as a camera lens would be) to prevent any artifacts or distortion. Against the inside surface of the face plate is a metal plate commonly referred to as the target, and it is coated with a special photoconducting material on its inside surface. This is also known as the signal plate or target because it plays a primary role in producing the T.V. signal. There are a number of photoconductive materials available but the

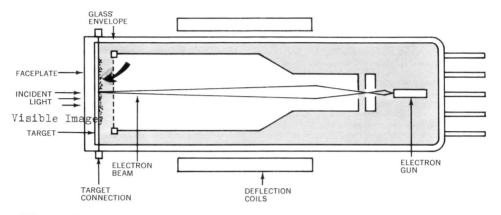

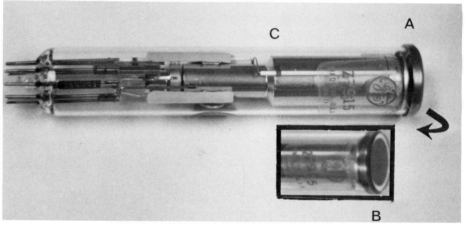

Figure 265.   As the visible image passes through the lens, it strikes the photoconductive surface of the target and produces corresponding electrical charges indicated by the bold arrow. These are scanned by the electron gun. The deflection coils control movement of the electron stream as well as focus them to a pinpoint beam. The photograph shows an actual T.V. pickup tube with its photoconductive face. *A* is the target connection, *B* is the photoconductive surface, *C* is the deflection coils. *Courtesy of* General Electric.

one most commonly used today is antimony trisulfide. Photoconductive means that this particular material produces an electric charge when stimulated by visible light. In other words, this material will conduct electricity at a rate that is controlled or tempered by the amount of visible light striking the material over any given area. The photoconductive plate contains the same image that passed through the T.V. camera lenses, and the image simply is made up of a distribution of negative electrical charges—the amount of which is commensurate with the degree and pattern of light incident upon it from the output phosphor of the I.A. tube. At this particular stage the image is invisible to us. Thus, the entire photoconductive layer converts the visible image and holds it in the form of millions of tiny negative charges (in a remote way it is analogous to the latent image in the emulsion of x-ray film with its pattern of ionized and non-ionized silver bromide crystals). The electronic image is therefore constantly changing with the scene or object.

The Electron Gun of the T.V. Camera

At this point you are probably wondering what the millions of electrical resistance elements are actually resisting. The answer lies in the component of the camera known as the electron gun, which is the heart of the T.V. camera itself. Its purpose is to project a single stream of electrons toward the photoconductive plate. The point to appreciate here is that the stream of electrons is not stationary but moves or scans back and forth from left to right beginning with the upper left corner of the image. When it completes each particular horizontal scan, the electron stream is redirected to the left side of the image and begins moving immediately to the right side to complete the second line. Each time the electron beam is directed to the left margin it indexes down to the next line until it reaches the bottom of the picture.

Progressive versus Interlaced Scanning

There are basically two scanning systems that T.V. cameras use. One is called a progressive type (which has just been described) in which the electron gun indexes down one line at a time after each horizontal scan. A second type of scanning system is known as the interlacing system. With the interlacing system (see Figure 266) the electron gun scans every other line until it

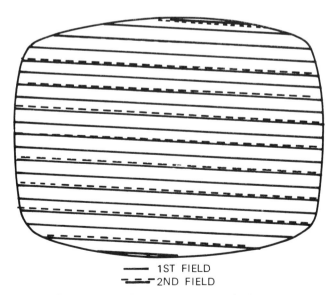

——— 1ST FIELD
⎯⎯⎯ 2ND FIELD

Figure 266. *Courtesy of* General Electric

reaches the bottom of the image. At that point it immediately jumps to the second line on the upper left hand corner and scans each line that was missed during the first scanning. Thus, with the first scanning, lines one, three, five, seven, etc., are scanned. When the gun reaches the bottom right hand corner of the image it retraces to the upper left side and begins lines two, four, six, eight, etc. The progressive system produces a complete image in successive or progressive scanning lines and this is accomplished in 1/30 second. The interlacing system, however, produces two "fields" which are woven or interlaced together, and each field is one-half of a frame. Thus the interlacing system produces two fields at 1/60 second each to produce a complete frame as compared to the progressive system, which produces one complete frame in 1/30 second.

The reader should now begin to realize that the images we see on a fluoroscopic monitor and on a home T.V. screen produce a complete new picture thirty times a second. We, of course, do not see the individual frames change simply because it happens too fast for our eyes to see and our minds to comprehend: the changing frames every 1/30 second appear with smooth motion and continuity to our eyes.

The electron gun used in closed circuit T.V. systems for fluoroscopy to perform G.I. procedures can produce 525 horizontal lines from the top of the screen to the bottom per frame in the progressive mode; in the interlace mode it produces two fields, each made of 262.5 horizontal lines. This means that the electron gun must complete each horizontal scan line across the T.V. tube's target every 1/15,750 second. The number of lines from top to bottom on a screen that can be produced is important because, as will be explained later, it relates to the resolution that can be produced by the T.V. system as a whole.

Deflector Coils

Obviously the electron beam must be controlled to a extreme degree as it traces back and forth across the photoconductive material to assure that its tracing pattern is precisely measured and consistent with each frame. Since the beam is a stream of electrons, it is susceptible to electrical flux. This characteristic is used to its optimal advantage by the deflector coils that surround the T.V. camera glass envelope as shown in Figure 265. These coils are energized with current by separate electric systems in such a way that they produce flux lines that push or pull the electron beam across the image 15,750 times a second. They must also align the electron beam to index down every line after completing each left-to-right horizontal scan. Needless to say, the calibration, timing, and synchronization of the coils' own electrical supply is very sensitive and must be extremely accurate to assure the path of the electron beam will exactly reproduce each horizontal scan, a field, and frame.

FORMATION OF THE T.V. SIGNAL

We can now discuss in general terms how the actual T.V. signal is produced by the T.V. camera and how it is transmitted to the various other components (such as the T.V. monitor) and ultimately converted into a digital signal for clinical application in digital subtraction fluoroscopy. The photoconductive plate or target is approximately .001 of an inch thick and is able to produce a kind of latent electrical image composed of millions of negative charges, each having its own specific strength to act as individual differential resistors to the electron stream, depending on the amount of light their minute area received.

We will now use a very simple example to illustrate how this system works to produce a video signal. Figure 267 shows the electron gun projecting a stream of electrons across a simple test pattern. As the light striking any point of the photoconductive material of the T.V. camera increases, the degree of electrical charge on that point increases; conversely, as the amount of light striking the photoconductive material decreases, the electrical charge over any minute area of the photoconductive material will produce a decrease in the degree of electron charges at that point. Thus, an increased accumulation of negative charges will resist the flow of electrons from the gun and vice versa.

In Figure 267, the electron beam is scanning three bars of different densities from white to black. We will say for the sake of discussion that these bars are actually produced on the output phosphor of the image intensifying tube, which is typically coupled to the T.V. camera by a very carefully matched

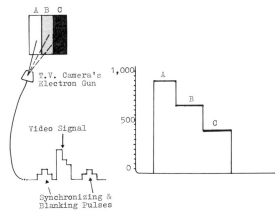

Figure 267. Using the test pattern, the electron gun scans one horizontal line of the image; the stream of electrons it produces change their rate of speed (voltage) as they move through the circuit to correlate with the degree of electrical resistance that was produced on the surface of the T.V. tube's target. Note the three density patterns portrayed by the video signal with the synchronizing and blanking phases separating each horizontal line.

pair of optical lenses. As the electron gun scans across the photoconductive material that makes up the electronic image on the target of the T.V. camera, it approaches the area at point A which offers relatively low electrical resistance to the flow of electrons from the gun. In effect, there is no change in the flow of electrons at this particular point as they come from the gun. However, as the electron stream continues to scan further across the image it comes into the area of a different density, which by virtue of its more intense light pattern, produced different electrical resistance in the photoconductive material. At this point, the electrical resistance to the stream of electrons coming from the gun will increase and therefore slow down the flow of electrons. In other words, the voltage of the current flowing from the electron gun will decrease commensurate with the strength of the resistance of the electron image on the T.V. target.

Looking at Figure 268 one should imagine the path of the electrons as they travel from the electron gun through the photoconductive material on to the target and pass behind and out through the target connection to complete their own circuit.

As the electron beam is driven across our test image (Fig. 267) it comes into the next area which is receiving somewhat less light from the output phosphor of the I.A. tube, so the electrical charge generated by the photoconductive material will be greater. With this set of circumstances, the voltage through the circuit of the electron gun will change due to the electrical resistance, and the amount of voltage exiting the tube through the target connection will be less. As the electron beam moves further to the right it reaches the next area, where the light is even less intense so offers more resistance to the voltage of the electron beam, and of course, the voltage exiting the target connection will be reduced even further. Voltage changes have been produced which are representative of light intensity patterns for just one horizontal scan line. In other words, the visible information as seen by

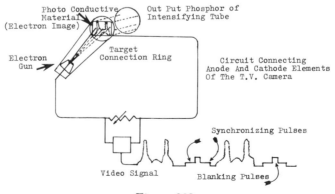

Figure 268.

the T.V. camera is converted to a pattern of various electrical resistance which serves to alter the voltage of the scanning beam producing a "video signal" consisting of fluctuating voltages.

THE DIGITAL IMAGE

To this point we have seen how the electron gun scans across the face of the target producing a fluctuating voltage we have referred to as the video signal. It is now appropriate to discuss how the video signal is converted into a digital signal, and we will continue using the example of the three density patterns discussed above. It is also necessary to introduce a device known as an A to D convertor (analog to digital). The continuous flow of fluctuating voltage exiting the T.V. camera is called an analog video signal. However, this signal must be converted to a format that can be handled by a computer. It is therefore necessary to quantify these voltage fluctuations into a format that a computer can accept and manipulate to produce the subtraction process. The A/D convertor's ultimate responsibility is to quantify ("digitize") this video signal. This is accomplished by producing something called pixels that form a kind of mosaic. These are tiny individual picture pieces or elements that together construct the image as seen on the T.V. monitor and hard copy. The number of pixels that form the digital image are usually in a matrix pattern of either 512 × 512 or 256 × 256. That is to say, the digital image is actually made of individual pixels that number 512 pixels across and down the image, totalling 262,144 pixels per image. The number of pixels used to produce the image is important because an image made up of smaller pixels will yield finer detail and resolution than if it had been made up of larger picture elements; however, there are some practical situations that make a 256 × 256 matrix advantageous.

The pixels or picture elements are produced by the A/D convertor in the following manner. The A/D convertor contains a type of subelectrical circuit that generates a signal which samples the video signal from the T.V. camera. In most digital subtraction systems, the matrix is used, and it is in fact common to hear that a particular image being viewed on a digital system is made up of a 512 × 512 matrix or a 256 × 256 matrix. This simply refers to the total number and size of the "pixels" that had been used to make up the image seen on the T.V. monitor, as well as the images that are produced by the multiformat cameras. A 512 × 512 matrix means, for example, that there are 512 individual points that are sampled in each horizontal scan of the electron gun and that there are 512 of these horizontal scan lines down the entire image, forming a matrix of 512 vertical columns and 512 horizontal rows. Thus a picture using a 512 × 512 matrix will contain a total of 262,144 individual picture elements. An image produced by a 256 × 256 matrix will produce only 65,536 individual picture elements.

The concept of picture elements or pixels is new and possibly confusing, but another analogy can be made with crystal or grain size on x-ray film. You will recall that the size of the crystal used in the emulsion of x-ray film affects the degree of detail and resolution that can be obtained on the radiographic image, other factors remaining constant. This is because smaller particles of silver bromide crystals can represent more accurately the very small finely detailed anatomical structures and can more easily represent them in density form on the radiologic image. With larger silver bromide crystals, each crystal will be responsible for representing a larger area of the anatomic structure and thus cannot specialize. This lack of "specialization" of each tiny segment of the anatomy results in decreased detail and sharpness.

The same concept then holds with picture elements, and we have already drawn the conclusion that a matrix size of 512 × 512 will produce an image with better detail compared to a 256 × 256 image. However, we have currently reached a point of diminished returns because the use of an image matrix greater than 512 × 512 will not yield noticeable increases in image sharpness, because of the chain of components that is necessary for digital subtraction fluoroscopy. Keeping in mind the weak link concept, an improvement in any one link of this imaging chain must be accompanied by a complimentary improvement in each of the other components. For example, if the T.V. camera and A/D convertor were able to produce a video signal that was made of up of a 1,024 × 1,024 matrix, very fine resolution and much more detail could be seen. However, this improvement would be unimportant and certainly unnecessary if a complimentary increase in the quality of the image produced by the image intensifying tube were not produced as well.

It is generally held at this point that the weak link in terms of producing ultra high detailed images lies with the image intensifier tube, and in making every attempt to minimize dose to the patient as much as possible (all other things being held constant, the more photons used to produce a digital subtraction fluoroscopic image the better the detail will improve). The term photon density is used to reference the number of x-ray photons within each pixel. In general the more photons used for a given digital exposure, the more patient information will be sent through the T.V. system; this increases the probability of better image detail and greatly reduces quantum size.

The Formation of Picture Elements (Pixels)

Let us examine for a moment how these pixels are actually formed. It has already been mentioned that the A to D convertor has a subelectrical circuit. This circuit generates a signal to turn on and off at a certain rate and in this manner acts like a high speed clock. In other words, 512 samples are taken during each horizontal scan; each time the video signal is sampled, a pixel is formed. In this way, 512 separate voltage samples are generated by the A/D convertor, and we obtain 512 voltage samples per horizontal scan. Each represents the degree of original resistance on the target of the T.V. camera, and each sample can be considered an element of the whole picture (pixel).

As mentioned earlier, the gun is constantly scanning from left to right in a horizontal pattern, and at the end of each horizontal scan the gun is turned off, is aimed back to the left side of the image, and drops down one line to make the next horizontal scan. In this manner, the electron gun is turned on and off 262,144 times per frame if a 512 × 512 matrix is used (recall that a complete frame is produced thirty times every second). One can easily calculate that when using a 512 × 512 matrix, the video signal will be sampled 7,864,320 times per second, and of course, in that one second time span the total number of pixels that will be produced will be 7,864,320.

It might be more desirable to use a 256 × 256 matrix, for reasons that will be explained later; in such an instance, the appropriate switch is thrown and the subcircuit in the A to D convertor will then sample the signal at a lower rate of speed—equaling 256 times with each horizontal scan, and 256 horizontal scans down the face of the image producing a total of 65,536 picture elements (or voltage samples) per 1/30 second. In viewing this image of 256 matrix for one second, a total of 1,966,080 pixels will be produced. See Figure 269.

Figure 271 illustrates a typical video signal as it applies to a bilateral carotid examination. The video signal is measured in millivolts and the useful amplitude or scale of the video signal ranges from zero millivolts to 700 millivolts. In other words, the voltage coming from the electron gun circuit, which constitutes the video signal, has a very low voltage measured in milli-

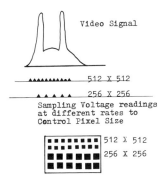

Figure 269. Note that the number of samples generated by the computer's A/D convertor can be changed. As the number of voltage samples increases per scan line, the relative area or pixel size decreases. Each sample of voltage constitutes one picture element or pixel.

volts. In other words, if an area on the T.V. camera's photoconductive material is receiving no light at all, the exiting voltage would be zero; conversely, the exiting voltage would be one thousand if the photoconductive material had received maximum light intensity.

Synchronizing and Blanking Pulses

You will see in Figure 268 that each horizontal scan line is separated by a synchronizing signal, a blanking pulse or signal, and a second synchronizing signal, and we will examine each of these briefly at this point. As the electron gun scans across a horizontal line and approaches the end of that line, another subcircuit generates an electrical current through the deflection coils that stops the scanning gun from continuing any further; therefore, it assures that each line will contain exactly the same number of picture elements and have a conforming width. A second electrical current generated by the same subcircuit is used to "blank" any flow of electrons through the gun of the T.V. camera; simultaneously, deflection coils drive and aim the electron stream back to the left side of the target and position it one line down from the line just scanned. The final pulse is another synchronizing pulse that starts the scanning gun across the second horizontal line and at the same time allows the electron gun to produce a stream of current across the image until it reaches the right side of the line when another synchronizing pulse stops the electron beam.

These synchronizing and blanking electrical signals are generated from a source independent of the electron gun; also, consistent precision and timing must occur to make this produce a recognizable video image. The full T.V. scan image (or frame) is produced every 1/30 second. This means that the electron gun must produce a single horizontal scan every 63 millionth of a second. Of this 63 millionth of a second, 13 millionth of a second is needed to generate and execute the two synchronization and one blanking signal which occurs at the end of each horizontal line, so the actual horizontal scan that produces the video image is accomplished in about 50 millionth of a second, or 50 microseconds. When we consider that this only accounts for the production of one frame (a frame is produced every 1/30 second) the precise repeatability of the T.V. system over many minutes of viewing time reaches staggering proportions.

The Digital Conversion

It is important at this point to remember the fluctuation of voltage obtained from a single horizontal scan, as shown in Figure 270. The first major function of this entire "digitizing" operation was to generate an independent electrical signal via the A/D convertor that produces the individual picture elements. At this point the sampled voltage is changed into a series of

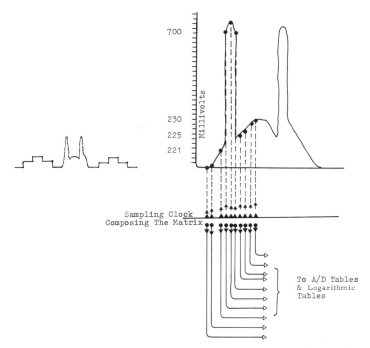

Figure 270. Shows a series of samples taken by the A/D convertor's sampling clock from the incoming video signal. The number of these samples taken per T.V. scanning line determines the matrix of the computerized image along with the number of lines down the face of the entire image.

binary numbers with each binary number equal to one pixel. The "numerized" pixel is then addressed (assigned) a place in the semiconductor memory of the computer.

Nearly all computers use this binary (base of two) system to handle the information and voltage sampling coming from the video signal.

However, before it is addressed and sent to its location in the memory, the voltage value that constitutes each pixel is channelled by the computer through a series of logarithmic tables that determine their final digital values. These tables are very important since this is what determines the relative density value of each pixel, and therefore establishes the degree of contrast of the final image. See Figure 271. After the logarithmic look up tables, the pixels are sent to the memory.

Once this pixel image information is properly addressed in the memory, the computer (upon appropriate instructions from the operator) can make a positive image from any one of the regular fluoroscopic images it receives. Upon further instructions from the operator, it can then go through a process that in essence compares and subtracts the negative fluoroscopic image from

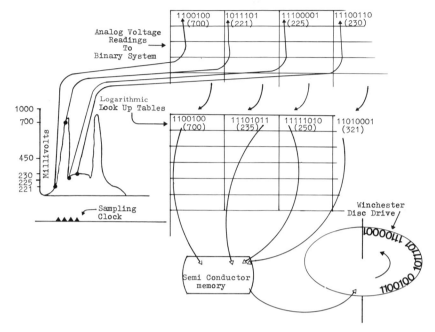

Figure 271. Here we see a typical video signal of a bilateral carotid arteriogram. Note the sequence that occurs to convert the numbers from millivolts to binary to logarithmic numbers, which in effect expand their original variances to produce contrast enhancement to produce grey tones. Once the final tone values are achieved, it is "addressed" and located in its assigned space in the computer's semiconductor memory. These do not usually provide limited space for the huge volume of pixel numbers, so they can only hold about eight complete images. Once these are filled, the first image stored is automatically transferred by the computer to the Winchester disc, which has sufficient space for about 64 images. Most Winchester discs used in digital subtraction systems can hold about 84,000,000 bytes of information or 672,000,000 bites of information. The computer can draw the pixel information from the disc almost instantaneously upon instruction from the operator to produce images of previously performed cases. The term *postprocessing* refers to the manipulation of images sometime after the examination has been completed for the purpose of improving contrast, edge enhancement, magnification, and other effects.

the positive image produced by the computer on a pixel by pixel basis to yield the final subtracted image. See Figure 263.

Contrast Enhancement

An important advantage offered by the digital subtraction technique is contrast enhancement. You will also recall the human eye can see two adjacent structures only if there is a 15 percent difference in density between them. We know the remnant radiation contains photon intensities that encompass diagnostic information but that vary much less than 15 percent and until now could not be appreciated or seen.

Once the video signal has been converted to pixels with a binary value by the A/D convertor, the operator can instruct the computer to use logarithmic look up tables, which are used to expand the relative differences between density values of each pixel. In this manner, the relative density of each picture element can be very much increased or enhanced by the computer to a point where it can be seen. Two structures that might originally have only a 2 percent difference, which is invisible to us, can be expanded by the computer and become visible.

The more direct example of contrast enhancement can be afforded by simply reviewing the number of perceptible densities on a routine radiographic image. In most instances one would find that a radiographic image will not usually differentiate more than twenty densities from white to the darkest tone. In comparison to this, the digital system—by virtue of the computer's enhancement capabilities using the logarithmic tables—can produce and yield information that is increased to as much as 1024 gray tones ranging from black to white. This expanded and much more refined contrast scale will obviously provide more diagnostic information than would have been possible on the radiographic image itself.

THE T.V. MONITOR

The monitor is a vital part of the imaging chain (which includes the I.A. tube, the T.V. camera lenses, and the T.V. camera). It is important to recall the scanning operation of the T.V. camera's electron gun and the various synchronizing and blanking pulses produced, as well as the electrical signal to the deflection coils of the T.V. camera. See Figure 272. The coils in turn produced appropriate flux lines that pushed and pulled, thereby controlling the movement of the electron stream.

The generating circuit that guided the electron gun also affects the picture tube in the T.V. monitor. A quick look at the back or neck of a T.V. picture tube will reveal a rather large set of electrical coils. These coils receive the same blanking and synchronizing signals as those of the T.V. camera and produce similar flux lines which push and pull a stream of electrons from its own gun, also located in the neck of the T.V. picture tube. The gun of the T.V. monitor receives the video signal after it exits the T.V. camera. Keep in mind the raw video signal has been modulated somewhat by a host of capacitors, transitors, amplifiers, etc. However, the video signal is eventually transferred to the gun of the T.V. tube in the form of voltage fluctuations. The gun in the picture tube receives and then projects the fluctuating voltage signal against the phosphor coating on the inside wall of the picture tube itself. As the varying voltage strikes the phosphorescent material, the phosphor material ionizes, causing a light emission with an intensity commensurate with the varying voltage as it is projected across the picture tube line by line. As the voltage increases it strikes phosphorescent material and more light is given off by the phosphors; as less voltage strikes the phosphor layer, it produces less visible light. This is similar to the effect increasing or decreasing KvP has on light output of a standard intensifying screen.

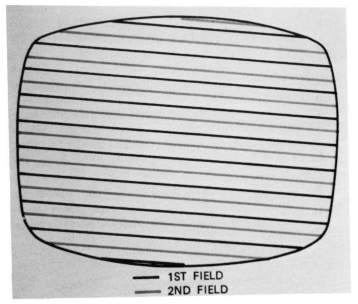

1ST FIELD
2ND FIELD

Figure 272A. T.V. raster lines.

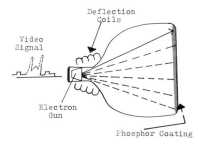

Figure 272B. Schematic of a T.V. picture tube.

You will hear the term *raster lines* mentioned during almost any discussion of the operation of a T.V. picture tube. This is a reference to the scanning pattern as the T.V. picture tube's gun projects its electron beam toward the phosphor material. Its own deflection coils push and pull the electron flow from left to right across and down each line until the entire picture is formed. Thus the T.V. raster is the pattern (Figure 272) of individual scanning lines traced on the T.V. picture tube. In this way the electron gun both in the camera tube and in the T.V. picture tube scans across the image left to right down each row in unison via the blanking and synchronizing phases produced by the subcircuit generator in the T.V. camera.

We have been talking about a digital image being made of a matrix consisting of 512 horizontal and 512 vertical rows of pixels, and of a matrix of

256 × 256. One might ask under these circumstances how these matrix patterns fit into the raster lines with 525 lines. The answer lies in the way the pixel addressing information is coded by the computer; a complex explanation would be necessary but is not terribly important to the scope of this chapter. It is only important to keep in mind that some alignment modifications are made via the computer to compensate for this disparity in matrix and raster line patterns. Actually, even though all of the technical and printed sales specifications of the commercial companies regarding digital subtraction fluoroscopic equipment talk extensively about a 512 matrix or 256 matrix image, in reality because of some technical problems the 512 × 512 matrix is in its truest sense not possible. In reality, a 512 × 512 matrix is actually closer to a 480 × 480 matrix. A 256 × 256 matrix in reality would be approximately 450 × 450.

Image Registration

Registration brings us back to the immediate subject of image subtraction and to a concept that we discussed earlier in this chapter. You will recall that the basic idea is to sandwich a positive (mask) image with a negative image containing opacified vessels. When these two films are sandwiched properly, the anatomic structures representing the area of interest are precisely superimposed or registered. With conventional subtraction methods this registration was accomplished as the technologist slides the negative over the mask film until optimal registration or superimposition of the images was visually achieved. A digital system must certainly perform this task as well, and it is accomplished by the computer operating on the command of the technologist or radiologist. A schematic of the digital computer's memory can be seen in Figure 263 to provide some insight as to how the subtracted image is accomplished.

Image Reregistration

Any technologist who works with subtraction technique will emphasize without hesitation that the quality of the subtracted image is very much dependent upon immobilizing the patient. The reason for this, of course, is if the patient moves between exposures he will be in a different position as various exposures are taken so the mask film and the negative film of the opacified vessels will not exactly superimpose. This lack of superimposition is commonly referred to as misregistration, the result of which is a totally unsuccessful and nondiagnostic subtraction study. A digitally subtracted image, however, provides the opportunity to correct patient motion caused by this particular situation, thus providing a subtraction study with optimal result. This is accomplished once again by the computer on command from the operator, and the process is known as reregistration. A moment of thought relative to the picture elements will come in handy at this point. You will recall each frame is made of a composite of tiny picture elements called pixels. If the patient did move between the positive and negative images, the computer has each pixel addressed and "knows" the "address" of each pixel in the computer's chip memory. The computer upon command will then sort out and compare the density values of each picture element of both the negative and the positive images and in effect reorganize the pixels until the atomic structures in each image are reregistered and in alignment, yielding optimal superimposition. From this point, the computer goes through its regular process of subtracting densities to produce the final subtraction film. The accuracy of the reregistration technique is quite high since computers can move these pixels around in increments as small as 1/10 pixel or .02 mm. Figure 273 shows two images of the same examination before and after registration. Note the visibility of the smaller vessels.

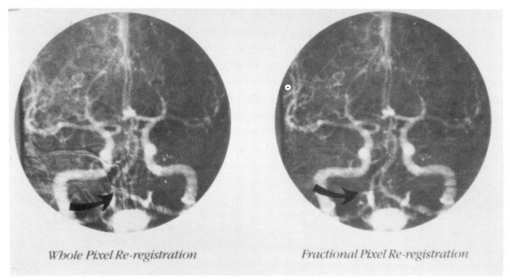

Whole Pixel Re-registration　　　　　　　　*Fractional Pixel Re-registration*

Figure 273.　Example of image registration.

ENERGY SUBTRACTION

Certainly one of the key factors in producing high quality angiographic work is to maximize contrast. The probability of contrast improvement is great, since the computer can capture as many as 256 gray tones. However, some additional development of digital subtraction technique is being appreciated by the time-honored method of adjusting kilovoltage to improve subject contrast. This makes use of a special generator that alternately produces one exposure at a higher Kv and a second exposure at a preestablished lower setting. The combination of the two exposures with different Kv and related subject contrast with the 1024 gray tone capability works to improve the resulting contrast to a greater point than has been thus appreciated, even with conventional digital subtraction fluoroscopic equipment. The disadvantage in this system is the development and eventual cost of a generator with the capability of alternating from one Kv to another many times a second without fluctuating from the desired Kv by more than 1 percent. In addition to these problems, the computer will be required to handle twice the amount of digital information as is necessary with the second lower Kv exposure and would conceivably be 50 percent to 100 percent more expensive. Despite these problems, the development and manufacturing of x-ray generators and computers to accomplish energy subtraction studies is under way, and only time with the critical evaluation of the degree or extent of the improvement will determine its future in the marketplace as a viable adjunct to the basic digital subtraction system.

IMAGE QUALITY PARAMETERS

As was mentioned earlier, digital subtraction fluoroscopy's strong suit is contrast; its weak point is in resolution or sharpness. We will take time now to discuss the various factors that interlace and eventually affect the contrast and resolution of the digital image.

Spacial Resolution

The spacial resolution, or simply resolution, is measured by the number of line pairs that can be seen per millimeter when a line pair is considered to be one line of subject information next to one space of equal width as the subject line. See Figure 274. The final image is very much dependent on the individual quality of the x-ray tube's focal spot, the intensifying tube, the T.V. camera, and the computer. We will discuss each of these very briefly. However, it is helpful to keep in mind that current resolution is about 2 line pairs per millimeter, compared to about 9 line pairs per mm common with conventional x-ray film, but this will certainly increase as technical improvements are built into the systems of the future.

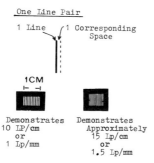

Figure 274.   Example of line pairs per cm and mm.

Focal Spot Size and Digital Subtraction Fluoroscopy

As with conventional radiography, the focal spot affects the sharpness of the radiographic image. The geometric pattern of the remnant beam is not projected on the film but directly to the input phosphor of the I.A. tube. In conventional radiography as the focal spot size decreases, the geometric image becomes sharper; if all other things are held constant, as the focal size decreases the digital image will yield increased sharpness. However, the chain of components making up today's digital systems requires nothing smaller than a 0.6 F.F.S. Generally the stated specifications call for a 0.6 to 1.2 focal spot. When other components can make better use of smaller focal spot sizes they will certainly be added into future installations, keeping in mind the well-known limitations on anode heat that are related to smaller focal spot sizes.

The Image Intensifier

The image intensifier can produce about four line pairs per millimeter on its output phosphor. To date this is more than adequate. There is, as you would expect, much activity aimed at improving this for the near future. However, an improvement beyond 4 lp/mm currently would yield no noticeable improvement in the image because of the T.V. and the image matrix produced by the computer.

The T.V. Pickup Camera

You will recall the T.V. pickup camera produces a video signal composed of fluctuating voltages. The camera currently in use cannot produce resolution quality much over 2 lp/mm. This is an important limiting factor since the 0.6 mm F.F.S. produces a geometric image of much finer quality, and the I.A. tube can produce about 4 lp/mm. Special high resolution cameras are available but their inclusion alone, without making improvements in the other components by increasing matrix and patient dose, *would have little affect* on the final image.

The Matrix of the Computerized Image

Each pixel of information represents a binary number and is stored in the computer and, therefore, must be manipulated by the computer. If the A/D convertor produces a 512 × 512 matrix image, the computer must handle 262,144 pixels every 1/30 second or 7,864,320 pixels every second of actual fluoroscopic time. To make use of a special high resolution T.V. camera, the A/D convertor would have to increase the matrix of the image. This would, of course, yield a smaller pixel size and would therefore potentially produce an image with about twice the lines per millimeter. As it turns out, the next largest matrix size would be a 1024 × 1024, and this would produce four times the number of pixels. Thus a 1024 × 1024 matrix would produce 1,048,576 pixels every 1/30 second, or 31,457,280 per second of actual fluoroscopic time. While the fluorosope is on, the T.V. camera produces an entirely new and separate image every 1/30 second with either a 256 × 256 or 512 × 512 matrix. Thus a T.V. frame is composed of 1,048,576 pixels, and computers are being developed to handle and manipulate that much computer information for the necessary framing speeds in angiography and heart catheterization work.

Framing Speed

When doing conventional serial filming for arteriograms, the radiologist typically chooses a program for the films to be taken. Usually a cerebral arteriogram can be accomplished at two films per second for two seconds immediately followed by one film per second for two additional seconds, or some variation thereof depending on the radiologist.

With digital angiography, the same concept in filming is used; however, there are some limitations imposed as to the number of images that can be captured by the digital system per second. The primary reason for this limitation lies in the volume of computer data, which are expressed in binary pixel density numbers that make up each individual T.V. picture or frame. Even the fastest computers available for medical application require time (1) for their A/D convertors to digitize each pixel from the video signal, (2) for the pixel to be given a density number according to its logarithmic table, (3) to address the pixel, (4) and finally to send the addressed pixel to the correct place in the semiconductor memory. In addition, after perhaps eight complete digital images have been placed in these semiconductor memory banks, they become full and then all the images (pixel by pixel) are automatically transferred to disc drive for longer term storage. Thus, when the radiologist wishes to take more than the semiconductors can hold during the injection function of the examination, the computer must add these images to the front end of the semiconductor memory, and must simultaneously transfer images to the disc from the back end to make room in the entry end of the semiconductor for new images.

With all this in mind considerations must be made as to how fast the computer can produce images and process them through its internal system. However, most systems can produce four images per second in a 512 × 512 matrix and about eight images per second when using a 256 × 256 matrix. Obviously, the lower the matrix size the less pixels are present in the image and the more images per second can be processed during the serial filming part of the examination.

Noise

Noise—as it relates to intensifying screens and x-ray film and as we see it on fluoroscopic T.V. monitors—will now be reviewed in terms of the digital system. There are two primary sources of noise seen in the digital image. The first and most important is quantum noise, which is caused by not enough photons in the beam as they strike the input phosphor of the I.A. tube. Thus as the concentration of photons increases per square inch, for example, the I.A. tube simply receives more diagnostic x-ray information. When the concentration of remnant photons on the I.A. tube is low, "blank" or "photon starved" areas are present. These become part of the image as a whole and are ultimately projected to the output phosphor and scanned by the T.V. camera's electron gun. These tiny diagnostically useless blank areas are then given some pixel density value by the computer and stored along with the other pixels that do contain diagnostic information. The fewer the photons per pixel area that fall on the input phosphor of the I.A. tube, the more blank areas will appear and the image thus increases in quantum noise.

Certainly the higher the concentration of x-ray photons in the remnant beam the "smoother" (less noisy) the image will be. There is a noise ratio that seems to be acceptable, and in brief a concentration of 6^{10} photons per pixel is considered necessary to maintain an acceptable noise level. An important drawback in obtaining ultra low noise in the digital image lies in the area of patient dose, because to obtain a less noisy image the exposure values would have to be set significantly higher.

Another source of noise that contributes to the digital image is commonly called electronic noise. Each electronic component in the digital system produces signals that appear on the film as noise. The amount of electronic noise that is typically produced is significant to the final image.

Image Contrast

Whatever might be lost in spacial resolution is very much compensated for by the system's ability to enhance contrast. This is to say that when it comes to producing a diagnostic image there is legitimate room for trading resolution (sharpness) for contrast. Basically, the contrast has about 256 separate tones and can demonstrate as little as 1 percent variances in tissue or structure absorption. This means that any anatomic structure that causes as little as 1 percent variance in x-ray absorption can be distinguished by the digital system and reproduced on the final image, whether it be the T.V. monitor or the hard copy. This enormous contrast advantage offered by today's digital systems goes a long way to make up for the loss in resolution.

In summary, the digital image quality parameters normally fall in the area of 1 percent contrast differentiation with sufficient sharpness to resolve structures as small as one millimeter.

GLOSSARY

ABSORBED DOSE: The energy imparted to matter by ionizing radiation per unit mass of irradiated material at the place of interest. The special unit of absorbed dose is the rad. One rad equals 100 ergs per gram.

ABSORPTION: The *total* transfer of energy (x-rays) to a substance, which results in the removal of x-ray photons from the beam.

ACCELERATOR (agent): *See* Activator (agent).

ACETATE (film base): A substance formerly used in making the base material of x-ray film (now replaced by polyester.)

ACIDIFIER: An additive to the fixing solution. It stops the developing action and neutralizes the developing chemicals instantly when the film is placed in the fixer solution.

ACTIVATOR (agent): A chemical additive to the developer solution for the purpose of increasing the swelling and softening of gelatin during the developing cycle so the reducing agents can more easily penetrate the emulsion and reduce the exposed silver bromide crystals.

ACTIVE LAYER: That part of the intensifying screen that contains the phosphorus emission materials used to expose the film.

ACTIVITY: The number of nuclear disintegrations occurring in a given quantity of material per unit time. *See* Curie.

ADDRESSING (computer system): The process of assigning individual bites of computer information to a specific hold place in the computer, where it can be retrieved by the computer later to perform various calculations and functions.

A/D CONVERTOR: That part of the computer system that samples the voltage coming from a video analog type signal and assigns it a number.

AFTERGLOW: The momentary continuation of intensifying screen emissions after the exposure had been terminated. (*See also* Phosphorescence.)

AGITATION (film): A light, but definite shaking or flexing of the x-ray film while in the developing solution for the primary purpose of loosening unwanted reaction particles from the film surface, thereby allowing more even development. Accomplished by hand in manual processing and by the transport system in automatic processing.

AGITATION (solution): The mixing or stirring of various processing solutions (developer and fixer) to assure even distribution of temperature and chemical activity. Accomplished by hand in manual processing and by pumps in automatic processing.

AIR GAP TECHNIQUE: A specific radiographic technique that requires a space (gap) of approximately 6 inches between the film and the patient to reduce scatter reaching the film. This (gap) allows the tangent scatter photon to stray away from the area of the x-ray film, thus reducing S/S fog. Grids are usually not used with this technique and the focal film distance is approximately 10 feet to compensate for the increased O.F.D.

ALUMINUM EQUIVALENT: The thickness of aluminum affording the same attenuation, under specified conditions, as the material in question.

AMPERE: An electrical current of one coulomb per second. It is also the steady current produced by one volt applied across a resistance of one ohm.

AMPLITUDE: Usually, amplitude refers to the arcing distance over which the tube travels during the exposure. As the amplitude increases, the objective plane becomes thinner.

ANODE: An electrode of the x-ray tube that is given a positive charge during the exposure.

ANODE—HEEL EFFECT: An uneven distribution of x-ray photons along the long axis of the x-ray tube caused by the angle of the target. This distribution of photons has little or no practical effect on the radiographic image (when a 14 × 17 or smaller film is used at 40 inches.)

ARC-TO-ARC MOTION: A particular type of tomographic linkage that allows both the x-ray tube and film to simultaneously move in an arcing pattern during the exposure. This type of linkage keeps a consistent relationship between the tube, object, and film thus producing a clearer image of the objective plane compared to line-to-line or arc-to-line linkages.

ARTIFACT: An unwanted mark or blemish on the film (usually caused by roughly handling film, improper processing, dirty screens, and static).

ATTENUATION: The reduction of exposure rate upon passage of radiation through matter.

AUTOMATIC TIMING: *See* Phototiming.

AUTOTRANSFORMER: A self-induction transformer with a wide variety of taps from which contracts are made to vary the voltage to the H.T. transformer.

AVERAGE GRADIENT: A line that measures the "average" slope of the diagnostic portion of an H & D curve for the purpose of indicating film contrast characteristics. It is a straight line drawn from a density of .25 above base plus fog to 2.0 above base plus fog and is meant to represent the "average" of all the various bends along the H & D curve between .25 and 2.5.

BACK SCATTER: Scatter photons resulting from interactions between primary photons and various structures behind the film. After passing through the cassette, the photons interact with structures behind the cassette and the resulting scatter photons reexpose the film by entering through the back of the cassette.

BALANCED RADIOGRAPHIC IMAGE: An image that possesses the optimal contrast and density for the particular body region being radiographed and the purpose of the examination. These characteristics are accomplished by properly matching the KvP, subject contrast, part thickness, MaS, and the imaging system.

BASE KvP: The lowest kilovoltage possible that can be expected to penetrate a body part effectively and produce a good quality radiographic image. The formula for finding base KvP for single phase is the centimeter thickness times two, plus 30 KvP. For three phase equipment, 20 KvP should be sufficient.

BASE PLUS FOG: A low level density that is present in all diagnostic radiographs after processing without exposure to light or x-rays. In general, normal base fog readings for diagnostic radiographs should not exceed 0.2 D.

BEAM (x-ray): The emission of x-ray photons from the x-ray tube. The beam is not a continual emission, but rather is emitted in spurts (pulses) dependent on each electrical phase of tube current—120 spurts (pulses) per second for single phase full wave rectification.

BEAM EFFICIENCY: *See* Exposure Efficiency.

BEAM FREQUENCY: The rate at which x-ray quanta vibrate (waves cycles per second) while in route from the x-ray source.

BEAM-LIMITING DEVICE: Any device or system that restricts or confines an x-ray beam to a desired area or shape.

BEAM QUALITY: The penetrating power (frequency of a given x-ray beam). As beam quality (frequency) increases, its penetrating power increases.

BEAM QUANTITY: The amount of x-ray photons contained in a given beam moving from the target. The term is sometimes used in place of the more precise term *exposure rate.*

BINARY NUMBERING SYSTEM: A base two numbering system, as opposed to the more commonly used numbering system that has a base ten. The binary system only uses 0 and 1. Computers use this system because they can only recognize "on" and "off" type of instructions for computing functions: 1 means on, and 0 means off to a computer.

BODY HABITUS: The general composition or body makeup with respect to such things as age, fat content, muscle content and tone, and water. Body habitus changes with age and pathology.

BODY SECTION: Tomography.

BOOK CASSETTE: *See* Multiplanography.

BOOSTER CIRCUIT: A special circuit that increases the temperature of the x-ray filament to the required point to obtain the desired Ma values for a given exposure.

BRACHYTHERAPY: A method of radiation therapy in which an encapsulated source is utilized to deliver gamma or beta radiation at a distance up to a few centimeters either by surface, intracavitary, or interstitial application.

BRIGHTNESS (image): The general appearance of the radiograph as measured by the amount of illuminator light that can pass through the film to the viewer. Usually a bright image is the product of high contrast scale.

CALCIUM TUNGSTATE: A particular type of phosphor most commonly used in the active layer of intensifying screen. Its purpose is to convert x-ray energy into a low energy (usually of visible light) emission that exposes the x-ray film. Calcium tungstate produces a blue-violet emission.

CALIPER: A device with a sliding arm that is used to measure thickness of the body parts in centimeters.

CARDBOARD IMAGING: A radiograph obtained with film exposed in a cardboard or plastic holder that does not contain intensifying screens.

CASSETTE: A hard-backed light-tight device made of metal and plastic used for handling x-ray film and protection from light exposure. Cassettes usually contain a pair of screens that expose the film.

CASSETTELESS RADIOGRAPHY: A system in which the x-ray equipment used employs its own pair of screens and film transport system, which eliminates the use of conventional cassettes. Such types of radiographic equipment have their own film supply bins internally and a conveyor system that transports the film directly to a processor.

CATHODE: The electrode of the x-ray tube that is given a negative charge and contains an element known as the filament. (*See also* Filament.)

CENTRALIZED PROCESSING: A type of arrangement or location of automatic processors whereby almost all are placed in one specific central area of the x-ray department.

CENTRAL RAY: An imaginary photon that is located in the central portion of the beam that moves *perpendicular* to the plane of the film. Proper positioning in most cases requires that the central beam pass directly through the part being radiographed to reduce shape distortion.

CHARACTERISTIC CURVE: A plotting of film density versus exposure that is used to quantify the film characteristics of latitude, speed, and contrast.

CHEMICAL ACTIVITY: The level at which the developer and fixer chemicals work at the film's emulsion to produce a visible diagnostic x-ray image. Chemical activity is affected by replenishment, solution temperature, dilution with water, and oxidation.

CHEMICAL FOG: A particular type of fog caused by processing chemicals that have gone out of control (increased activity). The appearance of a chemically fogged radiographic image is that of increased density and very low contrast.

CIRCUIT BREAKER: A device that senses sharp increases in current and reacts by opening the circuit. These are generally to protect electrical components from damage.

CIRCULAR TOMOGRAPHY: A type of tomographic motion during which the tube and film move in a circular pattern during the exposure.

CLEAN-UP (grid): The degree to which a grid eliminates excessive scatter from the remnant beam.

COATING (the film): The process of putting the emulsion material on a base during manufacturing. The film is coated with an emulsion approximately one-thousandth of an inch thick. Film coating is one of the most important procedures in modern film manufacturing.

COLLIMATOR: A type of beam-limiting device that employs a high intensity light source and movable lead diaphragms. The light source is used to outline the field of exposure and movable lead diaphragms are used to adjust the size and the shape of the field to be exposed.

COMPENSATING SCREENS: A type of intensifying screen that gradually produces increased emissions toward one end of the screen. Often these screens are used in 14 × 36 inch cassettes and are used to compensate for radiographing body parts with different absorption rates. A compensating screen mounted in a 14 × 36 inch cassette may yield good radiographic results from the knee to the pelvis using a single exposure, with the part of the screen producing higher emissions placed under the pelvis and the low emission under the knees. Also called gradient screens.

COMPTON INTERACTION: A specific type of tissue-photon interaction in which an outer orbital electron of a tissue's atom is removed by an incident x-ray photon. The ejected electron, which is removed from the tissue's atom, is called a compton electron. The balance of the energy between the incident photon and what was carried by the ejected electron is known as scatter radiation.

CONE: A general term that refers to a beam-limiting device usually having a fixed field size. These devices can be a true cylinder style or a flared cylinder.

CONSTANT POTENTIAL: In radiological practice, this term is applied to a unidirectional potential (or voltage) which has little, or no, periodic variation. The periodic component is called the ripple potential (or ripple voltage).

CONTACT FOCAL FILM DISTANCE: This refers to a circumstance when the focal film distance is reduced to a point where the x-ray tube or collimator touches the patient's skin. Usually used under situations where certain structures are to be blurred from the image—as in the case of the sternum, a. c. joints, t. m. joints, and sternoclavicular joints, etc.

CONTACT THERAPY APPARATUS: X-ray therapy apparatus designed for very short treatment distances (S.S.D. of 5 cm or less) usually employing tube potentials in the range of 20 to 50 KvP.

CONTAMINATION (developer): When fixer solution mixes with the developer. The developer can become contaminated by only 1½ ounces of fixer.

CONTAMINATION (radioactive): A radioactive substance dispersed in materials or places where it is undesirable.

CONTRAST: The difference between radiographic tones.

CONTRAST (low or long): A given distribution of densities in which various tones show subtle (low percentage) differences between each other.

CONTRAST (short or high scale): A distribution of densities (gray scale from white to black in which the various tones show obvious (high percentage) differences. (*See also* Short Scale Contrast.)

CONTRAST ENHANCEMENT (digital subtraction): A computer function by which logarithmic numbers are assigned by the computer which produces a higher percentage variation than what was originally obtained.

CONTRAST SCALE: A distribution of radiographic tones ranging from black to white (gray scale).

CONTROLLED AREA: A defined area in which the occupational exposure of personnel to radiation is under the supervision of the radiation protection supervisor.

CONTROL PANEL (x ray): Generator.

CONVERGENT POINT: The relationship between the general alignment of primary x-rays to the angle of the lead strips in a particular grid.

COOLIDGE TUBE: The first hot cathode tube that requires a complete vacuum; the tube current is produced by heating the filament to a point of incandescence.

CROOKS TUBE: The earliest type of x-ray tube that utilizes a partial vacuum and a cold cathode. The tube current is obtained from ionizing the gas within the envelope.

CROSS GRID: A particular type of grid that is made by superimposing two linear grids over each other so that the lead strips from tiny squares on the film as the x-ray beam passes through.

CROWN STATIC: A particular type of static characterized by a short horizontal black line across the film from which numerous smaller wavy vertical lines emanate, usually caused by improper separation of film and packaging material.

CURIE (Ci): (1) The special unit of activity equal to 3.7×10^{10} disintegrations per second. (2) By popular usage, the quantity of any radioactive material having an activity of one curie.

CYLINDER CONE: A particular type of beam-limiting device that has true cylinder form with a plate at one end for fastening to the x-ray tube housing.

DEAD-MAN SWITCH: A switch so constructed that a circuit closing contract can be maintained only by continuous pressure on the switch.

DECENTERING: When the central ray is not centered to the grid. Decentering almost always produces grid cutoff.

DEFINITION: How sharply and accurately the borders of a structure appear on the radiographic image.

DEFINITION OF DETAIL: That part of the radiographic image that relates to how well borders of a structure are defined.

DEFLECTION COILS: An assembly of electronic coils carefully calibrated and located around a T.V. camera pickup tube that produces flux lines to control the speed and motion of the stream of electrons coming from the gun.

DEGREE OF DEVELOPMENT: A degree to which a latent image is transfered into a visible image. As the degree of development increases, the overall radiographic image becomes more dense (as more black metallic silver crystals are produced). Degree of development is primarily dependent upon replenishment, temperature, oxidation, time, and dilution with water.

DENSITOMETER: A device used to quantify radiographic tones.

DENSITY (radiographic): Refers to the amount of blackness seen in the radiographic image in a specific or over a general part of the film.

DENSITY (tissue): The concentration of various molecules (mass) composing a given body part or tissue within a body. As the density of a body part increases, its absorption to x-rays increases. Also, body or tissue density is an important factor in producing scatter.

DETAIL: The ability to see small structures in the radiographic image.

DEVELOPER ACTIVITY: The rate at which the various chemicals in the developing solution act upon the exposed image. Developer activity is greatly affected by solution temperature, dilution with water, replenishment, and oxidation.

DEVELOPER SOLUTION: A combination of various chemical additives that together with water will reduce a latent image of exposed silver bromide crystals to black metallic silver to form a visible image.

DEVELOPING TIME: The prescribed time during which an x-ray film is bathed in developing solution.

DIAGNOSTIC QUALITY: A term used to refer to the amount of "diagnostic patient information" that can be seen on a radiograph. As the amount of patient information increases on a radiographic image, the diagnostic quality of that image increases.

DIAGNOSTIC-TYPE PROTECTIVE TUBE HOUSING: An x-ray tube housing so constructed that the leakage radiation measured at a distance of 1 meter from the source does not exceed 100 mr in 1 hour when the tube is operated at its maximum continuous rated current for the maximum rated tube potential.

DIAPHRAGM: A beam-limiting device usually in the form of a lead plate in which an opening has been cut out to any shape or size desired so x-rays can pass through to expose the film. A diaphragm with its fixed opening is usually used to accommodate the specific needs of the examination, such as key-hole shape for carotid arteriograms.

DIGITAL IMAGE: An image composed of a matrix of pixels which contain densities assigned by a computer.

DIGITAL SUBTRACTION FLUOROSCOPE: A type of x-ray equipment that utilizes fluoroscopy to produce a remnant beam that is later processed by a computer to form a digital matrix image composed of pixels.

DILUTION: The mixing of water into a solution such as the developer or fixer.

DISPERSAL PROCESSING: An arrangement whereby film processors are located throughout a department to service specifically prescribed regions. Dispersal processing is the opposite of central processing and is generally considered to be more efficient.

DISTORTION (shape): When there is a difference between the shape of a structure as it appears on the radiographic image and its actual anatomic shape.

DISTORTION (size): When there is a difference between the size of the structure as it appears on the radiographic image and its actual anatomic size in the body.

DOUBLE COATING (film): Coating emulsion on both sides of the film base.

DRYING SYSTEM: The mechanism of an automatic processor where x-ray film is dried to complete the processing cycle.

DRY TO DRY: A general term used to refer to total processing time of automatic processing.

EDDIE CURRENTS: Low energy electrical currents that are produced as a kind of byproduct resulting from rapidly changing flux lines in a magnet.

EDGE SEAL: A material used to coat the four edges of an intensifying screen to prevent moisture from entering between the screen's component parts and causing their separation. Such separation could ultimately lead to permanent screen damage.

ELECTRICAL PRESSURE: The electrical force generated by an imbalance of positive and negative charges in a circuit. This imbalance drives electrons through the circuit. Electrical pressure is measured in volts. As pressure increases, voltage increases as the electrons move more quickly to try to equalize the imbalance.

ELECTRON GUN (T.V. camera): This component of the T.V. camera pickup tube projects an extremely fine stream of electrons toward the target of the T.V. camera tube.

ELECTRON IMAGE: An accumulation of electrons typically produced by a photoconductive material whereby each degree of visible light produces a corresponding number of electrons.

ELLIPTICAL TOMOGRAPHY: A particular type of tomographic motion in which the tube and film move in an elliptical pattern during the exposure.

EMULSION (film): The component of x-ray film that contains the mixture of gelatin and silver bromide crystals.

ENERGY SUBTRACTION: The use of rapidly alternating exposure of very different KvP levels to enhance subject contrast.

EXHAUSTED SOLUTION: A solution, developer or fixer, that is weakened by a lack of replenishment solution or by an increase in oxidation to a point where it cannot produce optimal contrast and density.

EXPOSURE: A measure of the ionization produced in air by x or gamma radiation. It is the sum of the electrical charges on all of the ions of one sign produced in air when all electrons liberated by photons in a volume element of air are completely stopped in air divided by the mass of the air in the volume element. The special unit of exposure is the roentgen. (For radiation protection purposes, the number of roentgens may be considered to be numerically equivalent to the number of rads or rems.)

EXPOSURE ANGLE: The actual distance the tube travels during a tomographic exposure while the x-rays are being emitted.

EXPOSURE ARC: The distance the x-ray tube travels during a tomographic sweep while the exposure is on.

EXPOSURE EFFICIENCY: The relationship between the quantity of x-ray photons and the resulting radiographic density. Exposure efficiency is primarily a function of KvP; as higher KvP exposures are formulated, exposure efficiency increases.

EXPOSURE RATE: The concentration of x-ray photons in a specified area over a specified time. Primary factors effecting exposure rate are Ma, F.F.D., and KvP.

EXPOSURE TIME: The actual period during which x-rays are being emitted.

FIELD: One-half of a complete T.V. image which is typically obtained in 1/60 second.

FIFTEEN PERCENT RULE: A system by which radiographic density can be expected to double by increasing the kilovoltage 15 percent, or by expecting the radiographic density to be cut in half by reducing the Kv 15 percent.

FILAMENT: A fine coil of tungsten wire attached to the cathode portion of the x-ray tube. The filament is heated and through thermionic emission provides a cloud of electrons that ultimately are driven toward the anode.

FILAMENT CURRENT: The current that supplies the x-ray tube's filament with sufficient heat to cause incandescence.

FILM (radiographic): A permanent image composed of various individual gray, black, and white tones that form a pattern representing body structures.

FILM CRITIQUE: A three-step process of viewing a film and evaluating its diagnostic quality, determining the problem, and formulating the correction.

FILTER: A plate, usually of aluminum, that is placed in the x-ray beam for the purpose of greatly reducing unnecessary soft rays to the patient. In most cases, the filter is mounted to the port of the x-ray tube housing.

FILTRATION (added): The amount of filtration placed in the primary beam (not including inherent filtration).

FILTRATION (inherent): Filtration provided by glass envelope of the x-ray tube. Inherent filtration is approximately equal to 0.5 mm of lead.

FILTRATION (total): An accumulation of filtering effect provided by the sum of inherent and added filtration.

FINE-GRAINED FILM: An x-ray film (almost always a slow film) made with very small silver bromide crystals. A film using very small or very fine grains of silver bromide crystals can be expected to produce a very fine textured image yielding high detail.

FINE-LINE GRID: A grid having a minimum of 100 lead strips per inch. The lead strips used are much thinner than conventional grids. Depending on the manufacturer, the lines per inch range from 105 to 110 including interspace material. Because the lead strips are so thin, fine-line grids are used as stationary grids when a bucky is too slow for the required exposure time, i.e. chest and pediatric radiography.

FIXED KvP: A type of technique chart that provides for a single KvP for a given body part or examination with MaS used for compensating body thickness.

FIXING AGENT: A chemical added to the fixing solution for the primary purpose of dissolving the unexposed silver bromide crystals and thus producing the low density areas of the radiographic image.

FLUORESCENCE: The emission of visible light resulting from ionizing fluorescent phosphors. For diagnostic purposes, this emission is produced by calcium tungstate crystals and has the characteristic of terminating simultaneously with the exposure.

FOCAL FILM DISTANCE: The distance span between the focal spot of x-ray tube and the surface of the film.

FOCAL PLANE: The specific level "slice" of the body seen most clearly on a tomographic examination. The location (or level) of the focal plane is determined by the fulcrum, and the thickness is determined by the exposure angle. (*See also* Objective Plane.)

FOCAL SPOT: The specific area on the anode from which a primary beam is emitted.

FOCAL SPOT SIZE: The actual dimensions of the focal spot producing the useful primary beams. Focal spot is generally measured in millimeters. Focal spots in a diagnostic tube commonly range in size from .5 millimeter to 2.0 millimeters depending on specific purpose and function of the tube. Unless otherwise stated refers to the effective focal spot.

FOCAL TRACK: The rim or bevelled area around the anode disc upon which the tube current strikes.

FOCUSED GRID: A type of grid made in such a way that its lead strips are vertical in the center but gradually become tilted toward the center ray at an increasing angle toward the center of the grid as they are located along the side of the grid.

FOCUSING CUP: A cup-shaped device in which the x-ray tube filament is placed, the purpose of which is to aim the space charge directly at the focal spot area of the anode.

FOG (base plus): *See* Base Plus Fog.

FOG (chemical): *See* Chemical Fog.

FOG (scatter): A radiographic fog specifically caused by an excessive ratio of scatter photons compared to primary radiation. Fog caused by excessive scatter can be improved greatly with the use of radiographic grid. (These scatter photons are actually stray or "renegade" photons that are emitted by the body in indiscriminate patterns and expose the film in areas that would otherwise have lower density, thus diminishing the desired balance of light and dark areas.)

FORCED EXTINCTION: A system of timing an x-ray exposure that can terminate the time in segments smaller than a pulse or 1/120 second.

FRACTIONAL FOCAL SPOT: A focal spot size measuring less than .3 millimeter.

FRAME: One complete picture as scanned by a T.V. camera, typically obtained in 1/30 second.

FRAMING SPEED: The number of complete computerized images that a system is capable of, whether it be in conventional serial x-ray studies or digital.

FULL-WAVE RECTIFICATION: The process of producing direct current by changing the negative phase to a positive phase.

GELATIN (film): A material that resembles a gel during x-ray film manufacturing. Gelatin is a basic component of the x-ray film emulsion and has such important properties as being chemically inert, transparent, and able to swell and contract during various processing cycles. The purpose of the gelatin is to hold the silver bromide crystals in place.

GENERATOR (x-ray): The x-ray control panel. Its purpose is to "generate" the proper amount of current in the primary circuit from the incoming line and deliver that current to the high tension transformer. It has the various meters and controls for selecting voltage, amperage, and time to the x-ray tube.

GEOMETRIC IMAGE: That component of the x-ray beam that controls the degree of sharpness and distortion seen in the radiographic image.

GHOST SHADOW: Blurred and vaguely visible structure seen in the tomographic image. These are anatomic structures lying just superior and inferior to the desired objective plane. These ghost shadows are blurred anatomy lying close enough in location to the actual objective plane so that total blurring is not possible.

GRADIENT SCREENS: *See* Compensating Screens.

GRAIN (coarse grain): X-ray film manufactured with relatively large silver bromide crystals. Coarse grain x-ray usually yields less detail but has a higher speed than a fine grain film.

GRAIN (film): Silver bromide crystal.

GRAIN (fine grain): *See* Grain (film).

GRID: A device made with alternating lead strips and interspace material (plastic, aluminum, or cardboard). A grid is used to absorb large quantities of excess scatter radiation from the remnant beam. As this excess scatter is absorbed, a lower ratio of scatter to primary photons is created in the remnant beam, thereby increasing radiographic contrast.

GRID CUT-OFF: Excessive absorption of the primary beam by a radiographic grid resulting from improper grid alignment or grid positioning with respect to the x-ray beam.

HALF VALUE LAYER: The thickness of aluminum or lead equivalent that when placed in an x-ray beam will reduce the exposure rate of the beam to one-half of its original value.

HALF-WAVE RECTIFICATION: The process of producing direct current by suppressing the negative phase.

HARDENING AGENT: A chemical additive put in the fixer solution to produce the required shrinking and hardening of the film's emulsion for long-term storage.

HEAT EXCHANGER (processing): A device placed in the recirculating system of most automatic processors that utilizes conduction between hot and cold water lines for the purpose of moderating developer solution temperature.

HEAT LOADING (tube): This refers to the amount of heat imposed on the anode of the x-ray tube during a given exposure or series of exposures.

HEAT UNITS: The amount of heat specified as "units" that accumulates in the x-ray tube resulting from a given exposure or an accumulation of heat units resulting from a series of exposures. The number of heat units generated is calculated by KvP × Ma × Time.

H & D CURVE: *See* Characteristic Curve.

HETEROENERGETIC BEAM: A beam composed of x-ray photons with moderate to widely variant wavelengths.

HIGH KILOVOLTAGE TECHNIQUE: A technique that utilizes KvP values between 100 and 150.

HIGH SPEED ANODE: An anode that spins at approximately three times the rate of a conventional anode. High speed anodes generally turn at approximately 10,000 revolutions per minute.

HOMOENERGETIC BEAM: A beam composed of x-ray photons with similar wave lengths.

HYPOCYCLOIDAL TOMOGRAPHY: A complex motion of the x-ray tube during a planographic examination, which resembles a three-leaf clover pattern.

HYPO SOLUTION: *See* Fixer Solution.

IMAGING FILM SYSTEM: Any combination of film with or without intensifying screens used to produce a permanent radiographic image, including Xerox® and Polaroid® recording media.

IMAGING SYSTEM (fluoroscopic): Any workable combination of television camera, mirrors, lenses, image intensifiers, and related electronic circuitry used in conjunction with fluoroscopic viewing.

IMPULSE TIMER (x-ray): A timing device that can differentiate segments of time into individual pulses of 1/20 second.

INCANDESCENCE: The emission of visible light caused by excessive heating. A light bulb incandesces when its filament is heated by electrical current passing through it, as does the filament of the x-ray tube.

INPUT PHOSPHOR (I.A. tube): The part of the I.A. tube that the remnant beam first strikes to produce a visible image on the inner surface of the glass envelope.

INTENSIFICATION FACTOR: The radiographic effect produced by x-ray screen emissions as opposed to density produced by the x-ray beam itself. Screens amplify radiographic density over x-rays by 10 to 60 times.

INTENSIFYING SCREENS: A device primarily consisting of a backing material coated with an active layer, usually of calcium tungstate phosphors. The phosphors are excited by an x-ray exposure and produces an emission (usually of blue-violet) to expose the x-ray film. The purpose of screens is to reduce x-ray exposure to patients, to make possible shorter exposure times than would otherwise be possible.

INTENSITY (beam): The concentration of photons in a specified measured area during a specified time. Exposure rate is the beam intensity given in quantitative form.

INTERLACED SCANNING: A method of producing a complete T.V. picture by the T.V. camera by scanning every other line then going back to scan the lines in between. In this fashion two separate fields are interlaced to form one complete frame.

INTERSPACE MATERIAL (grid): A space or filler (of aluminum or cardboard) put between a grid's lead strips. It separates the lead strips, helps establish grid ratio, and permits primary radiation to pass through to expose the film. Some scatter traveling at a similar angle to the primary photons will also pass through the interspace material.

INVERSE SQUARE LAW: Describes the relationship between beam intensity and distance over which x-rays travel. The intensity of the x-ray beam increases by a factor of four each time the distance decreases one-half and decreases in intensity by a factor of four each time the distance is doubled.

ION: An electron that has been separated from its atom or an atom itself that has lost one of its electrons, or an atom that has gained another electron.

IONIZATION: The addition or deletion of an electron of an atom, usually resulting from an atom interacting with an external energy, such as an atom of body tissue with an x-ray photon. An atom is left with either a positive or negative charge.

KILOVOLTAGE: One thousand volts. Kv regulates the speed of tube current that, in turn, determines the resulting x-ray beam's penetrating power.

KILOVOLTAGE PEAK (KvP): The kilovoltage value at the peak (top) of the electrical phase of current moving through the tube.

KvP METER (prereading meter): A device connected in parallel that displays the voltage anticipated or expected to go through the x-ray tube.

LAMINOGRAPHY (body section): *See* Tomography.

LATENT IMAGE: The distribution of ionized (exposed) and nonionized (unexposed) silver bromide crystals in a sheet of film. When the film with a latent image is placed in the developer solution, that distribution of ionized crystals is converted into various gray and black tones and the unexposed crystals yield the clear areas—forming a composite of gray, black, and white areas representing body detail.

LATITUDE (exposure): The amount of error or deviation from an optimal exposure that can be used before image quality is diminished. Exposure latitude is primarily regulated by KvP. Wide exposure latitude is desirable for the technologist because changes in exposure are less critical.

LATITUDE (film): The ability of x-ray film to produce patient information at the toe and shoulder portions of the H & D curve. A film with wide latitude expands the diagnostic range beyond a film with narrow latitude characteristics. Often, wide latitude film is more useful to the radiologist.

LATTICE (crystal): The molecular arrangement or structure of the atoms in a silver bromide crystal.

LEAD EQUIVALENT: The thickness of lead affording the same attenuation, under specified conditions, as the material in question.

LEAKAGE RADIATION: All radiation coming from within the source housing except the useful beam. (Note: leakage radiation includes the portion of the radiation coming directly from the source and not absorbed by the source housing as well as the scattered radiation produced within the source housing.)

LINEAR GRID: A grid made with all its lead strips placed in the same direction, but not necessarily at right angles to the plane of the film.

LINEAR TOMOGRAPHY: The simplest type of tomographic motion, in which the tube and film move only in a back and forth opposing motion.

LINE COMPENSATOR: A device that corrects manually or automatically incoming line voltage fluctuations which could affect x-ray tube output.

LINE FOCUS PRINCIPLE: The difference between the "actual" focal spot size of the tube and the focal spot as "seen by the film" because of the bevel (angle) of the anode. As the angle increases, the focal spot projected to the film decreases without lessening the actual focal area.

LINE PAIRS: The system used to quantify sharpness where one line pair is a line of subject information next to a space of equal width as the subject line.

LINES PER INCH: Refers to the actual number of grid strips present per inch in a radiographic grid. Conventional grids usually have 60 to 80 lines per inch, whereas micro line grids have up to 110 lead strips per inch of grid.

LINE SPREAD FUNCTION: A method of quantifying the degree of unsharpness. Often used to evaluate screen unsharpness.

LINE TO ARC: A type of tomographic motion in which the film is moved horizontally, but the tube moves in an arcing pattern as it crosses over the film.

LINKAGE: The type of mechanical connections in a tomographic unit that synchronizes the x-ray tube and bucky movement during a tomographic exposure.

LUMINESCENCE: A general term referring to the emission of light from phosphor materials. The luminescent material could be phosphorescent or fluorescent.

MAGNIFICATION: An enlargement of the structure size as seen on the radiograph compared to the body part's actual size.

MASK FILM: A positive image produced from a negative x-ray film using photographic image reversal techniques.

MATRIX: A geometric arrangement of pixels which makes up the computerized image.

MAXIMUM PERMISSIBLE DOSE EQUIVALENT (MPD): For radiation protection purposes, the maximum dose equivalent that a person or specified parts thereof shall be allowed to receive in a stated period of time. For radiation protection purposes, the dose equivalent in rems may be considered numerically equal to the absorbed dose in rads and the exposure in roentgens.

MICRO LINE GRID: *See* Fine Line Grid.

MICROSECOND: 1/1,000,000 of a second.

MILLIAMPERE (x-ray): The number of electrons moving from and to the cathode during an x-ray exposure.

MILLIAMPERE SECONDS: The total quantity of electrons moving from cathode to anode during the exposure. MaS is the primary factor in regulating radiographic density. MaS = Exposure Time × Milliamperage.

MILLROENTGEN (mr): One-thousandth of a roentgen.

MILLISECOND: 1/1,000 of a second.

MILLION ELECTRON VOLTS (MeV): Energy equal to that acquired by a particle with one electronic charge in passing through a potential difference of one million volts (one MV).

MODULATION TRANSFER FUNCTION: A concept that puts a quantitative value on the amount of edge blurring of a specific structure in a radiographic image.

MONOCHROMATIC BEAM: A beam containing photons of equal or nearly equal energy.

MOTION (involuntary): Movement generated from the parasympathetic nervous system, i.e. heart peristalsis.

MOTION (voluntary): Movement of the body part being radiographed, due to the central nervous system, i.e. breathing, unconscious movement of the hand and foot.

MOTTLE: An uneven (gritty salt and pepper) look in the image most noticeable at the low density areas. Mottle is caused by an insufficient number of primary photons in the remnant beam (producing a sprinkly effect instead of an evenly textured density). High speed screens are one of the primary causes of mottle.

MOTTLE (screen): Mottled appearance in the radiographic image that is specifically caused by intensifying screens.

MULTIPLANOGRAPHY: A technique used to produce more than one objective plane with a single exposure and sweep of the x-ray tube. A specially designed cassette with from five to seven pairs of screens is used. When loaded properly with films, one tomo sweep yields many radiographs, each with a different body level corresponding to the film's position in the cassette. (*See also* Plesiotomography.)

MUTUAL INDUCTION TRANSFORMER: A transformer that has two separate sets of primary and secondary coils to produce changes in the voltage via electrical flux and magnetic domains.

NOISE (screen): Very subtle irregular density pattern over an area of the radiographic image that should otherwise have a very smooth appearing density. Noise is primarily caused by an inadequate supply of remnant x-ray photons carrying patient information to the film, i.e. the remnant beam is not concentrated enough to expose the film evenly, thus making an almost salt and pepper type appearance on the film. (*See also* Mottle (screen).) Nondiagnostic signals are included in the T.V. or digital produced image.

NONCONTROLLED AREA: Any space not meeting the definition of a controlled area.

NONSCREEN EXPOSURE SYSTEM: The film is exposed totally by the x-ray beam.

NONSCREEN FILM: X-ray film that is designed specifically to yield optimal radiographic characteristics when exposed directly by the x-ray beam as opposed to intensifying screens. It is usually a fine grain film, high resolution film, with high silver content.

OBJECT FILM DISTANCE: The distance between the specific part or section of the body being radiographed and the x-ray film.

OBJECTIVE PLANE: The plane "slice" of the body that is least blurred on the tomographic image. The thickness of the objective plane is primarily a function of exposure angle, and the level is established by the position of the fulcrum.

OHM'S LAW: A method of determining the degree of electrical resistance that will be achieved in a circuit or by an electrical component.

OPTIC SYSTEM: The assembly of photographic lenses that focus and transmit the visible image from the output phosphor of the I.A. tube to the target of the T.V. camera lens.

OPTIMAL KVP TECHNIQUE: A format of establishing exposure techniques in which the KvP is held constant for a given body part regardless of its thickness, and MaS (usually time) is changed to accommodate varying body thicknesses. This technique tends to reduce patient exposure and produce radiographs with more consistent contrast. Also known as fixed KvP technique.

OUTPUT PHOSPHOR (I.A. tube): The part of the I.A. tube that receives the electron image and is coated with a phosphorescent material. This converts the electron image into a visible image.

OVEREXPOSURE: The exposure used is too great for the body part being radiographed and the imaging system being used, resulting in a radiograph with too much density.

OXIDATION (chemical): A mixing of air molecules with developer or fixers. This almost always results in a depletion of the solution's strength. If unchecked, the strength of the solution will be diminished to a point where it becomes unusable.

PANEL, FLUOROSCOPIC: Surface of a vertical fluoroscope analogous to the tabletop of a tilting table fluoroscope.

PARALLAX DISTORTION: Blurring on a radiographic image resulting from incomplete superimposition of the images on each side of an x-ray film. Usually the blurring is so minimal that it has no practical effect on image quality.

PARALLEL GRID: A radiographic grid constructed with all the lead strips parallel to each other and standing 90 degrees to the plane of the film.

PENETRAMETER: A device usually made of aluminum or plastic that is often cut in steps of increasing thicknesses. The device produces changes in attenuation to the x-ray beam, which produces a series of density steps that correspond to the amount of beam penetration of each step.

PENUMBRA: Blurring of structures in the radiographic image caused by x-ray photons originating from slightly different areas of the focal spot, causing each photon to transcend a body part from slightly different angles producing an ill-defined (blurred) structure. Penumbra is always present to varying degrees in the radiographic image. Although it is a function of focal spot area, penumbra can be compensated for by other geometric adjustments.

PERSONNEL MONITOR: An appropriately sensitive device used to estimate the radiation exposure to an individual.

PHANTOM: For radiation protection purposes, a tissue-equivalent object used to simulate the absorption and scatter characteristics of the patient's body.

PHOSPHORS (intensifying screens): Substances that upon exposure to x-rays emit visible or ultraviolet light.

PHOSPHORESCENCE: Phosphorescence is the emission of light by a phosphor in which the emission continues for a short time after the stimulus to the phosphor material has been removed. Also known as screen lag or afterglow. The afterglow lasts about 10^{-3} to 10^{-6} seconds.

PHOTOCONDUCTIVE MATERIAL: A material that is used to coat the target of the T.V. camera which has the property of producing electrons when excited by incident light.

PHOTOELECTRIC (interaction): A specific turn of events that involve the interacting of x-rays and body tissue in which the incident photon is totally absorbed by the atom and a low energy electron is ejected from the atom K or L shell. The vacancy is filled by a free floating electron, and low energy characteristic (secondary) emissions result. This interaction is common in diagnostic exposures between 30 and 75 KvP.

PHOTOELECTRIC CELL: A device used in automatically timed radiographic systems, which upon being exposed to light (usually from a small piece of intensifying screen) produces a low level of electrical current that is channeled to the timing circuit in the primary circuit and terminates the exposure. These photo cells' sensitivity to light can be calibrated and adjusted to produce electrical current more quickly or slowly to the timing circuit, thus regulating radiographic density.

PHOTON DENSITY: The concentration of x-ray photons making up any given exposure which can be measured by the number of photons per square millimeter.

PHOTOTIMING: A method in which the x-ray exposure is automatically terminated to provide a desirable predetermined density on the radiograph. This is accomplished by the use of a photoelectric cell or an ionization chamber.

PICTURE ELEMENTS: *See* Pixel.

PIN HOLE DEVICE: A device used to measure the focal spot. It consists of a plate of metal that is positioned a few inches from above the film and a tiny hole through which the cathode ray is centered, resulting in a radiographic density that simulates the size of the effective focal spot.

PIXEL: Tiny parts of a computer produced image. The picture elements' size and number in a given image is determined by the computer.

PLANIGRAPHY: Tomography.

PLEISOTOMOGRAPHY: A tomographic technique in which a special cassette is used. The cassette has approximately four pairs of screens, and when loaded properly and exposed during a tomographic procedure produces four tomographs with approximately 1 mm increments. This application of multiplanography is primarily used for inner ear studies.

PLURIDIRECTIONAL TOMOGRAPHY: A tomographic x-ray machine that is designed to tomograph in a number of tube motions. The most common motions are circular, elliptical, hypocycloidal.

POINT SOURCE: An imaginary point from which all photons would emanate. Such a point source of x-rays would produce a perfectly sharp image on the films no matter what O.F.D. or F.F.D. (within reason) would be used. A true point source beam would have no crossover and ultimately no penumbra.

POLYCHROMATIC BEAM: An x-ray beam that contains photons of different wavelengths.

POLYESTER (film base): The plastic material that is used in manufacturing an x-ray film's base, replacing cellulose acetate.

POSTEXPOSURE: Fogging of the radiographic image caused by the x-ray film being overexposed to the safe light.

PREREADING VOLTMETER: A meter that displays the voltage actually going into the control panel from the hospital's supply.

PRESERVATIVE AGENT: An ingredient in the developer or fixer solution that retards the effect of oxidation, thus prolonging its chemical activity.

PRESSURE MARKS: Very small dark spots in the film resulting from excess pressure against the surface of the film, usually by transport rollers or by film stored in boxes placed horizontally as opposed to vertically on a shelf.

PRIMARY BEAM: *See* Useful beam.

PRIMARY CIRCUIT: A general reference to the route through which the electricity travels before reaching the high tension transformer. Generally this includes all the wiring in the x-ray control panel and the oil-emersed floor tanks, which house the rectifying system.

PRIMARY PROTECTIVE BARRIER: A barrier sufficient to attenuate the primary beam to the required degree.

PRIMARY RADIATION: That part of the x-ray beam that emanates from the target area of the anode and exits the collimator.

PROCESSING (film): A general term which refers to the developing, fixing, washing, and drying of the radiograph.

PROGRESSIVE SCANNING: A method of producing a complete T.V. picture by the T.V. camera as it scans line by line, as opposed to interlaced scanning.

PROJECTED IMAGE: The geometric pattern by which the x-ray photons travel from the focal spot to the film. The projected image controls sharpness, size distortion, and shape distortion.

PROTECTIVE BARRIER: A barrier of radiation absorbing material(s) used to reduce radiation exposure. (*See also* Primary protective barrier *and* Secondary protective barrier.)

PROTECTIVE SOURCE HOUSING: An enclosure, for a gamma-beam therapy source, so constructed that the leakage radiation does not exceed specified limits with the source in the "ON" and "OFF" positions.

QUALIFIED EXPERT: With reference to *radiation protection*, a person having the knowledge and training to measure ionizing radiation, to evaluate safety techniques, and to advise regarding radiation protection needs (for example, persons certified in this field by the American Board of Radiology, or the American Board of Health Physics, or those having equivalent qualifications). With reference to the *calibration of radiation therapy equipment*, a person having, in addition to the above qualifications, training and experience in the clinical applications of radiation physics to radiation therapy (for example, persons certified in Radiological Physics or X-ray and Radium Physics by the American Board of Radiology, or those having equivalent qualifications).

QUALITY (radiographic): The amount of usable diagnostic information contained by the radiographic image in relation to the amount of information available within the subject.

QUALITY (x-ray): The ability of an x-ray beam or photons to penetrate the subject being radiographed. Beam quality is primarily controlled by KvP.

QUANTITY (x-ray): The amount of x-ray photons contained in the primary beam. Quantity of the x-ray beam is primarily controlled by milliamperage.

QUANTUM: Tiny particles of x-ray energy that have no mass. The quantum theory identified x-ray photons as packets of energy commonly called photons that have the peculiar characteristic of being a high energy form, yet not having any mass. These packets of energy are produced as a result of a conversion of energy at the x-ray tube.

QUANTUM MOTTLE: A mottle specifically caused by decreasingly adequate numbers of primary photons used to produce a given radiographic image as a result of using high speed films and high-speed intensifying screens. (*See also* Mottle.) In most cases, high speed screens have a much greater influence on mottle than does film.

RAD: The special unit of absorbed dose equal to 100 ergs per gram.

RADIATION (ionizing): Any electromagnetic or particulate radiation capable of producing ions, directly or indirectly, by interaction with matter.

RADIATION PROTECTION SUPERVISOR: The person directly responsible for radiation protection.

RADIOGRAPHIC TONE: A density seen on the radiographic image.

RADIOLUCENT: Something that offers no attenuation or absorption to x-rays.

RADIOPAQUE: Something that offers total attenuation or absorption to x-rays.

RAPID PROCESSING: Ninety second processing, also known as dry-to-dry processing.

RARE EARTH SCREENS: An intensifying screen that has unusually high x-ray absorption characteristics, which allow the conversion of x-ray energy to light energy to be much more effective than in conventional screens. With a given x-ray exposure, rare earth screens are at least twice as fast as conventional high speed screens. Although the phosphors used in rare earth screens are not "rare," the process required to make them usable in radiography is quite complex, thereby increasing their market value.

RATIO (grid): The relationship between the height of the grid's lead strip and the space between them. A grid with lead strips five times higher than the space between them would be a 5 to 1 ratio. Grid ratio is a primary controlling factor in absorbing scatter radiation.

REACTION TIME: The period of time needed by all the component parts of a phototiming system to actually terminate the exposure after the photocell or ion chamber receives the critical dose for a body part.

RECIPROCITY LAW: When inverse but relative changes between milliamperage (the intensity of the beam) and changes in exposure time are made to maintain the original radiographic density. As milliamperage is increased and time is decreased proportionately, the original radiographic density will remain constant.

RECIPROCITY LAW FAILURE: When the relative or complementary changes in Ma and time do not maintain the original radiographic density. Usually the result of extremely long or short exposure times when intensifying screens are used.

RECIRCULATING SYSTEM: The component of an x-ray film processor that filters and agitates the developer and fixer solutions and helps maintain developer temperature.

RECTIFYING SYSTEM: Any device that converts alternating current to direct current.

REDUCING AGENTS: The specific additives in the developing solution that reduce the exposed silver bromide crystals to black metallic silver.

REDUNDANT SHADOW: Confusing shadow seen in a tomographic image in which those structures located immediately superior and inferior to the objective plane have not been totally blurred.

REFLECTIVE LAYER: The component of an intensifying screen that is responsible for bouncing stray light emitted by the crystals back toward the film, thereby increasing the screen's speed.

REGISTRATION: The process of matching the x-ray negative and the positive mask film to obtain nearly perfect superimposition. This can be done by a computer or manually.

REM: The unit of dose equivalent. For radiation protection purposes for only x and gamma radiation, the number of rems may be considered equal to the number of rads or the number of roentgens.

REMNANT BEAM: The beam of x-ray photons emitted (transmitted) by the patient that exposes the film. The remnant beam consists of primary, secondary, and scatter photons.

REPLENISHMENT (developer and fixer): A system by which a slightly fresh developer and fixer solution is added to the working solution. This additional replenishment supplements the strength of the working solutions already in use to produce the desired radiographic results.

REREGISTRATION: A computer manipulation of pixels to improve image misregistration of the positive and negative images.

RESOLUTION (film): Film resolution is primarily a function of grain size. As the x-ray film is able to reproduce more and more tiny structures as *separate radiographic images* no matter how close to each other they are, resolution increases. Resolution is measured by counting how many "line pairs" per millimeter are visible.

RESOLUTION (radiographic): The ability of a radiographic image to demonstrate two or more tiny structures, no matter how close, as separate structures in the radiographic image. Resolution is measured by counting number of line pairs that can be distinguished per millimeter.

RESTRAINER: A chemical additive given to the developing solution to moderate the reducing agents from developing unexposed silver bromide crystals. If the restraining additives were not used, overdevelopment of the radiographic image would result.

RHOMBIC GRID: A cross grid designed to have crossing lead strips in a diagonal orientation forming tiny diamonds rather than squares as in a conventional cross grid. The rhombic grid is a cross grid that may be used with some success in a bucky.

ROENTGEN (r): The amount of radiation that will produce a charge of 2.58×10^{-4} coulombs per kilogram of air. The special unit of exposure.

ROENTGENOGRAM: A permanent x-ray image.

ROTATING ANODE: A disc type structure mounted at midpoint to a shaft and motor enabling the disc to turn during the x-ray exposure. The advantage of a rotating anode is that a much more intense exposure can be made with it than if it were stationary.

SAFELIGHT: A device of various designs placed in a darkroom to provide suitable lumination by which one can function without the use of white lights.

SAFETY FILM: Film that has the quality of retarding combustion.

SATURATION POINT: A desirable situation by which increases in KvP will not cause increases in Ma.

SCATTER RADIATION: A kind of radiation "by-product" resulting from x-ray and tissue interactions (predominately from the Compton effect). The scatter rays resulting from the interaction are of moderate to high energy photons with many passing through the patient and having a very obvious effect on image quality. An excessive ratio of scatter to primary radiation reaching the film will cause film fogging.

SCATTERED RADIATION: Radiation that, during passage through matter, has been deviated in direction. (It may also have been modified by a decrease in energy.)

SCREEN CONTACT: The degree of surface-to-surface contact between the x-ray film and the intensifying screens. Under normal circumstances, the contact is a function of compression provided by the cassette and is vital to image sharpness.

SCREEN LAG: A continuation of intensifying screen emission beyond the time when the x-ray exposure had been terminated. Although once common, today's screens, for practical purposes, do not exhibit this characteristic.

SEALED SOURCE: A radioactive source sealed in a container or having a bonded cover, in which the container or cover has sufficient mechanical strength to prevent contact with and dispersion of the radioactive material under the conditions of use and wear for which it was designed.

SECONDARY CIRCUIT (x-ray): Any part of the circuit that is placed after the H.T. transformer.

SECONDARY PROTECTIVE BARRIER: A barrier sufficient to attenuate the stray radiation to the required degree.

SECONDARY RADIATION: A kind of radiation by-product resulting from interactions between tissue and x-rays (also known as characteristic radiation). The secondary radiation resulting from interactions are of low energy radiation, usually not traveling for more than a few centimeters of body tissue, thus contributing very little to exposing the film.

SELECTIVITY (grid): The ability of a grid to allow primary radiation to pass through to the film while restricting a certain percentage of S/S radiation. A high selectivity grid would allow many primary x-rays through while absorbing a great number of S/S ray photons (a most desirable effect).

SELF-INDUCTION TRANSFORMER: A type of transformer that utilizes one coil of wires to induce flux lines. These transformers usually have "taps" from which a variety of different voltages can be picked up on the secondary side.

SELF-RECTIFICATION: When the x-ray tube itself allows only direct current to pass through as it compresses the negative phase.

SENSITIVITY SPECK: An extremely small particle of silver to which negative and positive ions accumulate after the exposure, serving as the site of development.

SENSITOMETER: A device that sensitizes (exposes) a film, usually to produce a gray scale of known density values.

SENSITOMETRY: A technique by which the characteristics of speed, contrast, and latitude are evaluated in quantitative terms.

SHALL: *Shall* indicates a recommendation that is necessary or essential to meet the currently accepted standards of protection.

SHAPE DISTORTION: When the shape of the structure as seen in the radiographic image is not akin to the shape of the body part being radiographed. Although usually controlled to reasonable limits, shape distortion is often present.

SHARPNESS: The degree to which the borders of a given structure are defined in the radiographic image. A lack of sharpness would be present if the border of a given structure appeared ill defined (fuzzy). Edge definition can be given quantitative values by the concept of modulation transfer function.

SHOCK PROOF: Grounded properly to prevent electrical hazard (shock) to the operator.

SHORT SCALE CONTRAST: A distribution of densities (gray scale) in which obvious (high percentage) changes are seen in each progressive step. Also known as high scale contrast. The term "short scale" is used because the changes are required before a span of black to white can be accomplished.

SHOULD: *Should* indicates an advisory recommendation that is to be applied when practicable.

SHOULDER (H & D curve): The portion of an H & D curve that first shows only small increases in density while exposure values increase. In general, the shoulder portion of the curve is first noted at a density of 2.5 D.

SHUTTER: (1) In beam therapy equipment, a device, fixed to the x-ray or gamma-ray source housing to intercept the useful beam. (2) In diagnostic equipment, an adjustable device used to collimate the useful beam.

SILVER BROMIDE CRYSTAL: A complex combination of raw silver and bromide manufactured into a single substance with the ability to absorb radiation and intensifying screen emissions for the purpose of acquiring a latent image. The latent image can then be converted into a visible image during developing.

SINGLE COATING (film): X-ray film that has a coating of emulsion only on one side of its base.

SINGLE PHASE CURRENT: A type of electrical current that produces 60 complete cycles per second.

SIZE DISTORTION: When the size of a structure as seen on a radiographic image is not of the actual size of the body part. There is almost always some degree of size distortion, but in most cases, it is controlled to reasonable limits.

SOURCE-FILM DISTANCE (S.F.D.): The distance measured along the central ray from the center of the front surface of the source (x-ray focal spot or sealed radioactive source) to the surface of the x-ray film.

SOURCE-SURFACE DISTANCE (source-skin distance) (S.S.D.): The distance measured along the central ray from the center of the front surface of the source (x-ray focal spot or sealed radioactive source) to the surface of the irradiated object.

SPACE CHARGE: The bolus of electrons suspended immediately beyond the filament of an x-ray tube, which when put into motion becomes tube current.

SPACIAL RESOLUTION: A reference to how many line pairs can be seen.

SPEED (film): Relationship between film density and exposure (usually measured at 1^D, versus exposure); for example, a fast film will yield a density of 1.0^D with about 100 MaS as compared to a slower film that would need 200 MaS to yield a density of 1.0^D. Film speed is a function primarily of thickness of the emulsions and crystal size.

SPEED (screen): The ability of an intensifying screen to absorb x-rays and convert x-ray energy into an emission (usually visible light). The more effective or efficient the conversion from x-ray to visible light, the faster the screen will be.

SPIN TOP: A small shaft mounted to a small metal disc into which a hole is bored. As the disc spins on its shaft, the single hole produces small dots on the film as each burst of "phased" exposure is emitted from the tube. When this device is used with short exposures (1/120 to 1/10 second) of single phase equipment, the number of dots on the x-ray film indicates exposure times or problems with the rectifying system.

STATIC (crown): Black markings on a radiographic image usually seen as several tortuous strands of black lines rising vertically from a horizontal line about 1 inch long.

STATIC (smudge): Small individual black dots or clusters of black dots on the radiographic image most commonly caused by separating a sheet of film from a full film packette too quickly.

STATIC (tree): Black markings on a radiographic image usually seen as several tortuous strands or black lines emanating from a central point.

STATIONARY ANODE: An anode made of a small plate of tungsten embedded in a thick copper shaft that does not move during the exposure. Except for very specialized applications, stationary anodes have become obsolete.

STEP WEDGE: *See* Penetrometer.

STEREO RADIOGRAPHY: A technique by which two radiographs are produced with the x-ray tube shifted in position slightly in each exposure. When viewed simultaneously with a specially designed device, a three-dimensional image can be obtained of the body part.

STOP BATH: A solution used with hand processing made with plain water and small quantities of acetic acid. When the film is dipped into this solution immediately after the developing tank, the remaining developing solution on the film is neutralized instantly.

STRAY RADIATION: The sum of leakage and scattered radiation.

SUBJECT CONTRAST: Differences in x-ray absorption of adjacent body parts. It is the role of the technologist to formulate an exposure that compliments the subject contrast with the primary exposure factors the technologist has at his disposal to produce optimal radiographic results.

SUBTRACTED IMAGE: The final film obtained as a result of the subtraction of complimentary positive and negative images.

SUBTRACTION TECHNIQUE: A three-step photographic process that requires photographic image reversal to produce a positive image which is then superimposed with a negative x-ray film showing opacified vessel. These are sandwiched together and exposed to produce a subtracted image.

SYNCHRONIZING PULSE: An electronic signal that is produced by a special circuit to blank out the electron stream while the deflection coils position the electron stream to the next scan line.

SYNCHRONOUS TIMER (x-ray): An obsolete timer that uses a synchronous electric motor to determine exposure time.

TARGET: The beveled area of an anode that is intended to receive the tube current and from which primary photons are emitted.

TARGET FILM DISTANCE: The distance between the target of the x-ray tube and the film.

TECHNICAL IMAGE: That characteristic or component of the x-ray beam that affects radiographic contrast and density. The technical image is primarily controlled by KvP, time, and milliamperage.

TECHNIQUE CHART: An organized set of exposure factors designed to produce acceptable radiographic quality. The conventional chart should contain information about Ma, time, KvP, focal film distance, grid ratio, coning, and the film-screen combination used.

TELEROENTGENOGRAPHY: Radiography accomplished with a focal film distance of 6 feet or more.

THERAPEUTIC-TYPE PROTECTIVE TUBE HOUSING:

(a) For x-ray therapy equipment not capable of operating at 500 KvP or above, the following definition applies: An x-ray tube housing so constructed that the leakage radiation at a distance of one meter from the source does not exceed one roentgen in an hour when the tube is operated at its maximum rated continuous current for the maximum rated tube potential.

(b) For x-ray therapy equipment capable of operating at 500 KvP or above, the following definition applies: An x-ray tube housing so constructed that the leakage radiation at a distance of one meter from the source does not exceed 0.1 per cent of the useful beam dose rate at one meter from the source, for any of its operating conditions.

(c) In either case, small areas of reduced protection are acceptable providing the average reading over any 100 cm^2 area at one meter distance from the source does not exceed the values given above.

THERMIONIC EMISSION: When electrons are propelled (emitted) from the atom of a filament (usually tungsten) because of excess heating of the filament with electrical current. The electrons are propelled to high velocity when the filament is heated to an extremely high point, thus giving the electrons so much energy that they "stray" beyond the binding power of the tungsten atoms.

THREE-PHASE CURRENT: A type of electrical current that produces 360 complete phases per second.

TIME/TEMPERATURE PROCESSING: A system of establishing film developing techniques whereby a specific ratio of solution temperature and developing time is used to produce optimal radiographic results. As with Ma and time, there is a reciprocity between temperature and developing time; if used wisely, conservatively lowering developer solution temperature can be adjusted to compensate for increasing developing time.

TOE (H & D curve): The low density part of the characteristic curve that has a density less than .25 above base plus fog.

TOMOGRAPHY: A technique of moving the x-ray tube and film in opposing directions around a fixed pivot point (fulcrum). The movement of the tube and film in this fashion causes blurring of the body structures at all levels except at the specific level of the pivot point. Body structures lying within the level (objective plane) of the pivot point are reasonably clear and contain diagnostic information.

TOTAL FILTER: The sum of the inherent and added filters.

TRANSFORMER (auto): The autotransformer is directly connected to the KvP selectors that are used to regulate primary voltage (up or down), which ultimately feeds the high tension x-ray transformer.

TRANSPORT SYSTEM: The term that is usually used in reference to the series of rollers in an automatic processor that move the film from beginning to the end of processing, and is also responsible for flexing the film in such a way that reaction particles are shaken from the film surface.

TUBE COOLING CHART: A chart that uses heat units to indicate the total number of exposures made in sequence that a given x-ray tube can handle before going beyond the tube's heat loading capacity.

TUBE CURRENT: The bolus of electrons propelled to the target of the x-ray tube. Tube current is propelled to the anode in spurts (pulses) depending on phase and rectification of the equipment. Full wave single phase equipment will deliver 120 individual pulses of tube current toward the anode per second.

TUBE RATING CHART: A chart that indicates safe and unsafe single exposures in terms of maximum KvP, time, and Ma that can be used without causing damage to the tube. In order to use the chart correctly, one should know the phasing, rectification, and the focal spot size of the tube so that the proper chart can be used.

T.V. CAMERA: An electronic device that converts a visible image into a voltage pattern.

T.V. MONITOR: An electron device that contains a T.V. picture tube which receives the video signal and converts it to a visible image.

UMBRA: The main image. The umbra is not necessarily free of unsharpness.

UNDERDEVELOPED: When an insufficient number of black metallic silver crystals is produced on the film in relation to the number of silver bromide crystals that were exposed. The most common causes of underdevelopment are when the developing solution is too cool, inadequately replenished, overdiluted, or oxidized.

UNDEREXPOSED: When insufficient remnant radiation reaches the film because of either too little KvP or MaS for a given film-screen combination.

UNDERPENETRATED: When the energy of the x-ray beam used is not capable of adequately passing through the body part. When a body part is truly underpenetrated, it cannot be compensated for by inceasing MaS.

UNSHARPNESS (geometric): When the borders of a structure are not optimally seen (defined) on the radiographic image because of focal spot size, focal film distance, or object film distance (the poor projected image).

UNSHARPNESS (motion): When the borders of a structure are not optimally seen (defined) because of equipment motion or patient motion. Patient motion could be the result of either voluntary or involuntary motion.

UNSHARPNESS (radiographic): When the borders of a structure are not optimally defined on the radiographic image. Radiographic unsharpness will be the result of poor geometry of the beam, patient motion, or screen unsharpness.

UNSHARPNESS (screen): When the borders of a structure are not optimally seen (defined) because of the intensifying screen's phosphor emissions. (*See also* Modulation Transfer Function.)

USEFUL BEAM: Radiation which passes through the window, aperture, cone, or other collimating device of the source housing. Sometimes called "primary beam."

VARIABLE KvP TECHNIQUE: A formulation of exposure factors that requires the KvP to be adjusted to compensate for body thickness while the MaS remain constant. In the 70 and 80 range, two KvP increments are usually added for each 1 cm of body thickness. In the higher KvP ranges, greater KvP increments are required for each centimeter of body thickness.

VIDEO SIGNAL: The variation of voltage coming from the target of the T.V. camera.

VISIBILITY OF DETAIL: The ease with which "diagnostic information" can be seen in the radiographic image. Visibility of detail is affected by anything that alters contrast or density.

WASH CYCLE (processing): A phase in the processing cycle in which the film is bathed in fresh water to remove residual chemicals from the emulsion.

WETTING AGENT: *See* Stop Bath.

ZINC CADIUM SULPHIDE: A type of material that had been used to make the now obsolete conventional "direct view" fluoroscopic screens.

ZONOGRAPHY: A particular type of tomography where the exposure arc is 10 degrees or less. Because the exposure arc is so short, the objective plane is relatively thick (usually 10 millimeters) because there is less motion around the fulcrum. It is often used for roentgenogram examinations or when a relatively thin body part is to be viewed in total, such as a sternum, sternoclavicular joint, or temporomandibular joint.

BIBLIOGRAPHY

Cahoon: *Formulating X-Ray Techniques.* North Carolina, Duke Printing, 1970.

Cleare, Splettstosser, Seemann: Experimental Study of the Mottle Produced by X-Ray Intensifying Screens. *AJR*, No. 1:168-174, July 1962.

Cullinan: *Illustrated Guide to X-Ray Technics.* Philadelphia, Lippincott, 1972.

Curry III, Nunnally: *Introduction to the Physics of Diagnostic Radiography.* Philadelphia, Lea & Febiger, 1972.

Fuchs: *Principles of Radiographic Exposure Processing.* Springfield, Thomas, 1972.

Funke: Illumination. *Picker Corp.,* Vol. 18(3):2-22.

Hodes, Jacque DeMoor, Ernst: Body Section Radiography: Fundamentals. *Radiol. Clinic North Am.,* April, pp. 229-242.

Liebel, Flarsheim: *Characteristics and Applications of X-Ray Grids.* Cincinnati, 1968.

Lundh: Film Fogging by Radiation From Building Materials. *Photographic Science & Engineering.* Liebel, Flarsheim, July 9, 1974, Reprint No. 7468.

Meredith, Massey: *Fundamental Physics of Radiography.* Baltimore, Williams & Wilkins, 1968.

Miller: *Clinical Pathology.* Baltimore, Williams & Wilkins, 1966.

Rossmann: Image Quality in Medical Radiography. *The Journal Of Photographic Science,* Vol. 12:279-283, 1964.

Seemann: *Physical & Photographic Principles of Medical Radiography.* New York, John Wiley & Sons, Inc., 1968.

Selman: *Fundamentals of X-Ray and Radium Physics.* Springfield, Thomas, 1974.

Ter-Pogossian: *Physical Aspects of Diagnostic Radiography.* New York, Harper & Row, 1969.

Trout, Dahl: *Course Manual For X-Ray Applications.* Washington, D.C., Govt. Print. Office, 1973.

Tuddenham: Physiology of Roentgen Diagnosis. *AJAR,* Vol. LXXVIII, No. 1:116-123, July 1957.

Webb: Number of Quanta Required to Form the Photographic Latent Image. *Journal Optical Society of America.* Vol. 31, No. 9:559-569, Sept. 1941.

INDEX